Current Topics in Microbiology and Immunology

263

Editors

R.W. Compans, Atlanta/Georgia
M. Cooper, Birmingham/Alabama · Y. Ito, Kyoto
H. Koprowski, Philadelphia/Pennsylvania · F. Melchers, Basel
M. Oldstone, La Jolla/California · S. Olsnes, Oslo
M. Potter, Bethesda/Maryland
P.K. Vogt, La Jolla/California · H. Wagner, Munich

Springer
Berlin
Heidelberg
New York
Barcelona
Hong Kong
London
Milan
Paris
Tokyo

Arenaviruses II

The Molecular Pathogenesis of Arenavirus Infections

Edited by M.B.A. Oldstone

With 49 Figures and 10 Tables

 Springer

Professor Dr. MICHAEL B.A. OLDSTONE
Division of Virology
Department of Neuropharmacology
The Scripps Research Institute
10550 N. Torrey Pines Road
La Jolla, CA 92037
USA
e-mail: mbaobo@scripps.edu

Cover Illustration: The upper panel illustrates the distribution of lymphocytic choriomeningitis virus (LCMV) in brain during an acute intracerebral infection of 8-week-old adult mice. Six days following inoculation, LCMV (green) and cell nuclei (red) were visualized on a coronal brain section that was reconstructed using a Zeiss fluorescence microscope and image analysis software. LCMV is found in the cells of the meninges, ependyma, and choroid plexus. Photomicrograph courtesy of D. McGavern (The Scripps Research Institute).
  The lower panel illustrates the distribution of LCMV (red/yellow) throughout the body of a 8-week-old mouse that was persistently infected with virus since birth. Photomicrograph courtesy of W.I. Lipkin and M.B.A. Oldstone.

ISSN 0070-217X
ISBN 3-540-42705-8 Springer-Verlag Berlin Heidelberg New York

This work is subject to copyright. All rights are reserved, whether the whole or part of the material is concerned, specifically the rights of translation, reprinting, reuse of illustrations, recitation, broadcasting, reproduction on microfilm or in any other way, and storage in data banks. Duplication of this publication or parts thereof is permitted only under the provisions of the German Copyright Law of September 9, 1965, in its current version, and permission for use must always be obtained from Springer-Verlag. Violations are liable for prosecution under the German Copyright Law.

Springer-Verlag Berlin Heidelberg New York
a member of BertelsmannSpringer Science + Business Media GmbH

http://www.springer.de

© Springer-Verlag Berlin Heidelberg 2002
Library of Congress Catalog Card Number 15-12910
Printed in Germany

The use of general descriptive names, registered names, trademarks, etc. in this publication does not imply, even in the absence of a specific statement, that such names are exempt from the relevant protective laws and regulations and therefore free for general use.

Product liability: The publishers cannot guarantee the accuracy of any information about dosage and application contained in this book. In every individual case the user must check such information by consulting other relevant literature.

Cover Design: *design & production GmbH*, Heidelberg
Typesetting: Scientific Publishing Services (P) Ltd, Madras
Production Editor: Angélique Gcouta
Printed on acid-free paper   SPIN: 10763113   27/3020 5 4 3 2 1 0

Preface

Viruses are studied either because they cause significant human, animal or plant disease or because they are useful materials for probing basic phenomena in biology, chemistry, genetics and/or molecular biology. Arenaviruses are unusually interesting in that they occupy both categories. Arenaviruses cause several human diseases known primarily as the hemorrhagic fevers occurring in South and Latin America (Bolivia: Machupo, Argentine, Junin virus, and Brazil: Sabia virus) and in Africa (Lassa fever virus). Because such viruses produce profound disabilities and often kill the persons they infect, they are a source of health concern and economic hardship in the countries where they are prevalent. Further, they provide new problems for healthcare persons owing to the narrowing of the world as visitors from many countries travel increasingly to and from endemic areas and may incubate the infectious agent taking it from an endemic area into an area where the virus is not expected. Such cases are now being recorded with increasing frequency. In addition to these hemorrhagic fever viruses, the arenavirus lymphocytic choriomeningitis virus (LCMV) can infect humans worldwide, although the illness is most often less disabling and severe than those elicited by the other arenaviruses. Yet, LCMV is of greater concern to non-arenavirologists and experimentalists using tissue culture or animals, etc., because normal-appearing cultured cells or tissues from animals used for research may be persistently infected with LCMV without manifesting clinical disease or cytopathology and may transmit that infection to laboratory workers. For example, in 1975 Heinemann et al., recorded 48 cases of LCMV infection among personnel in the Radiotherapy Department and vivarium at the University of Rochester School of Medicine. These persons had contact with Syrian hamsters into which tumors, unknown to be infected with LCMV, were injected. The tumor cells were obtained from an outside, well-known, research supplier who distributed tumor cell lines to numerous laboratories primarily interested in SV40 and polyoma virus research. A subsequent investigation of the 22 tumor lines revealed that 19 yielded infectious LCMV. Thus, both production of SV40 and/or polyoma virus from these cells or the use of hamsters infected with such tumors provided an unanticipated human biohazard. In the Rochester scenario infection was spread from those doing basic research, presumably via air ducts to multiple individuals at the medical school/hospital including healthcare workers and patients undergoing radiologic procedures. Similarly, the Centers for Disease Control (CDC), USA, has recorded multiple cases of LCMV infection in families scattered from the northeast coast of the United States (New York) to the western region (Reno, Nevada) originating from hamsters sold by a single pet store supplier in the southeast (Florida). In addition, several investigators found that their hybridomas making monoclonal antibodies were infected with LCMV. The likely source was infected spleen feeder layers from clinically healthy but persistently infected mice purchased from a well-known commercial source which supplies mice to many universities and institutes in North America. Thus, researchers studying such diverse areas as the molecular biology of SV40, biological effects of *Chlamydia*, immune response of mice, and producing hybridoma cells or ascites fluids have found their preparations contaminated with LCMV. In these cases, the potentially dangerous effects of LCMV were not anticipated because the virus does not cause footprints associated with most acute viruses (i.e., necrosis, inflammatory response). The recognized arenaviruses of man, vectors and laboratory models used for their study are listed in Table 1.

LCMV has proven to be a Rosetta stone for uncovering numerous phenomena in virology and immunology. For example, research on both the acute and persistent infections of mice with LCMV led to the first description of virus-induced immunopathologic disease, of the role of the thymus in the immune response, of T cell-mediated killing and of the MHC restriction phenomenon associated with cytotoxic CD8 cells. In addition, recent studies with LCMV have detailed the kinetics of memory cells generated following an immune response for both CD8 and CD4 T cells and have defined the kinetics of B cell differentiation into plasma cells and the in situ location of such cells challenging previously held concepts. Lastly, LCMV has been instrumental in defining the concept of virus alterations of luxury or differentiated cells effecting differentiated products leading to disturbances in homeostasis and disease in the absence of observed cytolytic destruction of the involved tissues. These events are tabulated in Table 2.

Since the topic of arenaviruses was visited by *Current Topics in Microbiology and Immunology* 14 years ago, enormous advances have been made in this area. The receptor for several of the arenaviruses, alpha-dystroglycan has been uncovered, the cell biology and replication strategy has been decoded and a reverse genetics system for studying viral gene function and its application to study viral biology is well underway. These findings, in addition to the molecular phylogeny, discussion of rodent reservoirs and description of clinical diseases found in both new world and old world arenaviruses, are the topics included in the first volume. The second volume in this series deals with the biology and the pathogenesis of arenaviruses primarily through the study of LCMV. Interestingly and appropriately, the fundamental observation of MHC restriction and CD8 cytotoxic T lymphocyte killing derived initially from studies with LCMV in the mouse has been expanded to studies of most human pathogens, viral, bacterial, parasitic, as well as events in cancer. The scope and importance of this observation was recognized by awarding the Nobel Prize in 1996 to Rolf Zinkernagel and Peter Doherty, long-time workers in the field of LCMV and arenavirus biology.

Over the last 14 years many of the principles for understanding viral pathogenesis and the biology of animal viruses have been defined, in great part, from the lessons learned by studying LCMV. Those lessons and their implications are the subject of this second volume on the arenaviruses. In terms of virus-induced immune complex disease, the genetic and host control over such complexes has now been defined and extended to the injuries caused by immune complexes and the role of such complexes in persistent RNA and DNA viral infections of humans. In terms of T cell responses, a most remarkable finding is that over 60%, and frequently 80%, of a primary T cell response after viral infection reflect antigen-specific T cell expansion. A closely related discovery is the mechanisms by which apoptosis and other means cause such cells to contract numerically three to four days after the peak T cell response. In addition, the pool of virus-specific CD8 and CD4 memory T cells has been quantitated and their kinetics established. The constant level of antigen-specific CD8 T memory cells over an individual's lifetime contrasts with the gradual decline of CD4 T memory cells. This continuous decline of antigen-specific CD4 T memory cells may have broad implications with respect to the high incidence of cancers and enhanced susceptibility to infectious diseases in the aging human population as well as the potential need for revaccination of individuals every 10–15 years following primary immunization.

Recent understanding of persistent viral infection, as opposed to the acute infection that is cleared by antigen-specific immune responses, has now established the requirements for

Table 1. Recognized arenavirus diseases of humans

Group	Virus	Locality	Disease	Clinical Picture	Hemorrhagic fever syndrome in humans
Old World arenaviruses	Lymphocytic choriomeningitis virus (LCMV)	Probably originated in Europe, now worldwide	Lymphocytic choriomeningitis	Grippe and aseptic meningitis most common; more severe central nervous system disease of meningoencephalomyelitis occasionally occurs. May cause transient hydrocephalus during acute infection or congenital hydrocephalus and chorioretinitis after fetal infection.	No
	Lassa fever virus (LFV)	West Africa	Lassa fever	Severe systemic illness with changes in vascular permeability and vasoregulation. Worst cases often associated with bleeding.	Yes
New World arenaviruses	Junin virus	Argentine pampas	Argentine hemorrhagic fever (AHF)	Classical viral hemorrhagic fever. Similar to Lassa, except thrombocytopenia, florid bleeding, and neurological manifestations much more common.	Yes
	Machupo virus	Beni region of Bolivia	Bolivian hemorrhagic fever (BHF)	As in AHF	Yes
	Guanarito virus	Venezuela	Venezuelan hemorrhagic fever	Probably similar to AHF.	Yes
	Sabia	Brazil	Not yet named	Likely similar to AHF; extensive hepatic necrosis seen; only 3 cases.	Yes

Group	Person to person transmission	Expected mortality	Binding affinity to alpha-dystroglycan receptor	Laboratory model of human infection	Year isolated	Reference(s)
Old World arenaviruses	Never documented	<1%	High	Mouse	1933	Armstrong and Lillie 1934
	Frequently by blood contamination	15%	High	Monkey, guinea pig strain(s), LCMV WE in guinea pig	1969	Frame et al. 1970
New World arenaviruses	Occasionally	15–30%	Low	Guinea pig	1958	Molteni et al. 1961; Parodi et al. 1958
	Occasionally	25%	Low	Monkey	1963	Johnson et al. 1965
	Occasionally	25%	Low		1990	Salas et al. 1991
	Not known	33%	Low	Not known	1990	Clegg 1993

Table 2. Contributions to biology and medicine from the study of the LCMV model

Observation	Reference
Immune mediated injury in viral disease: role of the thymus	Rowe 1954
Definition of T cells in immune response disease	Cole et al. 1972
T cell killing + MHC restriction	Zinkernagel and Doherty 1974, 1996
Quantitation and kinetics of B and T cell memory regulation of T cell pool (apoptosis, clonal exhaustion)	Moskophidis et al. 1993; Razvi et al. 1995; Butz and Bevan 1998; Murali-Krishna et al. 1998; Slifka et al. 1998
Model of persistent viral infection/ tolerance/negative and positive T cell selection (thymic education)	Traub 1936; Burnet and Fenner 1949; Hotchin and Cintis 1958; Oldstone and Dixon 1967; Volkert and Larsen 1965; Zinkernagel et al. 1978
Etiology and pathogenesis of immune complex disease	Oldstone 1975; Oldstone and Dixon 1967
Cytopathology and disturbance of homeostasis in the absence of cytocidal injury	Oldstone et al. 1977, 1982
Immunocytotherapy to cure persistent viral infection	Oldstone et al. 1986; Ahmed et al. 1987; Berger et al. 2000

immunocytotherapy that can clear these persistent pathogens. Thus, the numbers of antigen-specific CD8 cells required to clear virus infection and the numbers of antigen-specific CD4 cells required to maintain those CD8 cells is now known to be 50 CD8 to 1 CD4 T cell. The clinical implication of these findings for adoptive immunotherapy for human diseases like those caused by cytomegalovirus, human immunodeficiency virus and Epstein-Barr virus are clear. Further, these results indicate the number of CD8 and CD4 T cells required following vaccination to prevent or control an acute infection, to cure persistent or latent infections and to shrink tumors.

The concept that virus infects differentiated cells in vitro and in vivo was developed during the study of persistent LCMV infection in the mouse. Such long-term infection disturbs cellular functions thereby altering the cells' homeostasis and causing disease without killing the cell. These well-established observations have now been extended to the understanding of neurologic and hormonal diseases. Related studies in vitro have determined both the host genes and the viral genes involved. Finally, several groups have applied the knowledge of T cell biology and LCMV to construct transgenic mice as experimental models. By cell-specific promoters, i.e., the rat insulin promoter, myelin basic protein promoter, neuron enolase-specific promoter, viral genes or host genes are expressed in specialized cells. For example, viral genes, MHC molecules and cytokines have been expressed in neurons, thymi, beta cells of the islets and oligodendrocytes. The use of such models is presented in this volume. These model systems have provided new information on how autoimmune diabetes and autoimmune CNS diseases occur, have defined the coactivating accessory molecules of the immune system that are required to potentiate or initiate disease and have addressed basic questions about thymic selection and T cell biology.

Other issues in this volume concern selection of viral variants. Those chapters were written by the authorities who initiated or developed the model systems involved.

MICHAEL B.A. OLDSTONE

References

Ahmed R, Jamieson BD, Porter DD (1987) Immune therapy of a persistent and disseminated viral infection. J Virol 61:3920–3929
Armstrong C, Lillie RD (1934) Experimental lymphocytic choriomeningitis of monkeys and mice produced by a virus encountered in studies of the 1933 St. Louis encephalitis epidemic. Public Health Report 49:1019–1027
Berger DP, Homann D, Oldstone MBA (2000) Defining parameters for successful immunocytotherapy of persistent viral infection. Virology 266:257–263
Burnet FM, Fenner F (1949) The production of antibodies. New York, MacMillan
Butz EA, Bevan MJ (1998) Massive expansion of antigen-specific CD8+ T cells during an acute virus infection. Immunity 8:167–175
Clegg JCS (1993) Molecular phylogeny of the arenaviruses and guide to published sequence data. New York, Plenum
Cole G, Nathanson N, Prendergast R (1972) Requirement for theta-bearing cells in lymphocytic choriomeningitis virus-induced central nervous system disease. Nature 238:335–337
Frame JD, Baldwin JM, Gocke DJ, Troup JM (1970) Lassa fever, a new virus disease of man from West Africa. I. Clinical description and pathological findings. Am J Trop Med Hyg 19:670–676
Hinman AR, Fraser DW, Douglas RG Jr, Bowen GS, Kraus AL, Winkler WG, Rhodes WW (1975) Outbreak of lymphatic choriomeningitis virus infections in medical center personnel. Am J Epidemiol 101:103–110
Hotchkin JE, Clintis M (1958) Lymphatic choriomeningitis infection of mice as a model for the study of latent virus infection. Canad J Microbiol 4:149–153
Johnson KM, Wiebenga NH, Mackenzie RB, Kuns ML, Tauraso NM, Shelokov A, Webb PA, Justines GJ, Beye HK (1965) Virus isolations from human cases of hemorrhagic fever in Bolivia. Proc Soc Exp Biol Med 118:113–118
Molteni HD, Guarinos HC, Petrillo CO, Jaschek FRJ (1961) Estudio clinico estadistico sobre 338 pacientes afectados por la fiebre hemorragica epidemica del Noroeste de la Provincia de Buenos Aires. Semana Med 118:838–855
Moskophidis D, Lechner F, Pircher H, Zinkernagel RM (1993) Virus persistence in acutely infected immunocompetent mice by exhaustion of antiviral cytotoxic effector T cells. Nature 362:758–761

Murali-Krishna K, Altman JD, Suresh M, Sourdive DJ, Zajac AJ, Miller JD, Slansky J, Ahmed R (1998) Counting antigen-specific CD8 T cells: a reevaluation of bystander activation during viral infection. Immunity 8:177–187

Oldstone MBA (1975) Virus neutralization and virus-induced immune complex disease: virus-antibody union resulting in immunoprotection or immunologic injury – two sides of the same coin. S. Karger, Basel

Oldstone MBA, Blount P, Southern P (1986) Cytoimmunotherapy for persistent virus infection reveals a unique clearance pattern from the central nervous system. Nature 321:239–243

Oldstone MBA, Dixon FJ (1967) Lymphocytic choriomeningitis: production of anti-LCM antibody by "tolerant" LCM-infected mice. Science 158:1193–1194

Oldstone MBA, Holmstoen J, Welsh R (1977) Alterations of acetylcholine enzymes in neuroblastoma cells persistently infected with lymphocytic choriomeningitis virus. J Cell Physiol 91:459–472

Oldstone MBA, Sinha YN, Blount P, Tishon A, Rodriguez M, von Wedel R, Lampert PW (1982) Virus-induced alterations in homeostasis: alterations in differentiated functions of infected cells in vivo. Science 218:1125–1127

Parodi AS, Greenway DJ, Ruggiero HR, Rivero E, Frigerio MJ, de la Barrera JM, Mettler NE, Garzon F, Boxaca MC, Guerrero LB, Nota NR (1958) Sobre la etiologia del brote epidemico de Junin (nota previa) Dia Med 30:62

Rowe WP (1954) Studies on pathogenesis and immunity in lymphocytic choriomeningitis infection of the mouse. Naval Med Res Inst Resp Rep 12:167–220

Salas R, de Manzione N, Tesh RB, Rico-Hesse R, Shope RE, Betaucourt A, Godoy O, Bruzual R, Pacheco DE, Ramos B, Taibo ME, Tamayo JG, James E, Vasquez C, Araoz F, Querales J (1991) Venezuelan hemorrhagic fever. Lancet 33:1033–1036

Slifka MK, Antia R, Whitmire JK, Ahmed R (1998) Humoral immunity due to long-lived plasma cells. Immunity 8:363–372

Traub E (1936) The epidemiology of lymphocytic choriomeningitis virus in white mice. J Exp Med 64:183–200

Razvi ES, Jiang Z, Woda BA, Welsh RM (1995) Lymphocyte apoptosis during the silencing of the immune response to acute viral infections in normal, Ipr, and Bcl-2-transgenic mice. Am J Pathol 147:79–91

Volkert M, Larsen JH (1965) Immunologic tolerance to viruses. Prog Med Virol 7:160

Zinkernagel ES, Callahan GN, Althage A, Cooper S, Klein PA, Klein J (1978) On the thymus in the differentiation of "H-2 self-recognition" by T cells: evidence for dual recognition? J Exp Med 147:882–896

Zinkernagel RM, Doherty P (1974) Restriction of in vitro T cell-mediated cytotoxicity in lymphocytic choriomeningitis within a syngeneic or semiallogeneic system. Nature 248:701–702

Zinkernagel RM, Doherty P (1979) MHC-restricted cytotoxic T cells: studies on the biological role of polymorphic major transplantation antigens determining T-cell restriction-specificity, function and responsiveness. Adv Immunol 27: 51–177

Zinkernagel RM, Doherty P (1996) Nobel Prize Lecture

List of Contents

R.M. ZINKERNAGEL
Lymphocytic Choriomeningitis Virus and Immunology ..   1

C.A. BIRON, K.B. NGUYEN, and G.C. PIEN
Innate Immune Responses to LCMV Infections:
Natural Killer Cells and Cytokines   7

J.M. MCNALLY and R.M. WELSH
Bystander T Cell Activation and Attrition   29

D. HOMANN
Immunocytotherapy   43

M.K. SLIFKA
Mechanisms of Humoral Immunity Explored
Through Studies of LCMV Infection   67

M.B.A. OLDSTONE
Biology and Pathogenesis of Lymphocytic
Choriomeningitis Virus Infection   83

L.T. NGUYEN, M.F. BACHMANN, and P.S. OHASHI
Contribution of LCMV Transgenic Models
to Understanding T Lymphocyte Development,
Activation, Tolerance, and Autoimmunity   119

M.G. VON HERRATH
Regulation of Virally Induced Autoimmunity
and Immunopathology: Contribution of LCMV
Transgenic Models to Understanding Autoimmune
Insulin-Dependent Diabetes Mellitus   145

C.F. EVANS, J.M. REDWINE, C.E. PATTERSON,
S. ASKOVIC, and G.F. RALL
LCMV and the Central Nervous System:
Uncovering Basic Principles of CNS Physiology
and Virus-Induced Disease   177

N. Sevilla, E. Domingo, and J.C. de la Torre
Contribution of LCMV Towards Deciphering Biology
of Quasispecies In Vivo.......................... 197

J.L. Whitton
Designing Arenaviral Vaccines.................... 221

D.A. Enria and J.G. Barrera Oro
Junin Virus Vaccines 239

Subject Index.................................. 263

List of Contents of Companion Volume I

J.C.S. Clegg
Molecular Phylogeny of the Arenaviruses   1

J. Salazar-Bravo, L.A. Ruedas, and T.L. Yates
Mammalian Reservoirs of Arenaviruses 25

C.J. Peters
Human Infection with Arenaviruses in the Americas 65

J.B. McCormick and S.P. Fisher-Hoch
Lassa Fever 75

S. Kunz, P. Borrow and M.B.A. Oldstone
Receptor Structure, Binding, and Cell Entry
of Arenaviruses 111

B.J. Meyer, J.C. de la Torre, and P.J. Southern
Arenaviruses: Genomic RNAs, Transcription,
and Replication 139

M.J. Buchmeier
Arenaviruses: Protein Structure and Function 159

K.-J. Lee and J.C. de la Torre
Reverse Genetics of Arenaviruses 175

Subject Index 195

List of Contributors

(Their addresses can be found at the beginning of their respective chapters.)

Askovic, S. 177
Bachmann, M.F. 119
Barrera Oro, J.G. 239
Biron, C.A. 7
de la Torre, J.C. 197
Domingo, E. 197
Enria, D.A. 239
Evans, C.F. 177
Homann, D. 43
McNally, J.M. 29
Nguyen, K.B. 7
Nguyen, L.T. 119

Ohashi, P.S. 119
Oldstone, M.B.A. 83
Patterson, C.E. 177
Pien, G.C. 7
Rall, G.F. 177
Redwine, J.M. 177
Sevilla, N. 197
Slifka, M.K. 67
von Herrath, M.G. 145
Welsh, R.M. 29
Whitton, J.L. 221
Zinkernagel, R.M. 1

Lymphocytic Choriomeningitis Virus and Immunology

R.M. ZINKERNAGEL

Lymphocytic choriomeningitis virus (LCMV), a non-cytopathic infection of mice, has revealed the following findings that have helped us to understand basic immune mechanisms:

1. Neonatal tolerance, absence of immunity after intra-uterine or neonatal infection (TRAUB 1936, 1938). The comprehensive view and interpretation was made by BURNET (1949).
2. Immunity can cause disease (ROWE 1954; HOTCHIN 1962).
3. Different LCMV isolates and mutants exhibit distinct tropism and distinct immunopathological diseases (HOTCHIN 1962; JACOBSON 1980; AHMED et al. 1984).
4. Immunity controls virus only partially, since virus survives at low levels (infection-immunity) (ROWE 1954; VOLKERT and LUNDSTEDT 1968; CIUREA et al. 1999), but adoptive cytotherapy of virus-carriers (OLDSTONE et al. 1986) is possible by $CD8^+$ T cells and/or $CD4^+$ T cells plus neutralizing antibody-producing B cells (VOLKERT 1963; PLANZ et al. 1997).
5. Immune complexes in neonatal virus-carriers question B cell tolerance (OLDSTONE and DIXON 1967).
6. Cytotoxic T cells (CTLs) against virus-infected cells act comparably to allo-reactive CTLs (OLDSTONE and DIXON 1970; MARKER and VOLKERT 1973).
7. CTLs cause lethal pathology in the form of T cell-mediated choriomeningitis (COLE et al. 1972).
8. T cells (MIMS and BLANDEN 1972), (in particular CTLs, DOHERTY et al. 1976; ZINKERNAGEL and WELSH 1976) also mediate anti-viral protection against LCMV, as had first been shown for ectromelia virus by R.V. Blanden.
9. MHC disease association of choriomeningitis (OLDSTONE et al. 1973; ZINKERNAGEL et al. 1985) and of aggressive hepatitis (LEIST et al. 1989).
10. CTLs recognize a combination of MHC plus viral antigen (MHC restriction) (ZINKERNAGEL and DOHERTY 1974). Heterozygous MHC expands T cell repertoire and response (hybrid vigor) (DOHERTY and ZINKERNAGEL 1975). MHC class I regulates cytotoxic T cell immune response (DOHERTY et al. 1978; ZINKERNAGEL et al. 1978).

University Hospital Zurich, Department of Pathology, Institute of Experimental Immunology, Schmelzbergstr. 12, 8091 Zurich, Switzerland

11. Interferons induced by virus infections may have detrimental effects (RIVIERE et al. 1977), particularly during early life.
12. Natural killer (NK) cell responses early after systemic virus infection (PFIZENMAIER et al. 1975; WELSH and ZINKERNAGEL 1977).
13. Primary anti-viral CTL responses in vitro (BLANDEN et al. 1977; DUNLOP and BLANDEN 1977).
14. Selective down-modulation of some viral transcripts during virus persistence in vitro and in vivo (OLDSTONE and BUCHMEIER 1982).
15. Immunosuppression by T cell-mediated anti-viral immunity indicates pathogeneic mechanisms that may apply to HIV (MIMS and WAINWRIGHT 1968; SILBERMAN et al. 1978; LEIST et al. 1988; ODERMATT et al. 1991). This includes destruction of antigen presentation (ALTHAGE et al. 1992; BORROW et al. 1995).
16. LCMV mutants that escape CTL activity may be selected (PIRCHER et al. 1990).
17. Viral antigen on cells strictly outside of lymphoid tissue, for example in pancreatic β-islet cells are immunologically ignored (OHASHI et al. 1991; OLDSTONE et al. 1991). In contrast to infection with LCMV, immunization with a vaccinia-recombinant expressing the LCMV-GP is not sufficient to cause disease, although it induces a potent, primed CTL response, Thus there is a high threshold for causing autoimmune disease (OHASHI et al. 1993).
18. Exhaustion of all available precursor T cells by virus that has spread throughout the lymphohemopoietic system, particularly LCMV-variants that replicate quickly and to high titers (MOSKOPHIDIS et al. 1993).
19. Perforin is the key for efficient, rapid $CD8^+$ T cell-dependent LCMV elimination (KAGI et al. 1994; WALSH et al. 1994).
20. Original antigenic sin at the $CD8^+$ T cells level. CTL escape of mutant virus may still induce a CTL response against the original wild-type virus that is not active, however, against the mutant (KLENERMAN et al. 1998).
21. Natural antibodies that are secreted and present in normal serum without immunization influence the initial virus distribution after LCMV infection (OCHSENBEIN et al. 1999a).
22. LCMV can persist by selection of mutant virus that escapes neutralizing antibody responses (CIUREA et al. 2000) and $CD4^+$ T helper cell responses (CIUREA et al. 2001).
23. Same open issues
    a) Delay of neutralizing anti-LCMV responses? Neutralizing antibody responses against LCMV are vastly delayed (ROWE 1954; HOCHIN 1962). The delay suggests that viral antigen exhibiting the correct neutralizing epitope must still be available or even increase after day 50–100 at the time when neutralizing antibodies appear. CD8 elimination during the early phase of virus infection enhances generation of neutralizing antibody responses (BATTEGAY et al. 1993; PLANZ et al. 1996). Both T cell-mediated general immunopathology and specific elimination of neutralizing antibody producing B cells may be involved.

b) Cytotoxic T cell memory? While it is generally accepted that increased T (or B) cell frequencies of primed mice are antigen-independent, protective immunity against challenge with viruses (or tumors) outside of lymphoid organs is antigen-dependent (Lau et al. 1994; Hou et al. 1994; Kündig et al. 1996; Bachmann et al. 1997; Ochsenbein et al. 1999b).

c) DNA forms of LCMV? Parts of LCMV being a negative-stranded virus with an ambisense genome were found in a DNA form in mouse cell-lines in vitro, but also in lymphohemopoietic cells in vivo. First experiments suggest episomal rather than integrated forms (Klenerman et al. 1997).

References

Ahmed R, Salmi A, Butler LD, Chiller JM, Oldstone MB (1984) Selection of genetic variants of lymphocytic choriomeningitis virus in spleens of persistently infected mice. Role in suppression of cytotoxic T lymphocyte response and viral persistence. J Exp Med 160:521–540

Althage A, Odermatt B, Moskophidis D, Kündig TM, Hoffman Rohrer U, Hengartner H, Zinkernagel RM (1992) Immunosuppression by lymphocytic choriomeningitis virus infection: competent effector T and B cells but impaired antigen presentation. Eur J Immunol 22:1803–1812

Bachmann MF, Kündig TM, Hengartner H, Zinkernagel RM (1997) Protection against immuno-pathological consequences of a viral infection by activated but not resting cytotoxic T cells: T cell memory without "memory T cells"? Proc Natl Acad Sci USA 94:640–645

Battegay M, Kyburz D, Hengartner H, Zinkernagel RM (1993) Enhancement of disease by neutralizing antiviral antibodies in the absence of primed antiviral cytotoxic T cells. Eur J Immunol 23:3236–3241

Blanden RV, Kees U, Dunlop MB (1977) In vitro primary induction of cytotoxic T cells against virus-infected syngeneic cells. J Immunol Methods 16:73–89

Borrow P, Evans CF, Oldstone MB (1995) Virus-induced immunosuppression: immune system-mediated destruction of virus-infected dendritic cells results in generalized immune suppression. J Virol 69:1059–1070

Burnet FM (1949) The production of antibodies. Macmillan Co, Melbourne

Ciurea A, Klenerman P, Hunziker L, Horvath E, Odermatt B, Ochsenbein AF, Hengartner H, Zinkernagel RM (1999) Persistence of lymphocytic choriomeningitis virus at very low levels in immune mice. Proc Natl Acad Sci USA 96:11964–11969

Ciurea A, Klenerman P, Hunziker L, Horvath E, Senn BM, Ochsenbein AF, Hengartner H, Zinkernagel RM (2000) Viral persistence in vivo through selection of neutralizing antibody-escape variants. Proc Natl Acad Sci USA 97:2749–2754

Ciurea A, Hunziker L, Martinic MM, Oxenius A, Hengartner H, Zinkernagel RM (2001) CD4$^+$ T-cell-epitope escape mutant virus selected in vivo. Nat Med 7:795–800

Cole GA, Nathanson N, Prendergast RA (1972) Requirement for theta-bearing cells in lymphocytic choriomeningitis virus-induced central nervous system disease. Nature 238:335–337

Doherty PC, Zinkernagel RM (1975) Enhanced immunological surveillance in mice heterozygous at the H-2 gene complex. Nature 256:50–52

Doherty PC, Dunlop MB, Parish CR, Zinkernagel RM (1976) Inflammatory process in murine lymphocytic choriomeningitis is maximal in H-2K or H-2D compatible interactions. J Immunol 117:187–190

Doherty PC, Biddison WE, Bennink JR, Knowles BB (1978) Cytotoxic T-cell responses in mice infected with influenza and vaccinia viruses vary in magnitude with H-2 genotype. J Exp Med 148:534–543

Dunlop MB, Blanden RV (1977) Induction of a primary cytotoxic T cell response to lymphocytic choriomeningitis virus-infected cells in vitro. I. Kinetics of response and nature of effector cells. J Immunol Methods 16:73–89

Hotchin J (1962) The biology of lymphocytic choriomeningitis infection: virus induced immune disease. Cold Spring Harbor Symp Quant Biol 27:479–499

Hou S, Hyland L, Ryan KW, Portner A, Doherty PC (1994) Virus-specific CD8+ T-cell memory determined by clonal burst size. Nature 369:652–654

Jacobson S, Pfau CJ (1980) Viral pathogenesis and resistance to defective interfering particles. Nature 283:311–313

Kagi D, Ledermann B, Bürki K, Seiler P, Odermatt B, Olsen KJ, Podack ER, Zinkernagel RM, Hengartner H (1994) Cytotoxicity mediated by T cells and natural killer cells is greatly impaired in perforin-deficient mice. Nature 369:31–37

Klenerman P, Hengartner H, Zinkernagel RM (1997) A non-retroviral RNA virus persists in DNA form. Nature 360:298–301

Klenerman P, Zinkernagel RM (1998) Original antigenic sin impairs cytotoxic T lymphocyte responses to viruses bearing variant epitopes. Nature 394:482–485

Kündig T.M, Bachmann MF, Oehen S, Hoffmann UW, Simard JJL, Kalberer CP, Pircher HP, Ohashi PS, Hengartner H, Zinkernagel RM (1996) On the role of antigen in maintaining cytotoxic T-cell memory. Proc Natl Acad Sci USA 93:9716–9723

Lau LL, Jamieson BD, Somasundaram T, Ahmed R (1994) Cytotoxic T-cell memory without antigen. Nature 369:648–652

Leist TP, Rüedi E, Zinkernagel RM (1988) Virus-triggered immune suppression in mice caused by virus-specific cytotoxic T cells. J Exp Med 167:1749–1754

Leist TP, Althage A, Haenseler E, Hengartner H, Zinkernagel RM (1989) Major histocompatibility complex-linked susceptibility or resistance to disease caused by a noncytopathic virus varies with the disease parameter evaluated. J Exp Med 170:269–277

Marker O, Volkert M (1973) Studies on cell-mediated immunity to lymphocytic choriomeningitis virus in mice. J Exp Med 137:1511–1525

Mims CA, Blanden RV (1972) Antiviral action of immune lymphocytes in mice infected with lymphocytic choriomeningitis virus. Infect Immun 6:695–698

Mims CA, Wainwright S (1968) The immunodepressive action of lymphocytic choriomeningitis virus in mice. J Immunol 101:717–724

Moskophidis D, Lechner F, Pircher HP, Zinkernagel RM (1993) Virus persistence in acutely infected immunocompetent mice by exhaustion of antiviral cytotoxic effector T cells. Nature 362:758–761

Ochsenbein AF, Fehr T, Lutz C, Suter M, Brombacher F, Hengartner H, Zinkernagel RM (1999a) Control of early viral and bacterial distribution and disease by natural antibodies. Science 286:2156–2159

Ochsenbein AF, Karrer U, Klenerman P, Althage A, Ciurea A, Shen H, Miller JF, Whitton JL, Hengartner H, Zinkernagel RM (1999b) A comparison of T cell memory against the same antigen induced by virus versus intracellular bacteria. Proc Natl Acad Sci USA 96:9293–9298

Odermatt B, Eppler M, Leist TP, Hengartner H, Zinkernagel RM (1991) Virus-triggered acquired immunodeficiency by cytotoxic T-cell-dependent destruction of antigen-presenting cells and lymph follicle structure. Proc Natl Acad Sci USA 88:8252–8256

Ohashi PS, Oehen S, Buerki K, Pircher H.P, Ohashi CT, Odermatt B, Malissen B, Zinkernagel RM, Hengartner H (1991) Ablation of "tolerance" and induction of diabetes by virus infection in viral antigen transgenic mice. Cell 65:305–317

Ohashi PS, Oehen S, Aichele P, Pircher HP, Odermatt B, Herrera P, Higuchi Y, Buerki K, Hengartner H, Zinkernagel RM (1993) Induction of diabetes is influenced by the infectious virus and local expression of MHC class I and tumor necrosis factor-alpha. J Immunol 150:5185–5194

Oldstone MB, Blount P, Southern PJ, Lampert PW (1986) Cytoimmunotherapy for persistent virus infection reveals a unique clearance pattern from the central nervous system. Nature 321:239–243

Oldstone MB, Dixon FJ, Mitchell GF, McDevitt HO (1973) Histocompatibility-linked genetic control of disease susceptibility. Murine lymphocytic choriomeningitis virus infection. J Exp Med 137:1201–1212

Oldstone MB, Nerenberg M, Southern P, Price J, Lewicki H (1991) Virus infection triggers insulin-dependent diabetes mellitus in a transgenic model: role of anti-self (virus) immune response. Cell 65:319–331

Oldstone MBA, Buchmeier MJ (1982) Restricted expression of viral glycoprotein in cells of persistently infected mice. Nature 300:360–362

Oldstone MBA, Dixon FJ (1967) Lymphocytic choriomeningitis: production of antibody by "tolerant" infected mice. Science 158:1193–1195

Oldstone MBA, Dixon FJ (1970) Tissue injury in lymphocytic choriomeningitis viral infection: virus-induced immunologically specific release of a cytotoxic factor from immune lymphoid cells. Virology 42:805–813

Pfizenmaier K, Trostmann H, Rollinghoff M, Wagner H (1975) Temporary presence of self-reactive cytotoxic T lymphocytes during murine lymphocytic choriomeningitis. Nature 258:238–240

Pircher HP, Moskophidis D, Rohrer U, Bürki K, Hengartner H, Zinkernagel RM (1990) Viral escape by selection of cytotoxic T cell-resistant virus variants in vivo. Nature 346:629–633

Planz O, Seiler P, Hengartner H, Zinkernagel RM (1996) Specific cytotoxic T cells eliminate cells producing neutralizing antibodies. Nature 382:726–729

Planz O, Ehl S, Furrer E, Horvath E, Brundler MA, Hengartner H, Zinkernagel RM (1997) A critical role for neutralizing antibody-producing B cells, CD4+ T cells, and interferons in persistent and acute infections of mice with lymphocytic choriomeningitis virus: implications for adoptive immunotherapy of virus carriers. Proc Natl Acad Sci USA 94:6874–6879

Riviere Y, Gresser I, Guillon JC, Tovey MG (1977) Inhibition by anti-interferon serum of lymphocytic choriomeningitis virus disease in suckling mice. Proc Natl Acad Sci USA 74:2135–2139

Rowe WP (1954) Studies on pathogenesis and immunity in lymphocytic choriomeningitis infection of the mouse. Navy Research Report 12:167–220

Silberman SL, Jacobs RP, Cole GA (1978) Mechanisms of hemopoietic and immunological dysfunction induced by lymphocytic choriomeningitis virus. Infect Immun 19:533–539

Traub E (1936) Persistence of lymphocytic choriomeningitis virus in immune animals and its relation to immunity. J Exp Med 63, 847–861

Traub E (1938) Factors influencing the persistence of choriomeningitis virus in the blood of mice after clinical recovery. J Exp Med 229–250

Volkert M (1963) Studies on immunological tolerance to LCM virus: 2. Treatment of virus carrier mice by adoptive immunization. Acta Pathol Microbiol Scand 57:465–487

Volkert M, Lundstedt C (1968) The provocation of latent lymphocytic choriomeningitis virus infections in mice by treatment with antilymphatic serum. J Exp Med 327–339

Walsh CM, Matloubian M, Liu CC, Ueda R, Kurahara CG, Christensen JL, Huang MT, Young JD, Ahmed R, Clark WR (1994) Immune function in mice lacking the perforin gene. Proc Natl Acad Sci USA 91:10854–10858

Welsh RM Jr, Zinkernagel RM (1977) Heterospecific cytotoxic cell activity induced during the first three days of acute lymphocytic choriomeningitis virus infection in mice. Nature 268:646–648

Zinkernagel RM, Doherty PC (1974) Restriction of in vitro T cell-mediated cytotoxicity in lymphocytic choriomeningitis within a syngeneic or semiallogeneic system. Nature 248:701–702

Zinkernagel RM, Welsh RM (1976) H-2 compatibility requirement for virus-specific T cell-mediated effector functions in vivo. I. Specificity of T cells conferring antiviral protection against lymphocytic choriomeningitis virus is associated with H-2K and H-2D. J Immunol 117:1495–1502

Zinkernagel RM, Althage A, Cooper S, Kreeb G, Klein PA, Sefton B, Flaherty L, Stimpfling J, Shreffler D, Klein J (1978) Ir-genes in H-2 regulate generation of anti-viral cytotoxic T cells. Mapping to K or D and dominance of unresponsiveness. J Exp Med 148:592–606

Zinkernagel RM, Pfau CJ, Hengartner H, Althage A (1985) Susceptibility to murine lymphocytic choriomeningitis maps to class I MHC genes: a model for MHC/disease associations. Nature 316:814–817

Innate Immune Responses to LCMV Infections: Natural Killer Cells and Cytokines

C.A. Biron, K.B. Nguyen, and G.C. Pien

1	Introduction: Innate Immunity	7
2	Natural Killer Cell Responses and Functions	9
2.1	Cytotoxicity and Blastogenesis Responses	10
2.2	Cytokine Production	10
2.3	Role in Defense or Lack Thereof	12
3	Endogenous Innate Cytokine Responses and Functions	13
3.1	Type 1 Interferon Effects on Interleukin 12 and Innate Interferon Gamma Production	13
3.2	Effects on Downstream T Cell Interferon Gamma Responses	15
4	Compartmental Issues	16
4.1	Natural Killer Cell Accumulation and Cytotoxicity in Peripheral Blood, Spleen, and Liver	16
4.2	Cytokine Responsiveness and Expression in Different Compartments	18
4.3	Exogenous Interleukin 12 to Induce Early Interferon Gamma for Antiviral Function	19
5	Signaling Pathways for IFNs	20
5.1	STAT 1 Effects on Natural Killer Cell Responses	22
6	Innate Responses and Disease	23
7	Summary	25
References		26

1 Introduction: Innate Immunity

There is a growing appreciation of the importance of innate immune responses to infections. These responses are produced by non-immune cells and cells of the innate immune system, including monocyte/macrophage populations, granulocytes, dendritic cells (DCs), and natural killer (NK) cells. Cytokines can be a part of the innate responses of non-immune cells to infections. The cell constituents of the innate immune system can respond with activation of other anti-microbial effector functions as well as cytokine production. Innate responses act both to mediate defense during periods of development of, and to direct the quality and nature of, downstream adaptive immune responses. The type 1 interferons,

Department of Molecular Microbiology and Immunology, Division of Biology and Medicine, Box G-B629, Brown University, Providence, RI 02912, USA

interferons alpha and beta (IFN-α/β), induced at early times during viral infections, are examples of innate cytokine responses to infections. Another is interleukin 12 (IL-12) induction during challenges with a variety of agents. Host pattern recognition molecules identifying the distinguishing chemical characteristics of infectious organisms induce initial innate responses (BIRON 1999). The best studied of these are cell surface receptors binding and stimulating responses to bacterial products, and associated with the induction of an innate pro-inflammatory cytokine cascade including IL-12 and accompanied by IL-12-induced NK cell type 2 interferon, interferon gamma (IFN-γ), production. In addition to promoting defense at early times of infection, these innate responses result in conditions preferentially driving T helper type 1 (Th1) CD4 T cells with their downstream production of IFN-γ. During infections with extracellular parasites, the earliest events are not well understood, but they result in conditions preferentially driving Th2 CD4 T cells producing IL-4 and IL-5. As IFN-γ is important in defense against bacteria, whereas IL-5 promotes the development of eosinophils important in defense against the extracellular parasites, the consequences of eliciting particular subset innate responses include the promotion of adaptive responses most effective in defense against the agent being encountered. Protective responses to intracellular protozoan infections are similar to those elicited during bacterial infections.

The understanding of innate NK cell and cytokine responses to viral infections is being advanced in the context of a variety of experimental infections of mice and natural infections of human (BIRON et al. 1999). Much of the work has focused on the herpes group viruses, cytomegalovirus (CMV) and herpes simplex virus (HSV), but responses to lymphocytic choriomeningitis virus (LCMV) infections of mice also are being studied. Although the understanding of the earliest signals induced by each of these viruses is still incomplete, it appears as though herpes virus infections elicit both early IFN-α/β production and a pro-inflammatory cytokine cascade with IL-12 expression and consequential induction of NK cell IFN-γ production. The IFN-α/β cytokines induce elevated NK cell cytotoxicity as well as NK cell blastogenesis and proliferation. In the case of certain viral infections, high circulating levels of the type 1 interferons, IFN-α/β, can be induced without the detectable induction of the pro-inflammatory cytokine cascade. LCMV infections fall into this class because they appear to predominantly induce IFN-α/β responses. The type 1 IFNs reach high levels, and their production is sustained for longer periods of time. Studies of innate responses to LCMV infections have been particularly important in revealing the role of IFN-α/β in regulating NK cell responses, and have provided insights into unique compartmental regulation of NK cell responses in vivo. Our group has also used these systems to ask if there is a particular innate to adaptive response cascade during viral infections inducing high levels of IFN-α/β. The suggestion is that the conditions which activate NK cell cytotoxicity negatively regulate the innate components of an alternate cascade associated with Th1 CD4 T cell responses, and preferentially promote CD8 T cell responses (Fig. 1). These topics are the major subjects reviewed here. The chapter concludes with a brief discussion of the

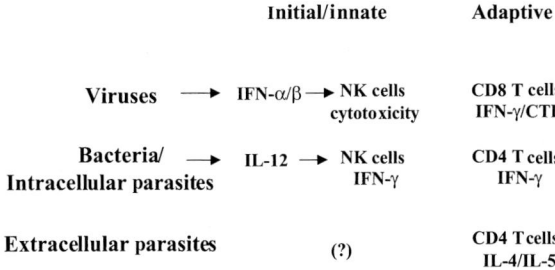

Fig. 1. Summary of protective initial/innate to adaptive immune responses during challenges with viruses eliciting high concentrations of type 1 IFNs as compared to those with bacteria or parasites. The cascade responses elicited during infections with bacteria and intracellular parasites result in type 1 CD4 T cell responses, whereas those elicited during extracellular parasitic infections lead to type 2 CD4 T cell responses. Certain viral infections induce high circulating levels of IFN-α/β and prominent CD8 T cell responses. LCMV infections are in this latter class

possible role of these innate responses in promoting or inhibiting disease during LCMV infections.

2 Natural Killer Cell Responses and Functions

NK cells were first defined by their spontaneous lysis of certain sensitive target cells (see BIRON et al. 1999). It is now clear that NK cells can be induced to mediate elevated levels of target cell killing and to kill a broader range of target cells including certain virus-infected cells, and that they can use perforin for this function. NK cells can also be induced to produce a variety of cytokines including IFN-γ. Classical NK cells mediating these functions are non-T cell lymphoid populations. They are CD3 negative, do not rearrange and express T cell antigen receptors (TCR), and develop in the absence of T and B lymphocytes. Mouse NK cells do express markers such as NK1.1, on restricted genetic backgrounds including C57BL/6, and DX5 on a broader range of genetic backgrounds. In the human and the mouse, they express high levels of asialo ganglio-N-tetraosylceramide (AGM1). They also express a complex family of receptors with activating or inhibiting functions. Although first identified for their interactions with class I MHC molecules, it is now known that certain of these NK cell receptors recognize structures expressed on particular target cells (BAUER et al. 1999; DIEFENBACH et al. 2000). None of these markers appears to be uniquely expressed by NK cells, however, as activated T cells may also be induced to express them. NK cells are equipped to respond to innate cytokines early during infections. Classical NK cell responses have been studied extensively during LCMV infections. There are also NKT cells expressing a restricted range of TCRs and certain NK cell markers, i.e., NK1.1, but little is known about the activation and function of these alternative NK cell populations during LCMV infections.

2.1 Cytotoxicity and Blastogenesis Responses

The first demonstration of the induction of NK cell cytotoxicity during viral infections was in the context of early infections of mice with LCMV. NK cell killing is induced at days 1–5, with peak responses observed at day 3, after LCMV infection (WELSH 1978). The kinetics of this response follows the induction of circulating levels of IFN-α/β during the infection. NK cells activated to mediate elevated killing under these conditions are also induced to undergo blastogenesis and proliferation. They can be characterized as blasting based on increased size after physical separation (BIRON and WELSH 1982) and increased forward scatter as evaluated by flow cytometric analysis (see below). They are proliferating based on DNA synthesis as evaluated by 3H-thymidine incorporation (BIRON and WELSH 1982), entry into S and G2 mitosis phases of the cell cycle as evaluated by propidium iodine (PI) binding (BIRON et al. 1990), and increased yields particularly in the liver (MCINTYRE and WELSH 1986; PIEN and BIRON 2000). The killing and blastogenic responses can be elicited in the absence of infection by treatments with IFN-α/β (BIRON et al. 1984, 1990).

Although a role for IFN-α/β in activating these NK cell responses can and has been defined in a variety of other in vitro and in vivo systems (BIRON et al. 1999), formal characterization of the requirement for IFN-α/β in driving NK cell cytotoxicity and blastogenesis during LCMV infection has never been reported. We have recently used mice blocked in their ability to respond to IFN-α/β as a result of being genetically mutated in their IFN-α/β receptor chain a, IFNRA1, to evaluate the requirement for these cytokines during this infection. As compared to infected wild-type mice, IFN-α/βR-deficient mice are dramatically inhibited in the infection-induced elevation of splenic NK cell-mediated lysis against YAC-1 target cells (Fig. 2A). As the genetic background of these mice is 129, a line not expressing NK1.1, flow cytometric analysis of NK cell subsets requires that the cells be identified as $DX5^+$ and $CD3^-$ or TCR^- cells. The blastogenesis of these populations, evaluated by flow cytometric analysis, is significantly reduced in mice lacking a functional IFN-α/β receptor (Fig. 2B). Thus, the endogenous IFN-α/β response to LCMV infection is required for activation of NK cell cytotoxicity and blastogenesis in vivo.

2.2 Cytokine Production

NK cell production of IFN-γ can be readily detected in serum, spleen and liver during infections of mice with murine CMV (MCMV) (ORANGE and BIRON 1996a; ORANGE et al. 1995b; RUZEK et al. 1997; SALAZAR-MATHER et al. 1998). Peak induction of the NK cell IFN-γ response during this infection is very sharp in serum and spleen, cresting at 36–40h, and are more extended in liver, with production from 36–48h, after infections of C57BL/6 mice (PIEN et al. 2000; SALAZAR-MATHER et al. 2000). The contribution of NK cells to cytokine production can be assessed by early IFN-γ responses in T and B cell-deficient but not NK cell-deficient mice (ORANGE and BIRON 1996b; ORANGE et al. 1995b), and by cytoplasmic staining for

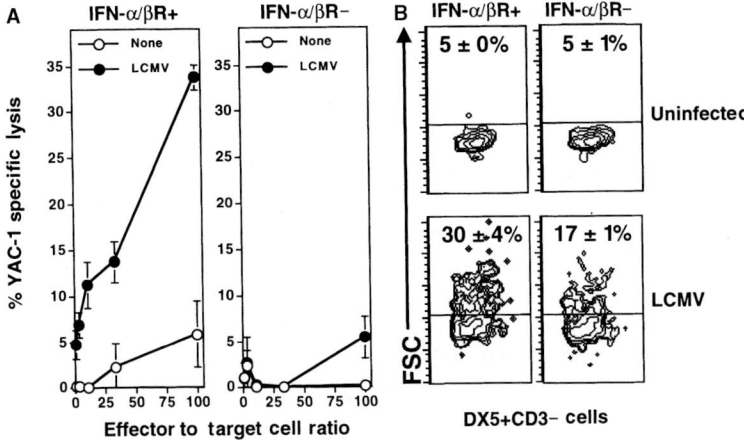

Fig. 2A,B. IFN-α/β requirements for induction of NK cell cytotoxicity and blastogenesis during LCMV infection. Immunocompetent or IFN-α/βR-deficient mice were either uninfected or infected with 2×10^4 plaque forming units (pfu) LCMV-Armstrong for 2 days. **A** Splenic leukocytes were obtained and evaluated for in vitro lysis of ^{51}Cr-labeled YAC-1 target cells at various effector to target cell ratios in a 4.5-h assay. Spontaneous release of 51Cr from target cells was subtracted from all samples prior to calculation of percent lysis relative to maximum release achieved by disruption of target cells with 1% Nonidet P-40. **B** Induction of NK cell blastogenesis was examined by first fluorescently labeling splenic leukocytes with the pan-NK cell marker DX5 and the T cell marker CD3ε. Classical NK cells are defined as $DX5^+CD3^-$. Blasting NK cells were evaluated by gating on NK cells and examining their FSC (size) profile. Results shown are means ± SEM of three mice per group

IFN-γ expression in $NK1.1^+$ and $CD3^-$ or TCR^- and $DX5^+$ and $CD3^-$ or TCR^- populations in immunocomplete and T and B cell-deficient SCID mice (PIEN et al. 2000). The MCMV infection induces expression of biologically active IL-12 which is required for the NK cell IFN-γ response in all three compartments (ORANGE and BIRON 1996a; PIEN et al. 2000). In contrast, it is extremely difficult to detect an NK cell IFN-γ response during LCMV infections. NK cells do not produce detectable levels of IFN-γ during LCMV infections of C57BL/6 mice (ORANGE and BIRON 1996a), and only extremely low levels of early IFN-γ production are detected during LCMV infections of 129 mice (NGUYEN et al. 2000). A low to undetectable IL-12 response accompanies the absence of strong NK cell IFN-γ responses, and poor IL-12 induction is one of the explanations for the lack of the NK cell IFN-γ production. A variety of isolates of LCMV have been examined in our laboratory for induction of early IL-12 and/or IFN-γ, i.e., LCMV-Armstong clone E350, LCMV-WE, LCMV-Armstrong clone 13, and mixed isolates from the tissues of mice failing to control LCMV replication on day 8 of infection as a result of being deficient in the functional IFN-α/βR. None of these has induced an innate IL-12 and NK cell IFN-γ response in immunocomplete mice equivalent to that observed during MCMV infections (Biron laboratory, unpublished results). Thus, although NK cell cytotoxicity and blastogenesis are induced during LCMV infections, there is little or no NK cell IFN-γ response under these conditions.

2.3 Role in Defense or Lack Thereof

Although there is clearly a profound NK cell cytotoxic response to LCMV infection, the cells do not appear to contribute to controlling viral replication in this system (BUKOWSKI et al. 1983). This lack of role is in dramatic contrast to the importance of NK cells in the early defense against MCMV (BIRON et al. 1999; BUKOWSKI et al. 1983; ORANGE et al. 1995b). The difference may be a result, at least in part, of the fact that NK cell IFN-γ production is not significantly induced during LCMV infection (ORANGE and BIRON 1996a). The NK cell IFN-γ response to MCMV is important in mediating defense against this virus, particularly in the liver where NK cells migrate to deliver the cytokine (ORANGE et al. 1995b; SALAZAR-MATHER et al. 1998, 2000). The antiviral effects can be mediated as a result of induction of nitric oxide synthase 2 (NOS2), also known as inducible nitric oxide synthase (iNOS) (TAY and WELSH 1997). The products of enzyme reactions driven by this molecule include NO, and the resulting NO can activate a variety of direct antiviral effects (MACMICKING et al. 1997). However, IFN-γ can also induce a number of immunoregulatory effects, which may contribute to defense. In particular, it is now clear that the local production of IFN-γ by NK cells in MCMV-infected livers promotes the expression of a chemokine identified as the monokine induced by IFN-γ (Mig) (SALAZAR-MATHER et al. 2000). Mig has potent chemotatic function for activated lymphocytes including T cells, and this chemokine is critical for antiviral defense following challenges with high dose MCMV (SALAZAR-MATHER et al. 2000). Thus, the trafficking of NK cell producing IFN-γ to MCMV-infected tissues acts to induce a chemokine, which can function to promote the migration of activated adaptive immune cells. NK cells also accumulate in the livers of LCMV-infected mice, but they are not induced to express IFN-γ under these conditions (ORANGE and BIRON 1996a; PIEN and BIRON 2000). Thus, NK cell migration to livers during LCMV infections is not likely to result in the same downstream consequences as their migration during MCMV infections.

The question remains, however, as to why the induction of NK cell cytotoxic function is not more important in the defense against LCMV. There is evidence in the MCMV system that NK cells may function through a perforin-dependent mechanism to mediate antiviral effects in the spleen (TAY and WELSH 1997). In order for NK cell killing to be effective in controlling a viral infection, virus-infected target cells would have to be sensitive to NK cell-mediated lysis. There is some evidence that MCMV-infected target cells may be more sensitive than LCMV-infected target cells to NK cell-mediated killing. This may be a result, in part, of the fact that LCMV-infected target cells can still access pathways inducing resistance to killing (BUKOWSKI and WELSH 1985). Alternatively or additionally, MCMV infection can result in down-modulation of class I molecules with the potential to deliver negative signals blocking induction of the release of NK cell killing machinery (WIERTZ et al. 1997). Thus, LCMV-infected target cells may escape from, but MCMV-infected target cells may have increased sensitivity to, NK cell lysis. Nevertheless, it is difficult to understand why, if NK cell cytotoxicity is not important in antiviral defense, the response appears to be induced under all conditions of IFN-α/β induction in vivo including LCMV infections. One possibility is that NK cell cytotoxicity has important immunoregulatory functions which remain to be defined. This would be consistent with the recently reported role for the perforin molecule, required for lytic function of both NK and T cells, in protection against the rare and devastating human disease associated with immune disregulation, familial hemophagocytic lymphohistosis (STEPP et al. 1999). Thus, although LCMV infections are potent inducers of NK cell killing, much remains to be learned about the biological importance of this response.

3 Endogenous Innate Cytokine Responses and Functions

As stated above, innate cytokine responses to infections can include the production of a variety of cytokines such as IFN-α/β, IL-12 and/or IFN-γ. In the case of LCMV infections, a prominent production of IFN-α/β is induced with little to no detectable IL-12 and IFN-γ. The type 1 IFNs are comprised of the products of multiple α genes, up to 14, and a single β gene (see BIRON and SEN 2001). This large number of genes is likely to be in place to allow access to type 1 IFNs in response to a variety of stimuli. Although pathways to IFN-α/β induction have not been defined in the LCMV system, those activated by a variety of other stimuli are being identified. They include, but are not limited to, induction by the presence of double stranded RNA in the cytoplasm and by stimulation through cell surface receptors. An amplification loop is in place by which the protein products of the first stimulated IFN-genes act through IFN-α/β receptors to prime neighboring cells to express a wider range of type 1 IFN genes upon infection with virus. Our group has been taking advantage of the characteristics of the high IFN-α/β but low IL-12 and innate IFN-γ responses during LCMV infection to use the system to ask if the type 1 IFNs mediate immunoregulatory effects to inhibit expression of the other cytokines.

3.1 Type 1 Interferon Effects on Interleukin 12 and Innate Interferon Gamma Production

IFN-α/β inhibit the induction of IL-12. The effect was originally tested because of the lack of induction of detectable IL-12 levels during LCMV infections. Studies carried out both in C57BL/6 mice treated with either control antibodies or antibodies neutralizing IFN-α/β functions, and in 129 wild-type (wt) and 129 IFN-α/βR-deficient mice, infected with LCMV for 2 or 3 days have been carried out (COUSENS et al. 1997). In contrast to the undetectable levels in infected normal mice, LCMV-infected mice blocked in IFN-α/β functions are induced to express low levels of circulating IL-12 soon after challenge. Both the inducible p40 chain (see Fig. 3A) and the biologically active p70 heterodimer of IL-12 are induced (COUSENS et al. 1997, 1999). Elevated levels of the cytokine are also observed in MCMV-infected mice having had IFN-α/β functions inhibited, and a replication-independent IL-12 response of murine splenic leukocytes to fixed *Staphylococcus aureus* Cowan strain is inhibited by exposure to exogenously added IFN-α or IFN-β (Fig. 3B) (COUSENS et al. 1997). The effects of IFN-α/β in these systems require high, but physiologically relevant, concentrations of IFN-α/β. They are somewhat "soft" in that the continued presence of the type 1 IFN is required for inhibition (COUSENS et al. 1997). Thus, part of the endogenous immune response to LCMV is the regulation of IL-12 expression. This immunoregulatory function of IFN-α/β has now been demonstrated in a number of different systems including IL-12 responses elicited with human cell populations (KARP et al. 2000; MCRAE et al. 1998).

There is also a second block in the IL-12 to NK cell IFN-γ production response induced during LCMV infections. This is a result of IFN-α/β-mediated inhibition of splenic NK cell responsiveness to IL-12 for IFN-γ production. It is not observed during LCMV infections of IFN-α/βR-deficient mice and can be mimicked in vivo or in vitro by exogenously added IFN-α (NGUYEN et al. 2000). Again, the effects are best observed at high, but physiologically relevant, concentrations of IFN-α/β. They are also apparent as the induction of a refractory response for IFN-γ production by T cells stimulated with antibodies directed against CD3 (NGUYEN et al. 2000). The type 1 IFN induction of unresponsiveness for IFN-γ production appears to be a "hard" block because the cells remain refractive even

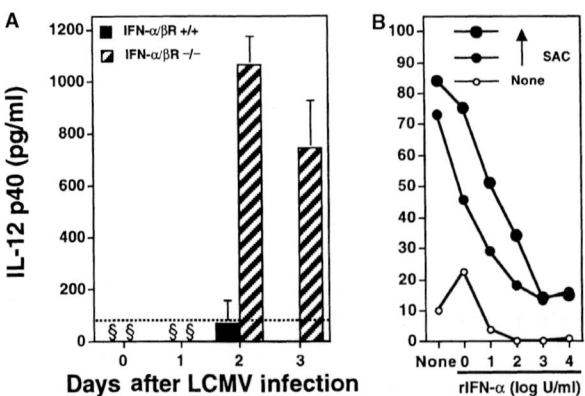

Fig. 3A,B. IFN-α/β-mediated inhibition of IL-12p40 expression during LCMV infection in vivo and during SAC stimulation in vitro. **A** Quantitation of IL-12p40 expression from sera of immunocompetent or IFN-α/βR-deficient mice uninfected or infected for 1, 2, or 3 days with 2×10^4 pfu LCMV-Armstrong. Results shown are means ± SEM of three mice per group. Symbol (§) denotes levels of IFN-γ below the detection limit. **B** Measurement of IL-12p40 after 24-h cocultures of splenic leukocytes left untreated or treated with increasing doses of SAC (indicated by increasing *circle* sizes) and increasing concentrations of rIFN-α A/D. IL-12 p40 was detected by cytokine-specific sandwich ELISAs. Results shown are from experiments with cells pooled from three mice

after they are taken out of the context of high IFN-α/β concentrations. A similar inhibition of T cell responsiveness for IFN-γ production can be observed with cells from patients receiving IFN-β (ROTHUIZEN et al. 1999). Thus, the presence of high levels of IFN-α/β negatively regulates both the induction of IL-12 and IFN-γ production in response to different stimuli including IL-12 during the early phases of LCMV infections. As the combination of these immunoregulatory effects acts to inhibit the alternative innate cytokine responses of IL-12 and NK cell IFN-γ production, the observations suggest that high levels of type 1 IFNs actively inhibit the innate to adaptive immune response linked with Th1 type CD4 T cell responses. Given that LCMV infections are strong inducers of prominent CD8 T cell responses, the high levels of IFN-α/β may set up conditions preferentially driving subset responses uniquely important in the context of viral infections. Thus, the IFN-α/β effects may be important under a variety of conditions associated with high level exposure to these factors, and should be considered in the development of therapeutic applications of the type 1 IFNs.

3.2 Effects on Downstream T Cell Interferon Gamma Responses

Although the accumulating evidence demonstrates that IFN-α/β can act to inhibit IFN-γ responses early during LCMV infections, these cytokines are known to enhance T cell IFN-γ responses, particularly in humans. Early reports from others examining cytotoxic T lymphocyte (CTL) responses in LCMV-infected mice suggested that the protective T cell responses to LCMV underwent exhaustion during IFN-α/βR-deficient mice (GALLIMORE et al. 1998; MULLER et al. 1994). The thought was that the virus failed to be controlled because antiviral effects mediated by the type 1 IFNs are missing in the mice, and that the resulting continuous stimulation of T cells lead to their deletion. However, LCMV is not particularly sensitive to IFN-α/β-mediated inhibition (MOSKOPHIDIS et al. 1994). Thus, the effects on the viral burden resulting from the lack of IFN-α/β function may have been downstream of their immunoregulatory rather than, or in addition to, their antiviral effects. Our group has evaluated the role of IFN-α/β in the development of CD8 T cell expansion and IFN-γ responses during infection with a relatively non-aggressive strain of LCMV, Armstrong clone E350 (COUSENS et al. 1999). These studies show that IFN-α/β functions are not required for the expansion of antigen-specific CD8 T cells, as evaluated by the binding of class I MHC tetramers presenting LCMV peptides or by LCMV peptide stimulation of cytoplasmic IFN-γ expression. However, the conditions are associated with the induction of the alternative innate cytokine, IL-12. If the functions of both innate cytokine responses are blocked, the antigen-specific T cell expansion is still induced during infections, but T cell IFN-γ production is reduced in vivo, as measured by levels in the serum, and in vitro, as measured by the spontaneous production in culture. Thus, two pathways of innate cytokine responses promoting T cell IFN-γ responses can be defined during LCMV infections. The first is in place during infections of normal mice and is dependent on IFN-α/β. In the absence of IFN-α/β functions, an IL-12 response is revealed which can substitute for the effect. The results begin to define the importance of innate cytokines in facilitating CD8 T cell responses to viral infections. It should be noted, however, that as the conditions of extreme antigen burden, resulting from infections with more "virulent" variants of LCMV, are indeed associated with the eventual exhaustion of antigen-specific CD8 T cells (GALLIMORE et al. 1998), the downstream consequences of failing to limit antigen can also result in T cell depletion.

4 Compartmental Issues

Viral infections elicit both systemic and local immune responses in vivo. The orchestration of innate and adaptive immunity across host compartments is crucial to protection, and requires both global and regional mechanisms of control. Evidence is accumulating suggesting the existence of divergent immunoregulation between lymphoid organs, like the spleen, and non-lymphoid organs, such as the liver. The mechanisms for the regulation are likely to benefit the host because they are in place to promote unique functions particular to individual sites while limiting potentially deleterious effects of such responses at other sites As examples, the spleen needs to fight off infections while providing a cytokine milieu capable of supporting the differentiation of the immune response most beneficial for controlling the agent being encountered. On the other hand, defense in solid organs such as the liver is dependent on the migration of activated effector cells induced in lymphoid compartments. The LCMV system has been, and is being, used to characterize certain of the regulatory pathways used in different compartments.

4.1 Natural Killer Cell Accumulation and Cytotoxicity in Peripheral Blood, Spleen, and Liver

LCMV infections induce accumulations of NK-phenotype cells in the liver (McINTYRE and WELSH 1986; PIEN and BIRON 2000). Flow cytometric analyses reveal that proportions of NK cells (defined as NK1.1+TCR-β−) increase up to fourfold in the liver and peripheral blood by day 3 of infection, whereas splenic percentages are relatively constant (Fig. 4A). In vivo treatment with anti-AGM1 antibodies depletes NK cells from these compartments in both uninfected and day 2 infected mice (Fig. 4B). Changes in lytic function can also be observed in the different compartments during LCMV infections. Two days after infection, cytolytic activity against NK sensitive YAC-1 target cells is enhanced by more than twofold in peripheral blood populations, fivefold in the spleen, and nearly tenfold in the liver (Fig. 5). These activities are also abrogated upon depletion with anti-AGM1. Thus, LCMV infections coordinately activate NK cells across a variety

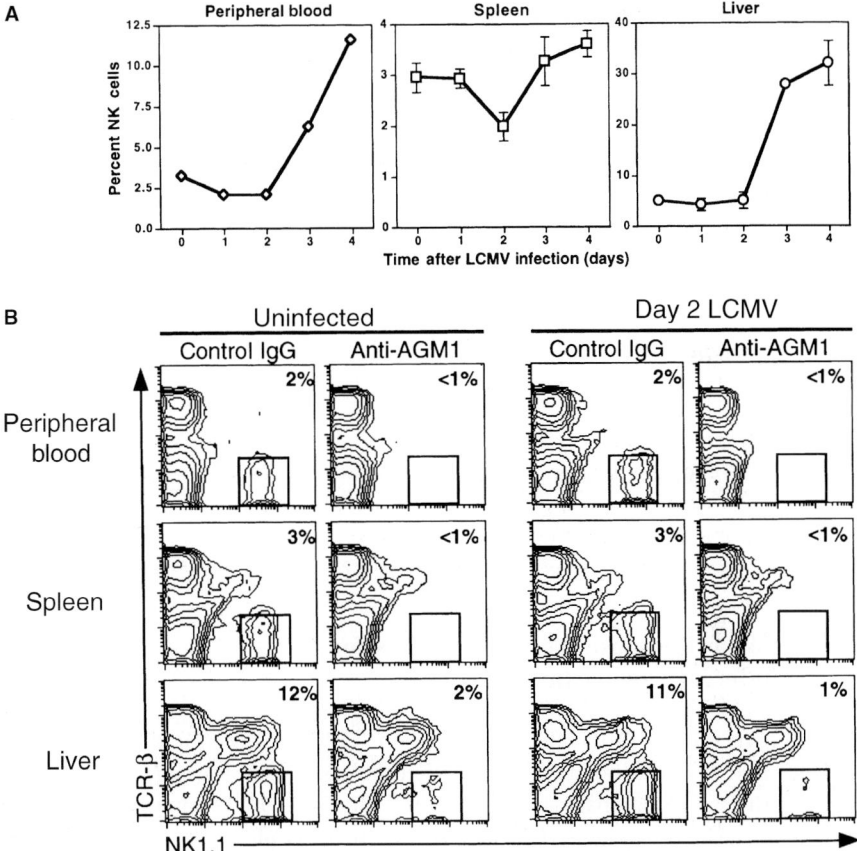

Fig. 4A,B. Compartmental distributions of NK cells during LCMV infection. C57BL/6 mice were infected i.p. with 2×10^4 pfu LCMV Armstrong Clone E350 for the indicated number of days. Peripheral blood mononuclear cells were harvested following density centrifugation on histopaque-1083 gradients. Spleen cells were obtained after mincing of organ and ammonium chloride lysis of erythrocytes. Liver leukocytes were obtained after mechanical and enzymatic (collagenase/DNAse) disruption of tissue, followed by ammonium chloride lysis and density centrifugation on percoll gradients. **A** Proportions of NK cells were determined by flow cytometry after labeling with anti NK1.1 (PK136) and anti TCR-β (H57–597) antibodies, and depicted as means with SEM of three mice per group. **B** To verify the identity of NK cell populations, mice were injected i.p. with control rabbit IgG or anti-AGM1 antibodies 12h prior to LCMV infection. Cells were isolated from the indicated compartments 2 days after infection for flow cytometric analyses. Values represent mean proportions of four mice per group

of host compartments, with pan-access to cytolytic functions. Interestingly, the increases in hepatic NK cell accumulation and cytotoxic activity occur in response to infection by either the hepatotropic variant WE, or the non-hepatotropic Armstrong strain (PIEN and BIRON 2000). Therefore, tropism of the virus does not appear to be a critical determinant in the recruitment of NK cells to the liver.

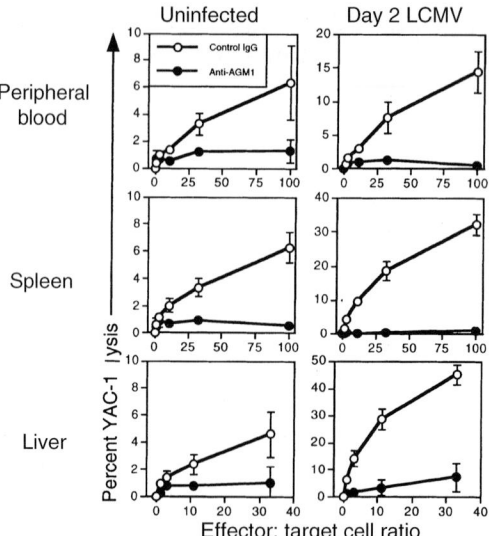

Fig. 5. Induction of NK cell cytotoxic activity after LCMV infection. As per Fig. 4, mice were treated with control or anti-AGM1 antibodies, then left uninfected or infected for 2 days before isolation of effector cells from the indicated compartments. YAC-1 target cells were labeled with ^{51}Cr for use in cytotoxic assays at indicated effector:target cell ratios. Following incubation at 37°C for 4.5h, cells were pelleted and supernatants harvested for determination of radioactivity. Values represent means ± SEM of four mice per group

4.2 Cytokine Responsiveness and Expression in Different Compartments

Despite activation of NK cell cytotoxicity in both the spleen and liver, there is evidence that the accessibility of other NK cell functions are disparately regulated between these two compartments (PIEN and BIRON 2000). Although IFN-γ can exert antiviral effects against LCMV (MOSKOPHIDIS et al. 1994), as stated above, infections by this virus do not elicit an early IL-12 or NK cell IFN-γ response in immunocompetent mice. In an attempt to access NK cell IFN-γ production for early defense against LCMV, our group has administered exogenous IL-12 during LCMV infections of immunocomplete mice (PIEN and BIRON 2000). Delivery of IL-12 after LCMV infection demonstrates the profound refractoriness in the ability of the factor to elicit IFN-γ from splenic leukocytes. However, hepatic leukocytes are not refractory and maintain their ability to produce IFN-γ upon IL-12 administration. These contrasting states between spleen and liver are also observable with in vitro IL-12 stimulation of leukocytes isolated after LCMV infection (PIEN and BIRON 2000). Splenic leukocytes are rapidly inhibited in their ability to respond by day 2 of infection, whereas hepatic leukocytes are not either in vivo (Fig. 6) or in vitro (Fig. 7). The primary IFN-γ producers under these conditions are NK cells. Direct intracellular staining for IFN-γ by flow cytometry demonstrates that both the proportions and intensities of IFN-γ expression by splenic NK cells are diminished in response to IL-12 administration during LCMV infection (PIEN and BIRON 2000). In contrast, hepatic NK cells maintain and/or enhance their level of expression. Taken together, the evidence demonstrates that although NK cells are activated for cytotoxic function in both the spleen and liver and that they are

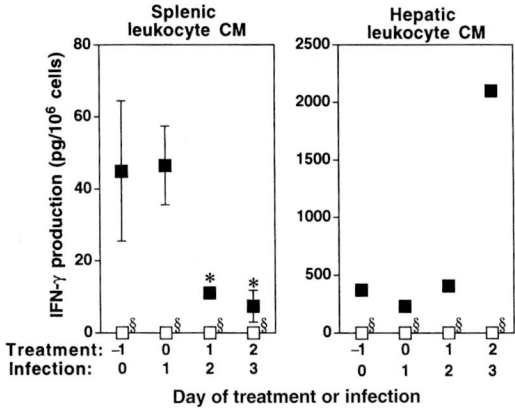

Fig. 6. Contrasting states of in vivo IL-12 responsiveness induced by LCMV infections. C57BL/6 mice were treated with a single i.p. injection of either vehicle control or 1μg IL-12 on indicated days relative to time of infection (denoted by *Treatment* on x-axis). All splenic and hepatic leukocytes were isolated 1 day following vehicle or IL-12 administration, at the indicated day post-infection (denoted by *Infection* on x-axis). Cells were cultured for 24h without additional exogenous stimulation. Conditioned media (*CM*) samples were harvested for determination of IFN-γ levels by sandwich ELISA. Values represent means ± SEM of three spleen samples per group or a single pool of three livers. Symbols denote: §, levels of IFN-γ below the detection limit of 2 and 10pg/10^6 cells for splenic and hepatic CM, respectively; *, statistically significant differences between uninfected and LCMV-infected mice receiving IL-12 ($p < 0.05$). (Adapted from data presented in PIEN and BIRON 2000)

differentially regulated at these sites in their ability to respond to IL-12 for IFN-γ production.

4.3 Exogenous Interleukin 12 to Induce Early Interferon Gamma for Antiviral Function

The virus-induced changes in local IL-12 sensitivity suggest downstream consequences for eliciting IFN-γ-dependent antiviral effects. Both the WE and Armstrong strains of LCMV induce splenic refractoriness with maintained hepatic responsiveness after infection (Fig. 7). Therefore, the contrasting virus-induced alterations in these compartments are not likely to be the result of virus tropism, just as the recruitment and induction of NK cell cytotoxic activity in the liver appear independent of tropism for that compartment. Experiments with Armstrong LCMV demonstrate that daily administration of IL-12 starting 1 day prior to infection could elicit a 2 log decrease in day 3 splenic viral titers as compared to controls. However, if daily IL-12 treatments are not begun until 1 day after infection, the antiviral efficacy of treatment is abrogated (Fig. 8A). Hepatic titers of virus are not detectable under these conditions. Studies with LCMV-WE yield similar results in the spleen, with IL-12 administration effective in reducing viral burden only when begun prior to infection. However, in the liver such treatments elicit a 2 log decrease in viral titers regardless of whether administration was begun

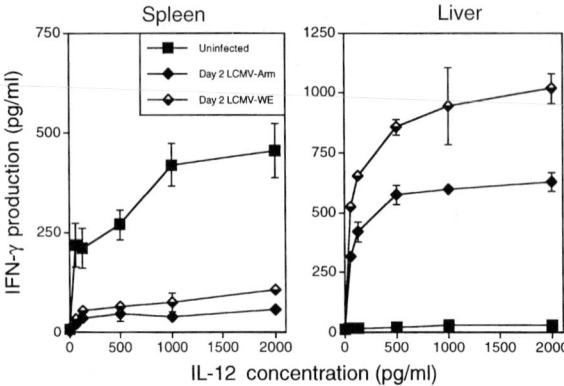

Fig. 7. Contrasting states of in vitro IL-12 responsiveness induced by LCMV infections. Spleen and liver leukocytes were isolated as per Fig. 4 from uninfected C57BL/6 mice or mice infected i.p. for 2 days with 2×10^4 pfu of either LCMV-Armstrong or the hepatotropic variant LCMV-WE. 10^6 cells per well were stimulated in 96-well microtiter plates with indicated concentrations of IL-12 in a final volume of 200µl for 24h at 37°C. Conditioned media samples were harvested for determination of IFN-γ levels by sandwich ELISA. Values represent means ± SEM of four mice per group

1 day prior or 1 day after infection (Fig. 8B). The reductions in viral burden with IL-12 therapy, when effective, are dependent upon IFN-γ in both the spleen and liver (Fig. 8C,D). Thus, the local effectiveness of therapy correlates with the compartmental responsiveness to IL-12 for IFN-γ production (Figs. 6, 7). The results therefore suggest that IL-12 intervention during an ongoing viral infection may be beneficial, but only if particular sites within the host retain sensitivity to IL-12 for mediating antiviral effects and are targets for infection.

Collectively, these and other studies demonstrate that: (1) early IFN-γ expression is negatively regulated at several levels in the spleen, as LCMV infections not only fail to induce an endogenous IL-12 response, but also decrease responsiveness to exogenous IL-12 in the spleen; (2) IL-12 responsiveness is maintained and/or enhanced in the liver, revealing the existence of differential mechanisms for regulating IFN-γ responses in specific compartments; and (3) these putative local regulatory mechanisms may promote beneficial responses at particular sites while limiting deleterious effects in other compartments.

5 Signaling Pathways for IFNs

IFN-α/β binding to their heterodimeric receptors results in rapid activation of the receptor-associated Janus kinase (JAK) 1 and tyrosine kinase (tyk) 2. These activated, i.e., phosphorylated, kinases then recruit the cytosolic factors signal transducer and activator of transcription (STAT) 1 and 2 to the receptor complex. STAT1 and STAT2 subsequently are activated by phosphorylation (see BIRON and

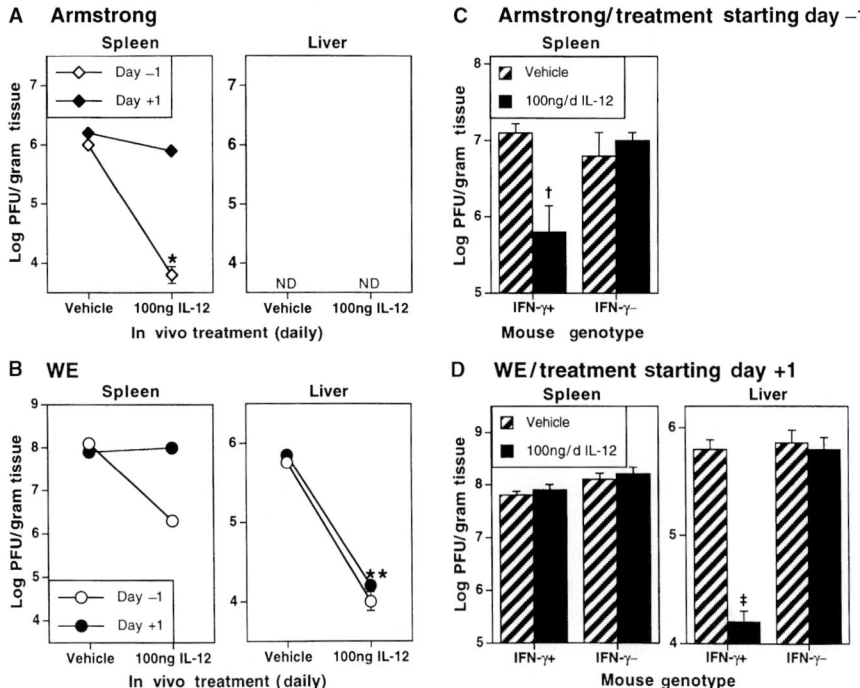

Fig. 8A–D. Effects of IL-12 administration on day 3 viral titers. **A,B** Mice received either vehicle or 100ng IL-12 i.p. daily starting 1 day prior (day −1) or 1 day after (day +1) LCMV-Armstrong (**A**) or LCMV-WE (**B**) infection with 2×10^2 pfu of virus. Three days after infection, spleen and liver samples were harvested for determination of viral titers by plaque assay on Veros cells. **C** Control C57BL/6 or mice genetically deficient in IFN-γ were injected daily with vehicle or 100ng IL-12 starting 1 day prior to LCMV-Armstrong infection with 2×10^2 pfu. Day 3 viral titers were determined in the spleen. Titers of virus in the liver were below the limits of detection by plaque assay. **D** Control or IFN-γ-deficient mice were infected with 2×10^2 pfu of hepatotropic LCMV-WE, and daily vehicle or IL-12 treatments were initiated 1 day after infection. Day 3 viral titers were determined in both spleen and liver. Symbols denote statistically significant differences, as determined by Student's t test, between control- and IL-12-treated groups: *$P = 0.001$, **$P = 0.001$, †$P = 0.02$, ‡$P = 0.001$. (Adapted from data presented in PIEN and BIRON 2000)

SEN 2001 for a detailed review). Activated STAT1/STAT2 heterodimers dissociate from the receptor complex and translocate to the nucleus, and associate with an interferon regulatory factor (IRF) family member, IRF-4/p48, to form a heterotrimeric complex referred to as the interferon stimulatory gene factor 3 (ISGF3). ISGF3 binds to regulatory DNA sequences known as interferon stimulated response elements (ISRE) to modulate gene transcription. In addition to the dominant ISGF3 complex formed in response to IFN-α/β activation, STAT1 homodimers can also be induced by type 1 IFNs. This complex, referred to as IFN-γ activated factor (GAF), binds to regulatory DNA elements known as IFN-γ activated sequences (GAS) to control transcription. Thus, in response to either IFN-α/β or IFN-γ, certain members of the JAK and STAT family are activated to regulate gene transcription. Although these pathways are generally thought to be the predominant factors induced in response to type 1 IFNs, other signaling molecules, including other STATs, have been reported to be induced by IFN-α/β. Moreover, non-STAT signaling pathways, including the phosphatidylionositol 3′-kinase and MAP kinase pathways, can be induced by type 1 IFNs. Thus, the IFN-α/β cytokines, upon engagement with their cellular receptors, have the potential to activate a large number of intracellular signal transduction pathways, and these may be differentially linked to antiviral and/or immunoregulatory functions. However, the contributions and the interactions of each of these pathways to the net IFN-α/β-mediated biological effects are still poorly understood.

The availability of STAT1-deficient cell lines as well as STAT1-deficient mice has allowed the characterization of the role of STAT1 in IFN-α/β-mediated gene regulation. STAT1 molecules, either as part of the ISGF3 or GAF complexes, are required for the transcription of a number genes, including IRF-1, 2′-5′ oligoadenyate (2′-5′A) synthetase, and MxA. There is accumulating evidence, however, that STAT1 also can serve as an inhibitor of gene expression. The IFN-α/β cytokines normally inhibit cellular proliferation in vitro in part by inhibiting the expression of certain genes such as the cellular proto-oncogene c-*myc*. In the absence of STAT1, the inhibitory effects are abrogated (RAMANA et al. 2000). Thus, the picture emerging is that STAT1 can have both positive and negative effects on gene transcription. Hence, IFN-α/β-mediated signal tranduction for biological effects can be mediated through STAT1, either positively or negatively, can be independent of STAT1, or is the net combinatorial result of these pathways. Our group has been evaluating the role of STAT1 in mediating IFN-α/β effects during LCMV infections.

5.1 STAT 1 Effects on Natural Killer Cell Responses

As discussed above, the IFN-α/β cytokines induced early during LCMV infection exert a number immunoregulatory effects on NK cells, including enhancing NK cell cytotoxicity and inducing NK cell blastogenesis and modest proliferation in vivo. As shown in Fig. 9A, STAT1 is required for the IFN-induced NK cell cytotoxicity during LCMV infections. Compared to wt mice, NK cell cytotoxicity against YAC-1 target cells is dramatically reduced in infected mice rendered STAT1 deficient by genetic mutation. The decreased NK cell cytotoxicity in STAT1-deficient mice is consistent with the observation that the perforin gene promoter contains STAT1 binding sites (YU et al. 1999). In contrast, STAT1 does not appear to be required for the induction of NK cell blastogenesis during LCMV infection, as NK cell blastogenesis is induced to similar extents in STAT1-deficient mice compared to their immunocompetent counterparts (Fig. 9B). Thus, during LCMV infections, STAT1 is required for induction of NK cell cytotoxicity but is dispensable for blastogenesis.

Studies of the requirement for STAT1 in the IFN-α/β effects for splenic NK cell IFN-γ production have revealed an interesting negative role for STAT1 as well as the importance of other pathways for type 1 IFN signaling in a biological

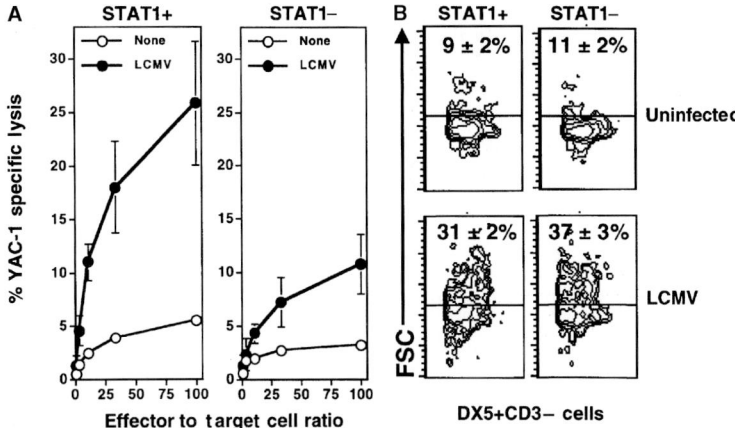

Fig. 9A,B. STAT1 requirements for the induction of NK cell cytotoxicity and blastogenesis during LCMV infection. Immunocompetent or STAT1-deficient mice were either uninfected or infected with 2×10^4 pfu LCMV-Armstrong for 2 days. **A** Splenic leukocytes were obtained and evaluated for in vitro lysis of ^{51}Cr-labeled YAC-1 target cells at various effector to target cell ratios in a 4.5-h assay. **B** Induction of NK cell blastogenesis was examined by first fluorescently labeling splenic leukocytes with the NK cell marker NK1.1 and the T cell marker CD3ε. Blasting classical NK cells were evaluated by gating on NK1.1$^+$CD3$^-$ NK cells and examining their FSC (size) profile. Results shown are means ± SEM of three mice per group

context. In contrast to the splenic leukocytes from day 2 LCMV-infected wt mice (Fig. 10A), which are induced to be refractory to IL-12 for IFN-γ production through IFN-α/β-dependent mechanisms, splenic leukocytes from day 2 LCMV-infect mice rendered STAT1 deficient as a result of genetic mutation are not only not inhibited but also hyper-responsive to IL-12 for IFN-γ production (Fig. 10A). Moreover, high levels of IFN-γ are detected in sera of day 2 LCMV infected STAT1-deficient mice (Fig. 10B), and NK cells are the major contributors to IFN-γ expression under these conditions (Fig. 10C) (NGUYEN et al. 2000). Furthermore, in the absence of STAT1, the IFN-α/β cytokines themselves can induce NK cells to express IFN-γ (NGUYEN et al. 2000). Thus, early during LCMV infection, type 1 IFNs regulate the NK cell response through STAT1-dependent and -independent mechanisms. A significant biological consequence of the presence of STAT1 early after infection is a block in a potential induction of IFN-γ by IFN-α/β. The results suggest that the consequences of cytokine exposure can vary greatly depending upon the ratio of intracellular signaling molecules available to induce the effects.

6 Innate Responses and Disease

There is no evidence to date that the induction of NK cell cytotoxicity during LCMV contributes to disease during LCMV infections. There are, however, indi-

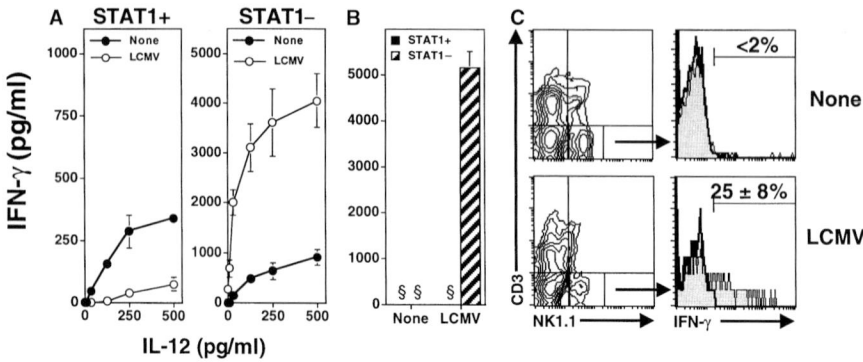

Fig. 10A–C. STAT1-dependent inhibition of NK cell IFN-γ responses in vitro and in vivo. **A** Splenic leukocyte responsiveness to in vitro IL-12 stimulation for IFN-γ production was evaluated in immunocompetent or STAT1-deficient mice uninfected or infected with 2 × 10⁴pfu LCMV-Armstrong for 2 days. Cells were cultured with varying concentrations of IL-12 for 24h, and IFN-γ levels were quantitated by ELISA. **B** Systemic IFN-γ responses following day 2 LCMV infection of immunocompetent or STAT1-deficient mice. Symbol (§) denotes levels of IFN-γ below the detection limit. **C** Intracellular NK cell expression of IFN-γ was evaluated following LCMV infection of STAT1-deficient mice. Splenic leukocytes from uninfected or day 2 LCMV-infected STAT1-deficient mice were obtained, fluorescent labeled with the NK cell marker NK1.1 and the T cell marker CD3ε, and followed by permeabilization and staining with fluorescent-conjugated anti-IFN-γ mAbs. Classical NK cells, e.g., NK1.1 + CD3⁻, were then gated and intracellular IFN-γ expression evaluated within the population and shown as histograms with frequencies of NK cells represented on a linear scale on the y-axes and IFN-γ fluorescence intensity on a log-scale on the x-axes. Results shown are means ± SEM in experiments with three mice per group

cations that the IFN-α/β responses may promote disease under particular conditions. The first indication of a role for type 1 IFNs in disease came from studies demonstrating that expression of these cytokines, during LCMV infections of suckling mice, is associated with the eventual inhibition of growth, liver necrosis, gromerulonephritis, and even death (GRESSER 1982; RIVIERE et al. 1980, 1977). The effects appear to be dependent on exposure during an early developmental stage. By resulting in the concentration of effector cells in organs essential for life, type 1 IFN regulation of viral spread has also been suggested to contribute to the CD8 T cell-dependent death following intra-cranial infections with LCMV (SANDBERG et al. 1994). LCMV-infected adult mice, after having had their IFN-α/β functions blocked by treatments with neutralizing antibodies, have virus dissipating at higher levels and throughout a broader range of target organs. Under these conditions, the mice are protected from the lethal consequences of the infections in the brain. One possible mechanism for the effect is dilution of effector CD8 T cells from the brain to the many other sites of viral infection. Another mechanism by which IFN-α/β may promote disease during viral infections, however, is to sensitize to the toxic effects of exposure to bacterial products such as endotoxin, lipopolysaccharide (LPS) (DOUGHTY et al. 2001). Mice infected with LCMV for 2 days are about twofold more sensitive to LPS-induced lethality, and two- to sixfold more sensitive to LPS for the induction of tumor necrosis factor alpha (TNF-α), than uninfected

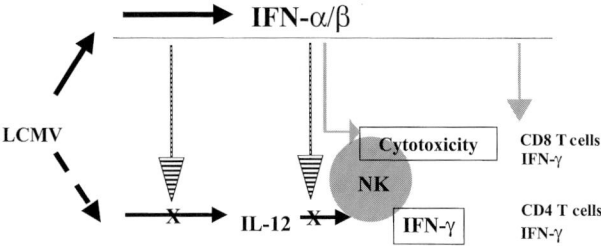

Fig. 11. Model for the mechanisms regulation initial/innate to adaptive immune responses during LCMV infections. *Solid arrows* denote positive effects, whereas *striped arrows* denote negative effects. The differential effects on NK cell responsiveness in the liver are not depicted

mice (DOUGHTY et al. 2001). The effects are dependent on IFN-α/β, but if IFN-α/β functions are blocked and IFN-γ is induced, this cytokine substitutes in sensitization. STAT1 is required for the enhanced responses (DOUGHTY et al. 2001). These observations suggest that the outcome of bacterial infections, in regards to the critical conditions resulting from induction of the pro-inflammatory cytokine cascade, including TNF-α, can be more severe in the context of IFN-α/β induction during a viral infection. Thus, the presence of IFN-α/β may result in increased illness and/or predispose to illness for a variety of reasons.

7 Summary

Although much remains to be learned, the study of early responses to LCMV infections of mice has contributed to the basic understanding of the regulation of a variety of important innate immune responses. Major discoveries have included the appreciation of the levels of type 1 IFNs induced during endogenous responses to viral infections, the importance of IFN-α/β for induction of NK cell cytotoxicity, and the roles for IFN-α/β in regulating the expression of other innate cytokines, i.e., IL-12 and IFN-γ produced by NK cells (Fig. 11). Taken together with the characterization of adaptive responses to LCMV, a paradigm is emerging for a possible initial to innate to adaptive response cascade during infections with viruses eliciting endogenous expression of high levels of IFN-α/β. The results not only advance the understanding of endogenous responses to viral infections and how they are balanced to achieve the best possible outcome for the host, but also give insights into possible consequences of therapeutic intervention with type 1 IFNs.

Acknowledgements. The authors thank their colleagues, past and present, for their research contributions to the material presented here. Work in our laboratory is supported by NIH grants CA41268, AI44644, and MH47674. G.C.P. is the recipient of a predoctoral fellowship from the Howard Hughes Medical Institute. K.B.N. is supported by the NIH training grant ES07272.

References

Bauer S, Groh V, Wu J, Steinle A, Phillips JH, Lanier LL, Spies T (1999) Activation of NK cells and T cells by NKG2D, a receptor for stress-inducible MICA. Science 285:727–729

Biron CA (1999) Initial and innate responses to viral infections–pattern setting in immunity or disease. Curr Opin Microbiol 2:374–381

Biron CA, Nguyen KB, Pien GC, Cousens LP, Salazar-Mather TP (1999) Natural killer cells in antiviral defense: function and regulation by innate cytokines. Annu Rev Immunol 17:189–220

Biron CA, Sen GC (2001) Interferons and other cytokines. In: Knipe D, Howley P, Griffin D, Lamb R, Martin M, Straus S (eds) Fields Virology. Lippincott, Williams & Wilkins, Philadelphia (in press)

Biron CA, Sonnenfeld G, Welsh RM (1984) Interferon induces natural killer cell blastogenesis in vivo. J Leukoc Biol 35:31–37

Biron CA, Welsh RM (1982) Blastogenesis of natural killer cells during viral infection in vivo. J Immunol 129:2788–2795

Biron CA, Young HA, Kasaian MT (1990) Interleukin 2-induced proliferation of murine natural killer cells in vivo. J Exp Med 171:173–188

Bukowski JF, Welsh RM (1985) Inability of interferon to protect virus-infected cells against lysis by natural killer (NK) cells correlates with NK cell-mediated antiviral effects in vivo. J Immunol 135:3537–3541

Bukowski JF, Woda BA, Habu S, Okumura K, Welsh RM (1983) Natural killer cell depletion enhances virus synthesis and virus-induced hepatitis in vivo. J Immunol 131:1531–1538

Cousens LP, Orange JS, Su HC, Biron CA (1997) Interferon-alpha/beta inhibition of interleukin 12 and interferon-gamma production in vitro and endogenously during viral infection. Proc Natl Acad Sci USA 94:634–639

Cousens LP, Peterson R, Hsu S, Dorner A, Altman JD, Ahmed R, Biron CA (1999) Two roads diverged: interferon alpha/beta- and interleukin 12-mediated pathways in promoting T cell interferon gamma responses during viral infection. J Exp Med 189:1315–1328

Diefenbach A, Jamieson AM, Liu SD, Shastri N, Raulet DH (2000) Ligands for the murine NKG2D receptor: expression by tumor cells and activation of NK cells and macrophages. Nat Immunol 1:119–126

Doughty LA, Nguyen KB, Durbin JE, Biron CA (2001) A role for IFN-alpha/beta in virus infection-induced sensitization to endotoxin. J Immunol 166:2658–2664

Gallimore A, Glithero A, Godkin A, Tissot AC, Pluckthun A, Elliott T, Hengartner H, Zinkernagel R (1998) Induction and exhaustion of lymphocytic choriomeningitis virus-specific cytotoxic T lymphocytes visualized using soluble tetrameric major histocompatibility complex class I-peptide complexes. J Exp Med 187:1383–1393

Gresser I (1982) Can interferon induce disease? Interferon 4:95–127

Karp CL, Biron CA, Irani DN (2000) Interferon beta in multiple sclerosis: is IL-12 suppression the key? Immunol Today 21:24–28

MacMicking J, Xie QW, Nathan C (1997) Nitric oxide and macrophage function. Annu Rev Immunol 15:323–350

McIntyre KW, Welsh RM (1986) Accumulation of natural killer and cytotoxic T large granular lymphocytes in the liver during virus infection. J Exp Med 164:1667–1681

McRae BL, Semnani RT, Hayes MP, van Seventer GA (1998) Type I IFNs inhibit human dendritic cell IL-12 production and Th1 cell development. J Immunol 160:4298–4304

Moskophidis D, Battegay M, Bruendler MA, Laine E, Gresser I, Zinkernagel RM (1994) Resistance of lymphocytic choriomeningitis virus to alpha/beta interferon and to gamma interferon. J Virol 68:1951–1955

Muller U, Steinhoff U, Reis LF, Hemmi S, Pavlovic J, Zinkernagel RM, Ague M (1994) Functional role of type I and type II interferons in antiviral defense. Science 264:1918–1921

Nguyen KB, Cousens LP, Doughty LA, Pien GC, Durbin JE, Biron CA (2000) Interferon α/β-mediated inhibition and promotion of interferon γ: STAT1 resolves a paradox. Nature Immunology 1:70–76

Orange JS, Biron CA (1996a) An absolute and restricted requirement for IL-12 in natural killer cell IFN-gamma production and antiviral defense. Studies of natural killer and T cell responses in contrasting viral infections. J Immunol 156:1138–1142

Orange JS, Biron CA (1996b) Characterization of early IL-12, IFN-alpha/beta, and TNF effects on antiviral state and NK cell responses during murine cytomegalovirus infection. J Immunol 156:4746–4756

Orange JS, Wang B, Terhorst C, Biron CA (1995b) Requirement for natural killer cell-produced interferon gamma in defense against murine cytomegalovirus infection and enhancement of this defense pathway by interleukin 12 administration. J Exp Med 182:1045–1056

Pien GC, Biron CA (2000) Compartmental differences in NK cell responsiveness to IL-12 during lymphocytic choriomeningitis virus infection. J Immunol 164:994–1001

Pien G, Satoskar AR, Takeda K, Akira S, Biron CA (2000) Cutting edge: selective IL-18 requirements for induction of compartmental IFN-gamma responses during viral infection. J Immunol 165:4787–4791

Ramana CV, Grammatikakis N, Chernov M, Nguyen H, Goh KC, Williams BR, Stark GR (2000) Regulation of c-myc expression by IFN-gamma through Stat1-dependent and -independent pathways. Embo J 19:263–272

Riviere Y, Gresser I, Guillon JC, Bandu MT, Ronco P, Morel-Maroger L, Verroust P (1980) Severity of lymphocytic choriomeningitis virus disease in different strains of suckling mice correlates with increasing amounts of endogenous interferon. J Exp Med 152:633–640

Riviere Y, Gresser I, Guillon JC, Tovey MG (1977) Inhibition by anti-interferon serum of lymphocytic choriomeningitis virus disease in suckling mice. Proc Natl Acad Sci USA 74:2135–2139

Rothuizen LE, Buclin T, Spertini F, Trinchard I, Munafo A, Buchwalder PA, Ythier A, Biollaz J (1999) Influence of interferon beta-1a dose frequency on PBMC cytokine secretion and biological effect markers. J Neuroimmunol 99:131–141

Ruzek MC, Miller AH, Opal SM, Pearce BD, Biron CA (1997) Characterization of early cytokine responses and an interleukin (IL)-6-dependent pathway of endogenous glucocorticoid induction during murine cytomegalovirus infection. J Exp Med 185:1185–1192

Salazar-Mather TP, Hamilton TA, Biron CA (2000) A chemokine-to-cytokine-to-chemokine cascade critical in antiviral defense. J Clin Invest 105:985–993

Salazar-Mather TP, Orange JS, Biron CA (1998) Early murine cytomegalovirus (MCMV) infection induces liver natural killer (NK) cell inflammation and protection through macrophage inflammatory protein 1alpha (MIP-1alpha)-dependent pathways. J Exp Med 187:1–14

Sandberg K, Kemper P, Stalder A, Zhang J, Hobbs MV, Whitton JL, Campbell IL (1994) Altered tissue distribution of viral replication and T cell spreading is pivotal in the protection against fatal lymphocytic choriomeningitis in mice after neutralization of IFN-alpha/beta. J Immunol 153:220–231

Stepp SE, Dufourcq-Lagelouse R, Le Deist F, Bhawan S, Certain S, Mathew PA, Henter JI, Bennett M, Fischer A, de Saint Basile G, Kumar V (1999) Perforin gene defects in familial hemophagocytic lymphohistiocytosis. Science 286:1957–1959

Tay CH, Welsh RM (1997) Distinct organ-dependent mechanisms for the control of murine cytomegalovirus infection by natural killer cells. J Virol 71:267–275

Welsh RM (1978) Cytotoxic cells induced during lymphocytic choriomeningitis virus infection of mice. I. Characterization of natural killer cell induction. J Exp Med 148:163–181

Wiertz E, Hill A, Tortorella D, Ploegh H (1997). Cytomegaloviruses use multiple mechanisms to elude the host immune response. Immunol Lett 57:213–216

Yu CR, Ortaldo JR, Curiel RE, Young HA, Anderson SK, Gosselin P (1999) Role of a STAT binding site in the regulation of the human perforin promoter. J Immunol 162:2785–2790

Bystander T Cell Activation and Attrition

J.M. McNally and R.M. Welsh

1	Introduction	29
2	Frequencies of Virus-Specific CD8 T Cells	30
3	Alloreactive CTL in Response to LCMV Infection	31
4	Reactivation of Memory CD8 T Cells by Heterologous Viruses	32
5	Cytokine-Induced Bystander Activation of CD8 T Cell Division	33
6	IFN-Induced Bystander T Cell Attrition	34
7	Minimal Proliferation of Bystander CD8 T Cells in Response to LCMV Infection	35
8	Consequences of Bystander Events on T Cells During Virus Infection	36
8.1	T Cell Homeostasis	36
8.2	Immune Deficiency	37
8.3	Autoimmunity and Transplantation Tolerance	37
9	Conclusions	39
	References	40

1 Introduction

Acquired immunity against viruses relies on the activation of virus-specific T cells to combat and clear the virus from the infected host. CD8 T cells are essential to this process, due to their ability to differentiate between self and non-self antigens presented by class I major histocompatibility antigens (MHC I) found on the surface of the majority of nucleated cells (MARGULIES 1999; GERMAIN 1999; AHMED and BIRON 1999). Virus-specific CD8 T cells that encounter viral antigen in the context of the proper MHC I must proliferate, migrate into areas of infection, and perform effector functions, which include the production of interferon gamma (IFN-γ) and the cytolysis of infected cells (AHMED and BIRON 1999). Much of our understanding of this process has been obtained using the lymphocytic choriomeningitis virus (LCMV) system. The immune response to an acute LCMV

Department of Pathology, Program in Immunology and Virology, University of Massachusetts Medical Center, Worcester, MA 01655, USA

infection of mice is associated with a massive expansion of CD8 T cells, fueling a debate about how much of this response is antigen-specific and how much might be a consequence of "bystander" T cell activation.

The issue of bystander activation is complicated in part because the term "bystander" is used to mean different things in different contexts. For the purpose of this review we will first define the term "bystander". We could apply the strict definition that bystander CD8 T cells are not stimulated through their T cell receptor (TCR) and, therefore, are not activated in an antigen-specific manner. The milieu of cytokines and pro-inflammatory signals present during a viral infection may be all that is required to stimulate T cells in an antigen-non-specific manner. It may, however, be best to limit the definition of bystander to T cells not specific for the virus that is driving the T cell response. This is because the term bystander has also been used to describe T cells whose TCRs are stimulated by third party antigens, such as auto-antigens, and in the context of a virus infection there is always a question of whether autologous MHC-peptide combinations can provide some TCR signaling in a virus-induced cytokine milieu. Further complicating this issue are the findings addressing the degeneracy of T cell recognition of peptide-MHC complexes. It has been estimated that a single TCR may have the capacity to recognize up to one million different peptide-MHC combinations, leading to potential cross-reactivity against multiple targets (MASON 1998). Since it is possible that T cells may recognize multiple antigens with differing affinities, cross-reactivities may be difficult to detect and complicate the ability to differentiate between antigen-specific and bystander events (SELIN et al. 1994; NAHILL and WELSH 1993). Recent work has demonstrated that as many as 5% of T cells may express two TCRs, with a common β chain paired with two different α chains (ALAM and GASCOIGNE 1998). These possibilities make the interpretation of whether a T cell is a true bystander or is receiving TCR stimulus extremely difficult.

The term "activation" must also be clarified. T cells can be activated to proliferate, perform effector functions, and eventually to undergo activation-induced cell death (AICD) as the immune response wanes. The term "proliferation" itself has caused much of the current controversy. It is apparent that a strong distinction between proliferation, an increase in cell number, and cell division must be made due to the fact that increased cell division can be offset by increased cell death. We will review much of the work that has looked at bystander activation and attempt to place this work in the context of what is currently known about the number and frequency of virus-specific and non-virus-specific CD8 T cells.

2 Frequencies of Virus-Specific CD8 T Cells

The swelling of the spleen and lymph nodes is a common event during viral infections and is associated with the proliferation of T lymphocytes. In the case of the LCMV infection, the CD8 T cell number can be five to 20 times higher 8 days postinfection than it was prior to infection (MURALI-KRISHNA et al. 1998). Associated with this increase in the number of CD8 T cells is a conversion of the CD4 to CD8 ratio from 2:1 before infection to 1:2–1:3 by 8 days postinfection (VARGA and WELSH 1998). This dramatic increase in the number of CD8 T cells led many to question how many of these cells were LCMV-specific. Early work to address this question utilized limiting dilution analysis (LDA) and single cell cytotoxicity assays to determine the frequency of virus-specific CTL and CTL precursors (CTLp). In many virus systems, very low frequencies (<1 in 1000) of virus-specific CTLp/CD8 cells were seen using these methods. However, in the LCMV system, where as many as 10% of the CD8 T cells were defined as virus-specific, it appeared that the techniques were more optimized for the analysis of antigen-specific T cells (MOSKOPHIDIS et al. 1987). Although studies using LCMV did yield higher frequencies of virus-specific CD8 T cells, the specificity of the other 90% of the CD8 T cells remained in question. This remaining 90% fueled speculation that they were present because of bystander proliferation of non-virus-specific CD8 T cells. However, the bystander proliferation hypothesis had to be reevaluated with the advent of sensitive technologies capable of enumerating antigen-specific T cells. More than 60% of the CD8 T cells present at the peak of an acute LCMV infection were identified as virus-specific by using peptide-loaded MHC tetramer staining and intracellular IFN-γ staining following peptide stimulation (MURALI-KRISHNA et al. 1998; BUTZ and BEVAN 1998; GALLIMORE et al. 1998). Peptide-loaded MHC-IgG dimers also revealed high proportions of viral peptide-specific CD8 T cells (SELIN et al. 1999; GRETEN et al. 1998). Each of these techniques has proven to be highly sensitive for the analysis of virus-specific T cells, and their use was instrumental in determining a more accurate frequency of LCMV-specific CD8 T cells during the acute virus infection. They also revealed that over 10% of the CD8 T cells in the LCMV-immune mouse remain virus-specific long after the resolution of the infection (MURALI-KRISHNA et al. 1998; SELIN et al. 1999). These studies did not, however, rule out the possibility that some of the undefined responding T cells were bystanders.

3 Alloreactive CTL in Response to LCMV Infection

Studies on the specificity of the virus-induced CTL response indicated that acute LCMV infection results not only in the activation of virus-specific CTL (SELIN et al. 1994) but also in the activation of CTL with specificities for other antigens. Infection with LCMV induced allospecific CTL that recognized and lysed targets expressing a wide variety of allogeneic but not syngeneic class I MHC antigens (NAHILL and WELSH 1993). LDA showed that these virus-induced allospecific CTL increased in number during LCMV infection and were found in lymphoblast fractions, suggesting cell division (NAHILL and WELSH 1993; YANG et al. 1989). This superficially may have suggested bystander activation of both their proliferation and effector function, but further study of those alloreactive CTL, using short-term cloning by limiting dilution assays, showed that many LCMV-specific clones isolated from C57BL/6 mice reacted with both virus-infected syngeneic targets and uninfected allogeneic targets (NAHILL and WELSH 1993). Hence, a great deal of the virus-induced allospecific CTL response could be accounted for by T cell cross reactivity, not bystander activation. This result did not preclude the possibility that there was still significant bystander activation of non-cross-reactive allospecific CTL, which are present in immunologically naive mice at relatively high CTLp frequencies. However, Balb/c and C3H mice, which have relatively high CTLp frequencies to H-2^b alloantigens, did not mount an anti-H-2^b allospecific CTL response as a result of an LCMV infection, suggesting that a high frequency of allospecific cells is not sufficient to account for the alloreactivity observed as a result of LCMV infection. C57BL/6 mice normally generate a wide variety of allospecific CTL responses during the LCMV infection, with particularly high responses against H-2^d and H-2^k alloantigens. However, C57BL/6 mice that harbored a restricted T cell repertoire, as a consequence of a HY TCR transgene, mounted an H-2^k but not an H-2^d allospecific response to LCMV, even though the mouse had comparable CTLp frequencies to each alloantigen prior to infection with LCMV (ZAROZINSKI and WELSH 1997). This means that it takes more than a high frequency of allospecific CTLp to activate alloreactivity at a detectable level, and it argues that cross reactivity, not bystander activation, is the major driving force in this process.

4 Reactivation of Memory CD8 T Cells by Heterologous Viruses

The bystander activation hypothesis was initially supported by the finding that heterologous viral infection of LCMV-immune mice with vaccinia virus (VV) resulted in the activation of LCMV-specific CTL effector function, which was found in the lymphoblast fractions (SELIN et al. 1994; YANG et al. 1989). Pichinde virus and murine cytomegalovirus could also activate LCMV-specific CTL in LCMV-immune mice. Similarly, putative bystander activation of effector function was observed using mice transgenic for a TCR specific for the LCMV peptide GP_{33-41}. Splenocytes from GP_{33-41} transgenic mice infected with VV exhibited cytolytic activity against target cells pulsed with the GP_{33-41} peptide (EHL et al. 1997).

This ability of heterologous viruses to reactivate LCMV-specific CTL in LCMV-immune mice was examined further by limiting dilution clonal strategies and unexpectedly came to the conclusion that many T cells cross react between heterologous viruses (SELIN et al. 1994). During viral infections the early phase of the T cell response is associated with the expansion of clones of T cells cross-reactive between the infecting virus and viruses previously encountered by the host. In both the two virus system and the allospecific CTL system, it was shown that the CTL were sometimes less effective at lysing targets infected with the stimulating virus than they were at lysing targets expressing the cross-reactive antigens (YANG and WELSH 1986; NAHILL and WELSH 1993; SELIN et al. 1994). As a consequence, it is easy to misinterpret results and conclude that a bystander event instead of a cross-reactive, antigen-specific event is occurring.

Although much of the seemingly aberrant cytotoxicity induced by a virus infection can be accounted for by cross-reactive responses, the degree of true bystander activation of cytolytic effector function remains unclear. Our preliminary data using the HY-TCR transgenic mouse system suggests that LCMV-infection can activate HY-specific cytolytic activity from T cells bearing the HY TCR transgenes under conditions where there is no proliferation of the transgenic T cells. This cytolytic activation may be independent of cross-reactive mechanisms and may be due to the presence of cytokines that can activate CTL function. This possibility complicates the analysis of bystander activation because the activation of effector function may take place independently of proliferation.

5 Cytokine-Induced Bystander Activation of CD8 T Cell Division

The proximity of bystander T cells to the milieu of cytokines that drive the proliferation of the virus-specific CD8 T cells offered a potential mechanism for their stimulation. Several cytokines, including interleukin 2 (IL-2), IL-15, and type I interferons α and β (IFN) have been suggested to non-specifically stimulate lymphocyte proliferation (BIRON 1998; KE et al. 1998; SUN et al. 1998; TOUGH et al. 1996; ZHANG et al. 1998; EHL et al. 1997). Memory CD8 T cells, defined by surface expression of CD44, can be stimulated in vivo to undergo cell division in response to IFN, which is induced at high levels during LCMV infection. $CD8^+CD44^{hi}$ cells were shown to incorporate bromodeoxyuridine (BrdU) 3 days following LCMV infection or injection with either purified IFN-β or poly I:C, a potent inducer of IFN [percent of $CD8^+CD44^{hi}$ cells that were $BrdU^+$ following poly I:C (80%) or purified IFN (40%)]. This incorporation of BrdU was interpreted to mean that the memory CD8 T cells were proliferating in response to IFN (TOUGH et al. 1996). It was subsequently shown that the effect of IFN may be to stimulate macrophages to produce IL-15, which in turn stimulates the division of memory CD8 T cells (ZHANG et al. 1998). However, in both cases there was little increase in the total number of CD8 T cells. Several studies have shown that memory CD8 T cells undergo a constant low level of cell division (MULLBACHER 1994; RAZVI et al. 1995b; SELIN and WELSH 1997; ZIMMERMAN et al. 1996) such that they maintain a constant frequency in the absence of antigenic challenge. It is thought that IL-15, which can be induced by IFN (ZHANG et al. 1998), maintains this low level of division. High levels of IFN and/or IL-15 induced during virus infection may enhance this cell division, but, surprisingly, do not necessarily result in an increase in the number of memory CD8 T cells. This is consistent with the fact that even though high levels of IFN are induced during LCMV infection, most of the CD8 T cells can be shown to be LCMV-specific (MURALI-KRISHNA et al. 1998; SELIN et al. 1999; BUTZ and BEVAN 1998).

6 IFN-Induced Bystander T Cell Attrition

Our own studies of the effect of IFN in vivo have paradoxically indicated that IFN is pro-apoptotic for CD8 T cells. During the early (day 2–4) phase of the acute LCMV infection of C57BL/6 mice there is a peak in type I IFN production which corresponds to a significant decrease in the number of CD8 T cells (WELSH 1978). At this peak in IFN during LCMV infection there is also a peak in the number of apoptotic cells in the spleen, as determined by in situ TUNEL staining (RAZVI et al. 1995a). Staining of spleen leukocytes with annexin V, an early indicator of apoptosis, reveals that many of the CD8 T cells, particularly those of the memory (CD44) phenotype, are dying. This decrease in cell number precedes the detection of virus-specific cells, suggesting that non-virus-specific "bystander" cells are being eliminated at this early time point. To determine if the CD8 T cell death observed during the early stages of the LCMV infection could be attributed to the effects of IFN, we injected mice with recombinant IFN-β or the IFN-inducer poly I:C and found decreases in the number of memory CD8 T cells in all tested sites (spleen, lymph node, peritoneal cavity, liver, blood, and bone marrow) (WELSH and MCNALLY 1999). CD8 T cells of the memory phenotype stained with annexin V, suggesting that the CD8 T cell loss due to IFN occurred via an apoptotic mechanism. Our hypothesis that the high systemic levels of IFN caused the decrease in CD8 T cell numbers and an increase in their apoptosis was supported by contrasting results with type I IFN receptor deficient mice, which showed an increase in the number of memory CD8 T cells with no increase in apoptosis in response to poly I:C injection (MCNALLY et al. 2001). Certain strains of LCMV can also lead to a more global hematopoetic deficiency by shutting down bone marrow function. This has recently been attributed to the effects of type I IFN (THOMSEN et al. 1986; BINDER et al. 1997).

Our results may appear to contradict earlier findings which concluded that IFN could induce the proliferation of memory CD8 T cells (SUN et al. 1998; TOUGH et al. 1996), but we believe that by way of IL-15, cell division and apoptosis of CD8 T cells are both occurring simultaneously. The consequence is that there is no "proliferation", i.e., an increase in cell number, in the true sense of the word. Rather there is an increase in cell turnover, with apoptotic events offsetting cell division. Many cytokines, including IL-2, have been shown to mediate either cell division or cell death, and the same may be true for IFN and IL-15. Alternatively, IFN may be stimulating apoptosis in bystander memory CD8 T cells, causing a loss in these cells. This cell loss, in the absence of antigen, may signal other homeostatic mechanisms (including IL-15) to repopulate the memory pool to the levels prior to IFN treatment, hence the incorporation of BrdU. Within the context of the LCMV infection, IFN-induced CD8 T cell loss may make room in the lymphoid compartments prior to the massive increase in the number of virus-specific cells. Consistent with this idea are the histological observations that IFN causes major changes in splenocyte architecture (ISHIKAWA and BIRON 1993). In this case, virus-specific antigens drive the repopulation of the T cell pool, favoring virus-specific and/or cross-reactive T cells that expand at the expense of the bystander cells.

7 Minimal Proliferation of Bystander CD8 T Cells in Response to LCMV Infection

What then do we really know about the degree of bystander T cell proliferation during the course of the LCMV infection? As described above, previous studies addressing the proliferation of bystander T cells were complicated in their interpretation by the inability to rule out cross-reactivity in the experimental systems. Zarozinski and Welsh used two systems, HY-transgenic mice and LCMV-carrier mice, to directly address whether non-virus-specific CD8 T cells were proliferating in response to the activation of virus-specific CD8 T cells in the absence of cross reactivity. First, HY mice contain transgenic T cells of a known specificity not cross reactive with LCMV antigens, but do contain a "leaky" population of CD8 T cells that express non-transgenic TCR α genes and proliferate in response to LCMV infection, resulting in the clearance of virus from infected host (ZAROZINSKI and WELSH 1997). Kinetic analysis of LCMV-infected HY mice showed no increase in the number of the HY-specific transgenic T cells, naive or memory phenotype, which express both α and β transgenes. This finding is evidence against bystander proliferation, as the "leaky" population underwent a fivefold increase in cell number, but did not stimulate an increase in the number of bystander transgenic T cells. Comparable results were observed in mice with adoptively transferred ovalbumin-specific transgenic T cells from OT-1 TCR transgenic mice and subsequently infected with LCMV (BUTZ and BEVAN 1998). Second, LCMV-carrier mice (Thy 1.2), persistently infected with the virus and lacking LCMV-specific CD8 T cells as a consequence of immunological tolerance (KING et al. 1992), received adoptive transfers of congenic (Thy 1.1) LCMV-immune splenocytes, providing them with a population of donor-derived, virus-specific $CD8^+Thy\ 1.1^+Thy\ 1.2^-$ cells, which proliferated in response to LCMV, and a well-defined, non-virus-specific population of host-derived $CD8^+Thy\ 1.1^-Thy\ 1.2^+$ cells that could be studied as bystander T cells. This system, examining a more heterogeneous population of bystander cells than the HY transgenic system, confirmed that there was no increase in the number of bystander cells while there was an increase in the number of donor-derived, virus-specific CD8 T cells. Because naive and memory T cells clearly defined as virus-non-specific did not increase in number during the LCMV-induced CD8 T cell response, it was postulated that those cells that had increased in number must be virus-specific (ZAROZINSKI and WELSH 1997).

What then do we know about the fate of clearly defined bystander cells during the course of an acute virus infection regarding their degree of expansion or loss? In the HY TCR transgenic and LCMV carrier models for bystander activation described above, a close scrutiny of the data revealed that bystander CD8 T cells actually decreased in total number (ZAROZINSKI and WELSH 1997). In both systems annexin V staining has revealed dramatic increases in apoptosis in bystander T cells during the peak of the LCMV-immune response. The bystander memory cells may experience an increase in their division rate during infection (TOUGH et al. 1996), but they also appear to experience an increase in their rate of apoptosis with an overall loss in the total number of bystander T cells in the wake of the viral infection (McNALLY et al. 2001).

8 Consequences of Bystander Events on T Cells During Virus Infection

It is clear that events happen to T cells that are not specific for the virus during an infection. As discussed above, the proximity of bystander T cells to the activated virus-specific T cells may lead to both increases in cell division and apoptosis, often causing a decline in number or attrition. The possible implications of these effects are discussed below.

8.1 T Cell Homeostasis

The consequences of attrition of bystander T cells may be twofold. First, bystander attrition may play a necessary role in eliminating cells from lymphoid compartments in order to make room for the expanding virus-specific T cell population. Two to 3 days following LCMV infection there are architectural changes in the spleen that precede the detection of virus-specific T cells which can also be induced by poly I:C and IFN (ISHIKAWA and BIRON 1993). These changes in splenic architecture result in reduced red pulp areas and increased white pulp areas. IFN-induced cell death and concomitant restructuring of the spleen may create the necessary physical space for the substantial increase in the number of CD8 T cells that takes place during the immune response to LCMV (McNALLY et al. 2001). Second, the loss, early in infection, of T cells, especially those of the memory phenotype, is compatible with the findings that a heterologous virus challenge of LCMV-immune mice results in a reduction in the number of LCMV-specific memory CD8 T cells (SELIN et al. 1996, 1999). The memory CD8 T cell response to a virus is very stable (SELIN et al. 1996), but other infections cause substantial losses in the memory pool, as T cells specific for the next virus compete for space with those from the earlier virus (SELIN et al. 1996, 1999). IFN-induced death of some of the original memory pool would allow for more successful competition by the newer memory cells. Because cross-reactive T cells may proliferate more efficiently than bystander cells, over time there may be a selection for CD8 T cells that cross-react with a wide variety of antigens. In both cases, bystander attrition may assist in the maintenance of homeostasis by preventing the memory pool from constantly increasing in number beyond the physical capacity of the host.

8.2 Immune Deficiency

During the course of the T cell response to many acute virus infections, including LCMV, the recall T cell response to other previously encountered antigens and T cell responses to T cell mitogens are greatly suppressed (RAZVI and WELSH 1995). This immune suppression is a transient phenomenon that resolves following the return to homeostasis after the infectious agent is eliminated. This immune suppression may in part be due to the observed attrition of bystander memory T cells, but it is more likely a consequence of increased sensitization to activation-induced cell death (AICD) (ZAROZINSKI et al. 2000). Highly activated T cells express Fas and Fas ligand (FasL), which is up regulated upon TCR stimulation (RUSSELL et al. 1991; RAZVI and WELSH 1995). As a consequence, strong stimulation of virus-induced T cell populations leads to apoptosis, rather than proliferation of T cells. Using the same HY-TCR transgenic and LCMV carrier systems described above, Zarozinski et al. examined the fate of bystander cells to AICD and found that bystander CD8 T cells exposed to the expanding LCMV-specific T cell population in vivo became sensitized to AICD upon triggering of their TCR in vitro (ZAROZINSKI et al. 2000). Transfer of LCMV-immune Thy 1.1^+ T cells into Thy 1.2^+ LCMV-carrier mice caused a sensitization of the Thy 1.2^+ host cells to undergo apoptosis (AICD) when they were stimulated through their TCR with anti-CD3. Immune T cells from mice lacking FasL did not sensitize the bystander cells to AICD, and bystander cells from mice lacking IFN-γ receptors resisted AICD (ZAROZINSKI et al. 2000). Of note is that IFN-γ causes an up regulation of Fas on T cell surfaces (OYAIZU et al. 1994). It therefore appears that FasL and IFN-γ, and probably other factors, sensitize bystander T cells to undergo AICD on antigen engagement. Such a mechanism may explain the immune deficiency seen during many viral infections and during other types of immune responses.

Hence, factors produced during a viral infection not only cause a direct attrition of T cells by apoptosis, but also sensitize the remaining T cells to undergo apoptosis if sufficiently stimulated through their TCR. After a virus infection has terminated there will likely be a reduction in the number of pre-existing bystander memory T cells, but their functions should return to normal preinfection levels after the decline in IFN-γ and FasL.

8.3 Autoimmunity and Transplantation Tolerance

Several studies have been performed to examine the ability of the LCMV infection to abrogate T cell tolerance, resulting in autoimmunity or rejection of tolerized skin allografts. Mice harboring LCMV transgenes in the islets of Langerhans β cells failed to develop diabetes, but on infection with LCMV did so (LEE et al. 1994, 1996; OEHEN et al. 1992). These mice had circulating peripheral T cells capable of responding to the LCMV antigens in the β cells, but because of "ignorance" these T cells remained unactivated and did not cause diabetes. In the presence of LCMV infection these antigen-specific T cells became activated and caused diabetes. This is used as a model to show that if there is some cross reactivity between a virus and a putative auto-antigen, the virus infection may be able to drive T cell responses specific to that antigen and cause an autoimmune disease. Expression of LCMV transgenes in the thymus reduced the available T cell pool and led, on LCMV infection, to diabetes with dramatically reduced kinetics (OEHEN et al. 1992; VON HERRATH et al. 1994). These experiments show that viruses can stimulate otherwise somnolent T cells to cause autoimmunity, but they rely on a cross reactive stimulation rather than an antigen non-specific bystander event to release the autoimmune T cells. Nevertheless, it remains possible that a completely non-cross-reactive T cell stimulated by virus-induced cytokines and third party auto-antigens might be triggered in some cases to cause autoimmune disease.

Virus infections have often been shown to precipitate rejection of allografts in transplant patients. This rejection could either be due to the generation of T cells cross-reactive between the virus and the allograft or a bystander event whereby the virus-induced T cell response elicits factors that activate T cells specific for a third party alloantigen. It has not been clarified which of these possible mechanisms is of greater importance. In recent studies, C57BL/6 mice were tolerized to $H-2^d$ allografts by infusions of Balb/c mouse allogeneic ($H-2^d$) lymphocytes in the presence of antibody to CD40 ligand, which regulates T cell costimulation (WELSH et al. 2000). This technique leads to a considerable cell loss in the number of antigen-specific T cells, as shown in a transgenic T cell system, and the remaining antigen-specific cells appear tolerized (WELSH et al. 2000). LCMV infection at the time of tolerization prevents this loss in transgenic T cells, which instead go up in number. Mice effectively tolerized by this process can then be transplanted with allogeneic Balb/c skin grafts, which will successfully implant and remain intact for many months. However, infection with LCMV during the first month after transplant will cause rejection of the allograft. This rejection is delayed in mice depleted of CD8 T cells (WELSH et al. 2000). Whether this is due to a cross reactive antigen-specific response or to virus-induced factors stimulating the expansion of the small number of T cells responding to third party antigens on the graft is not known. However, some viruses, like LCMV, uniformly cause rejection, whereas other viruses, such as VV, do not (WELSH et al. 2000). The mechanisms behind this need to be clarified, but what is clear is that as this transplantation technique enters the clinic, patients should be shielded from the possibility of acquiring infections with viruses like LCMV that might stimulate allograft-rejecting immune responses.

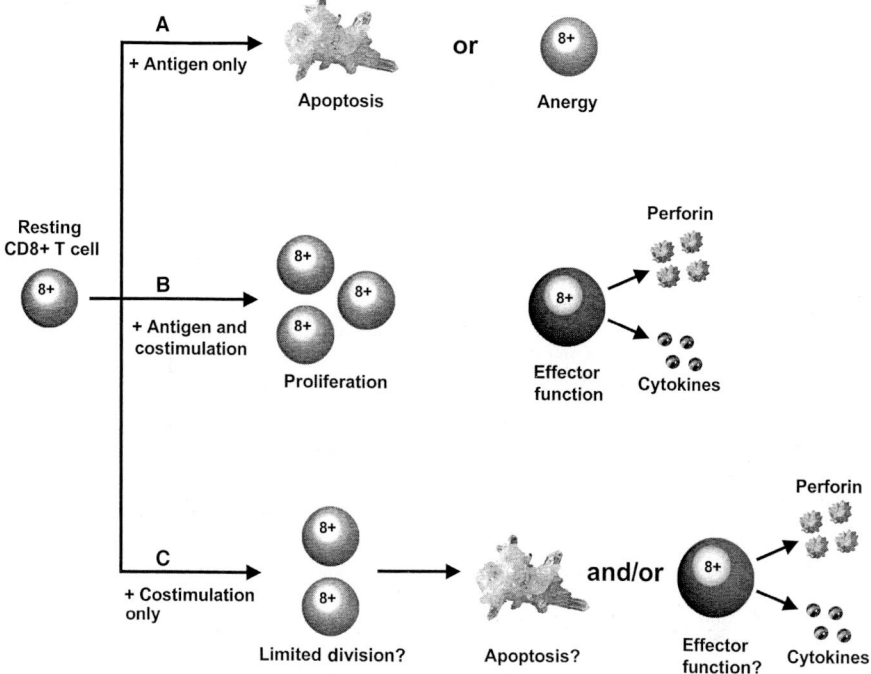

Fig. 1. Model of possible outcomes of T cell stimulation

9 Conclusions

Bystander T cells potentially receive a wealth of costimulatory and cytokine signals in the absence of triggering of their TCR by viral antigens. T cells that receive optimal signaling (antigen + costimulation) become effectors that can participate in the clearance of antigen from the system (Fig. 1B). However, T cells receiving only an antigen-derived signal through their TCR have been shown in other systems to become anergic or to be eliminated in the periphery (Fig. 1A). It is the third pathway (Fig. 1C) which we believe represents possible outcomes for bystander T cells. Bystander CD8 T cells encounter costimulatory signals in the absence of an antigen-derived signal through their TCR. A proportion of these cells are destined to be eliminated by apoptosis, but a limited amount of cell division and effector function may also take place. During the course of a viral infection, this apoptotic pathway may represent a mechanism for making space in lymphoid compartments prior to the expansion of antigen-specific cells and may help select for memory T cells that cross-react with the infectious agent. A memory pool that is highly cross-reactive allows for rapid, high-frequency responses against numerous antigens without the necessity of maintaining large numbers of memory CD8 T cells that would make the maintenance of homeostasis difficult.

Acknowledgements. This work was supported by United States National Institutes of Health training grant AI07272 to J.M.M. research grants AI17672 and AR35506 to R.M.W. and by the Diabetes and Endocrinology Research Core P30 DK32520. The contents of this publication are solely the responsibility of the authors and do not represent the official view of the National Institutes of Health.

References

Ahmed R, Biron CA (1999) Immunity to Viruses. In: Paul WE (ed) Fundamental Immunology. Lippincott-Raven Publishers, Philadelphia, pp 1295–1334

Alam SM, Gascoigne NR (1998) Posttranslational regulation of TCR Valpha allelic exclusion during T cell differentiation. J Immunol 160:3883–3890

Binder D, Fehr J, Hengartner H, Zinkernagel RM (1997) Virus-induced transient bone marrow aplasia: major role of interferon-alpha/beta during acute infection with the noncytopathic lymphocytic choriomeningitis virus. J Exp Med 185:517–530

Biron CA (1998) Role of early cytokines, including alpha and beta interferons (IFN-alpha/beta), in innate and adaptive immune responses to viral infections. Semin Immunol 10:383–390

Butz EA, Bevan MJ (1998) Massive expansion of antigen-specific CD8+ T cells during an acute virus infection. Immunity 8:167–175

Ehl S, Hombach J, Aichele P, Hengartner H, Zinkernagel RM (1997) Bystander activation of cytotoxic T cells: studies on the mechanism and evaluation of in vivo significance in a transgenic mouse model. J Exp Med 185:1241–1251

Gallimore A, Glithero A, Godkin A, Tissot AC, Pluckthun A, Elliott T, Hengartner H, Zinkernagel R (1998) Induction and exhaustion of lymphocytic choriomeningitis virus-specific cytotoxic T lymphocytes visualized using soluble tetrameric major histocompatibility complex class I-peptide complexes. J Exp Med 187:1383–1393

Germain RN (1999) Antigen Processing and Presentation. In: Paul WE (ed) Fundamental Immunology. Lippincott-Raven Publishers, Philadelphia, pp 287–340

Greten TF, Slansky JE, Kubota R, Soldan SS, Jaffee EM, Leist TP, Pardoll DM, Jacobson S, Schneck JP (1998) Direct visualization of antigen-specific T cells: HTLV-1 Tax11–19- specific CD8(+) T cells are activated in peripheral blood and accumulate in cerebrospinal fluid from HAM/TSP patients. Proc Natl Acad Sci USA 95:7568–7573

Ishikawa R, Biron CA (1993) IFN induction and associated changes in splenic leukocyte distribution. J Immunol 150:3713–3727

Ke Y, Ma H, Kapp JA (1998) Antigen is required for the activation of effector activities, whereas interleukin 2 is required for the maintenance of memory in ovalbumin-specific, CD8+ cytotoxic T lymphocytes. J Exp Med 187:49–57

King CC, Jamieson BD, Reddy K, Bali N, Concepcion RJ, Ahmed R (1992) Viral infection of the thymus. J Virol 66:3155–3160

Lee MS, Sawyer S, Arnush M, Krahl T, von Herrath M, Oldstone MA, Sarvetnick N (1996) Transforming growth factor-beta fails to inhibit allograft rejection or virus-induced autoimmune diabetes in transgenic mice. Transplantation 61:1112–1115

Lee MS, Wogensen L, Shizuru J, Oldstone MB, Sarvetnick N (1994) Pancreatic islet production of murine interleukin-10 does not inhibit immune-mediated tissue destruction. J Clin Invest 93:1332–1338

Margulies DH (1999) The Major Histocompatibility Complex. In: Paul WE (ed) Fundamental Immunology. Lippincott-Raven Publishers, Philadelphia, pp 263–286

Mason D (1998) A very high level of crossreactivity is an essential feature of the T-cell receptor. Immunol Today 19:395–404

McNally JM, Zarozinski CC, Lin MY, Brehm MA, Chen HD, Welsh RM (2001) Attrition of bystander CD8 T cells during virus-induced T-cell and interferon responses. J Virol 75:5965–5976

Moskophidis D, Assmann-Wischer U, Simon MM, Lehmann-Grube F (1987) The immune response of the mouse to lymphocytic choriomeningitis virus. V. High numbers of cytolytic T lymphocytes are generated in the spleen during acute infection. Eur J Immunol 17:937–942

Mullbacher A (1994) The long-term maintenance of cytotoxic T cell memory does not require persistence of antigen. J Exp Med 179:317–321

Murali-Krishna K, Altman JD, Suresh M, Sourdive DJ, Zajac AJ, Miller JD, Slansky J, Ahmed R (1998) Counting antigen-specific CD8 T cells: a reevaluation of bystander activation during viral infection. Immunity 8:177–187

Nahill SR, Welsh RM (1993) High frequency of cross-reactive cytotoxic T lymphocytes elicited during the virus-induced polyclonal cytotoxic T lymphocyte response. J Exp Med 177:317–327

Oehen S, Ohashi PS, Aichele P, Burki K, Hengartner H, Zinkernagel RM (1992) Vaccination or tolerance to prevent diabetes. Eur J Immunol 22:3149–3153

Oyaizu N, McCloskey TW, Than S, Hu R, Kalyanaraman VS, Pahwa S (1994) Cross-linking of CD4 molecules upregulates Fas antigen expression in lymphocytes by inducing interferon-gamma and tumor necrosis factor-alpha secretion. Blood 84:2622–2631

Razvi ES, Jiang Z, Woda BA, Welsh RM (1995a) Lymphocyte apoptosis during the silencing of the immune response to acute viral infections in normal, lpr, and Bcl-2-transgenic mice. Am J Pathol 147:79–91

Razvi ES, Welsh RM (1995) Apoptosis in viral infections. Adv Virus Res 45:1–60

Razvi ES, Welsh RM, McFarland HI (1995b) In vivo state of antiviral CTL precursors. Characterization of a cycling cell population containing CTL precursors in immune mice. J Immunol 154:620–632

Russell JH, White CL, Loh DY, Meleedy-Rey P (1991) Receptor-stimulated death pathway is opened by antigen in mature T cells. Proc Natl Acad Sci USA 88:2151–2155

Selin LK, Lin MY, Kraemer KA, Pardoll DM, Schneck JP, Varga SM, Santolucito PA, Pinto AK, Welsh RM (1999) Attrition of T Cell Memory: Selective loss of LCMV epitope-specific memory CD8 T cells following infections with heterologous viruses. Immunity 11:733–742

Selin LK, Nahill SR, Welsh RM (1994) Cross-reactivities in memory cytotoxic T lymphocyte recognition of heterologous viruses. J Exp Med 179:1933–1943

Selin LK, Vergilis K, Welsh RM, Nahill SR (1996) Reduction of otherwise remarkably stable virus-specific cytotoxic T lymphocyte memory by heterologous viral infections. J Exp Med 183:2489–2499

Selin LK, Welsh RM (1997) Cytolytically active memory CTL present in lymphocytic choriomeningitis virus-immune mice after clearance of virus infection. J Immunol 158:5366–5373

Sun S, Zhang X, Tough DF, Sprent J (1998) Type I interferon-mediated stimulation of T cells by CpG DNA. J Exp Med 188:2335–2342

Thomsen AR, Pisa P, Bro-Jorgensen K, Kiessling R (1986) Mechanisms of lymphocytic choriomeningitis virus-induced hemopoietic dysfunction. J Virol 59:428–433

Tough DF, Borrow P, Sprent J (1996) Induction of bystander T cell proliferation by viruses and type I interferon in vivo. Science 272:1947–1950

Varga SM, Welsh RM (1998) Detection of a high frequency of virus-specific CD4+ T cells during acute infection with lymphocytic choriomeningitis virus. J Immunol 161:3215–3218

von Herrath MG, Dockter J, Oldstone MB (1994) How virus induces a rapid or slow onset insulin-dependent diabetes mellitus in a transgenic model. Immunity 1:231–242

Welsh RM (1978) Cytotoxic cells induced during lymphocytic choriomeningitis virus infection of mice. I. Characterization of natural killer cell induction. J Exp Med 148:163–181

Welsh RM, Markees TG, Woda BA, Daniels KA, Brehm MA, Mordes JP, Greiner DL, Rossini AA (2000) Virus-induced abrogation of transplantation tolerance induced by donor-specific transfusion and anti-CD154 antibody. J Virol 74:2210–2218

Welsh RM, McNally JM (1999) Immune deficiency, immune silencing, and clonal exhaustion of T cell responses during viral infections. Current Opinion in Microbiology 2:382–387

Yang H, Welsh RM (1986) Induction of alloreactive cytotoxic T cells by acute virus infection of mice. J Immunol 136:1186–1193

Yang HY, Dundon PL, Nahill SR, Welsh RM (1989) Virus-induced polyclonal cytotoxic T lymphocyte stimulation. J Immunol 142:1710–1718

Zarozinski CC, McNally JM, Lohman BL, Daniels KA, Welsh RM (2000) Bystander sensitization to activation-induced cell death as a mechanism of virus-induced immune suppression. J Virol 74:3650–3658

Zarozinski CC, Welsh RM (1997) Minimal bystander activation of CD8 T cells during the virus-induced polyclonal T cell response. J Exp Med 185:1629–1639

Zhang X, Sun S, Hwang I, Tough DF, Sprent J (1998) Potent and selective stimulation of memory-phenotype CD8+ T cells in vivo by IL-15. Immunity 8:591–599

Zimmerman C, Brduscha-Riem K, Blaser C, Zinkernagel RM, Pircher H (1996) Visualization, characterization, and turnover of CD8+ memory T cells in virus-infected hosts. J Exp Med 183:1367–1375

Immunocytotherapy

D. HOMANN

1	Introduction	43
2	Balancing Antiviral Immunity and Virus Persistence	44
3	Defining Parameters of Acute Infection and its Control	46
3.1	The Role of CD4$^+$ T Cells in Acute LCMV Infection	47
3.2	The Role of B Cells in Acute LCMV Infection	49
4	Persistent LCMV Infection and Tolerance	49
5	Immunocytotherapy of Persistent Infection	50
5.1	Historical Background	50
5.2	Persistent LCMV Infection, Immunocytotherapy, and the CNS	53
5.3	The Limits of T Cell Tolerance in Persistent Infection	54
5.4	The Role of Virus-Specific CD8$^+$ T Cells and Clonal Diversity	54
5.5	The Role of Virus-Specific CD4$^+$ T Cells in Immunocytotherapy	55
5.6	The Role of B Cells and Antibodies in Immunocytotherapy	58
5.7	Other Factors Contributing to Virus Control After Immunocytotherapy	58
6	Other Applications of Immunocytotherapy	59
7	Conclusions	60
References		61

1 Introduction

Ever since the pioneering studies by Mogens Volkert and colleagues (VOLKERT et al. 1962–1967), lymphocytic choriomeningitis virus (LCMV), the prototypic member of the arenaviridae family, has become an important tool to evaluate the feasibility and efficacy of therapeutic immune cell grafting. Subsequent mechanistic and quantitative studies (OLDSTONE et al. 1986; JAMIESON et al. 1987; JAMIESON and AHMED 1988; TISHON et al. 1995; PLANZ et al. 1997; BERGER et al. 2000) have established the LCMV system as a primary immunocytotherapeutic model that continues to generate important insights for clinical applications in many human diseases. The remarkable flexibility of the LCMV system has also allowed for the

---

Division of Virology, Department of Neuropharmacology, The Scripps Research Institute, 10550 North Torrey Pines Road, La Jolla, CA 92037, USA

exploration of different experimental strategies generating a complex of answers that appears appropriate to the multifaceted and intricate nature of diseases whose prevention or treatment may be improved by the adoptive transfer of immunologically active cells.

In the LCMV system, immunocytotherapy, the adoptive transfer of immunocompetent cells into major histocompatibility complex (MHC)-matched hosts, has been used primarily for immune prophylaxis and the treatment of persistent virus infection. More recently, the transfer of dendritic cells intended to generate prophylactic or therapeutic virus-specific immune responses has also been explored. Lastly, immunocytotherapy in the LCMV system has been extended to tumor immunology by use of tumors that express LCMV epitopes and are thus sensitized against LCMV-specific T cells. For the purpose of this review, I will primarily focus on persistent LCMV infection and the adoptive transfer of lymphocytes with curative intent.

2 Balancing Antiviral Immunity and Virus Persistence

To reconcile the many observations made by several groups investigating the efficacy of immunocytotherapy in the LCMV system, it is critical to define the therapeutic parameters as much as the nature of the infection to be treated. Although persistent infection (a potentially profound disturbance of the immune status that may lead to increased morbidity and mortality) is generally considered to be a clinically and immunologically distinct state from immunity (the desired consequence of vaccination or sequela of resolved disease), both persistence and immunity may be regarded as quantitative opposites in the spectrum of possible interactions between virus and host. Even lifelong immunity is not necessarily sterile immunity. In fact, the very existence of sterile immunity to living pathogens may well be questioned. Treatment of LCMV-immune mice with "antilymphocytic serum" [generated by injecting rabbits with complete Freund's adjuvant (CFA) emulsified mouse spleen and lymph node cells] resulted in recrudescence of high virus titers for the duration of immunosuppressive intervention (VOLKERT and LUNDSTEDT 1968). More recently, Ciurea and colleagues have used highly sensitive techniques to detect viral nucleic acids and protein in fully immunocompetent LCMV-immune mice. Furthermore, infectious virus was detected by "exteriorizing" the immunosuppression experiment of Volkert and Lundstedt: adoptive transfer of immune splenocytes into highly susceptible interferon-(IFN) $\alpha/\beta/\gamma$-receptor-deficient mice demonstrated that in the absence of effective immune control in the hosts, infectious virus could replicate to high titers in several of these mice (CIUREA et al. 1999).

While immunity consequently does not rule out the possibility for viral persistence, the inverse conclusion is also valid since high titers of persisting virus do not necessarily lead to complete tolerance, i.e., the absence of virus-specific immune responses. Evidence for specific immunity in LCMV carrier mice comes from the observation that LCMV-specific antibodies, and by extension-specific B cells, are readily detectable in such mice (OLDSTONE and DIXON 1967). In addition, although considered "tolerant" at the T cell level, persistently infected mice do not lose their capacity to "break viral tolerance" and generate protective LCMV-specific $CD8^+$ T cell responses once virus has been cleared from the thymus by immunocytotherapy (JAMIESON and AHMED 1988). Furthermore, peptide immunization in combination with antiviral ribavirin therapy has documented the capacity of mice persistently infected since birth to generate functional $CD8^+$ cytotoxic T lymphocytes (CTL) (VON HERRATH et al. 2000). However, no reduction of virus titers accompanied this immune response, a likely consequence of low CTL numbers and avidities. Thus, quantitative as well qualitative considerations are crucial to define the nature of a viral infection. More precisely, it will be of importance to define the thresholds where quantitative changes will, phenotypically, turn into qualitative differences.

The outcome of a viral infection is the net result of host and viral factors in an equation that still contains poorly defined variables (Fig. 1). Recent technological breakthroughs have considerably advanced our understanding of virus-specific T and B cell responses (MCHEYZER-WILLIAMS and AHMED 1999; DOHERTY and CHRISTENSEN 2000; WHITTON and OLDSTONE 2001). The emphasis here lies on immune *response* since the initiating event, the immediate aftermath of the infectious insult that leads to induction of specific effector responses in the first place, is not yet understood in detail. One of the reasons for the powerful immune responses generated by LCMV may be its capacity to stimulate antigen presenting cells (APCs) by direct infection. However, this process can turn out to blunt or abrogate immune responses if virus-infected APC become targets for virus-specific CTL (LEIST et al. 1988; BORROW et al. 1995). Also, an LCMV strain-dependent differential tropism for particular APC subtypes may impact on the immune response (SEVILLA et al. 2000; SMELT et al. 2001), possibly by selective functional impairment

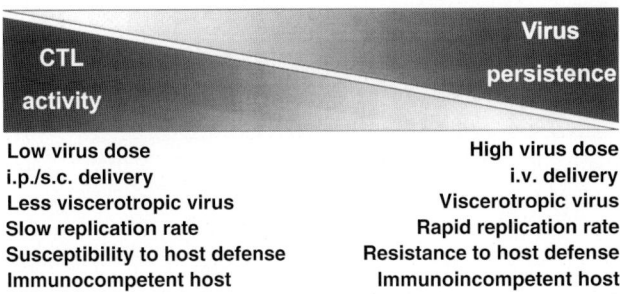

Fig. 1. The concept of virus-immune system balance in lymphocytic choriomeningits virus (LCMV) infected mice. Different factors may operate during acute infection of mice with LCMV. A continuum of host-virus interaction can lead to different states of coexistence between virus and anti-viral immune responses. Under many circumstances, e.g., acute LCMV Armstrong infection of adult immunocompetent mice, effective CTL responses and virus titers are inversely related. *CTL*, cytotoxic T lymphocyte; *i.p.*, intraperitoneal; *i.v.*, intravenous; *s.c.*, subcutaneous. (Modified from BORROW and OLDSTONE 1997)

of APC function. Finally, failure to remove infected APCs may lead to continuous stimulation which may drive immune responses into exhaustion.

Advancing our understanding of how viral tropism, virulence, degree and location of infectious insult (dose and route of inoculation) play a crucial role in shaping the ensuing immune response (whether successful or not), i.e., relating in detail the nature of pathology to the nature and scope of the generated immune response, remains an important goal (Table 1). As a consequence, we should eventually be placed in the position to predict the parameters required for virus control based on a detailed description of the pathology. The work with LCMV so far suggests, in simple terms, that control of a limited infection (e.g., LCMV Armstrong infection of adult immunocompetent mice) requires only limited resources, e.g., $CD8^+$ CTL and perforin (KAGI et al. 1994; WALSH et al. 1994). Resolving more pervasive, protracted or persistent infections, however, may require more complex immune responses including a multiplicity of cell types and effector molecules.

3 Defining Parameters of Acute Infection and its Control

Although the term "acute infection" is commonly used to imply virus exposure, infection and subsequent immunological control, the sequelae of acute virus-host interactions may be of varied nature. Some acute virus infections are effectively controlled by measurable immune responses and disappearing virus titers, others lead to pathology and even death. In the LCMV system, the essential role of $CD8^+$ T cells in both establishing immunity and mediating lethal immunopathology has been extensively reviewed (BUCHMEIER et al. 1980; BORROW and OLDSTONE 1997) and is discussed elsewhere (see Chap. 10). However, under certain circumstances virus clearance is delayed ("protracted infection"), only temporary ("recurrent" or

Table 1. Properties of some commonly used LCMV isolates

LCMV strain	Trophism for lymphohemopoietic tissues[a]	Interferon resistance or susceptibility[b]	Tendency to establish persistent infection in adult mice[c]	CTL exhaustion
Docile	High	R	Greatest	Greatest
Armstrong-clone 13	High	R	↓	↓
Traub	High	R	↓	↓
WE	Moderate	S	↓	↓
Aggressive	Moderate	S	↓	↓
Armstrong	Low (neurotrophic)	S	Least	Least

[a] In vivo trophism of different LCMV isolates for visceral organs, in particular lymphatic tissues vs. central nervous system is indicated.
[b] Replication of the different LCMV isolates is classified as relatively susceptible (S) or more resistant (R) to interferon-α, -β or -γ in vitro and in vivo.
[c] The relative ability of the different LCMV isolates to exhaust the virus specific cytotoxic T cell responses and establish a persistent infection following intravenous inoculation into adult immunocompetent mice is shown (from BORROW and OLDSTONE 1997).

"recrudescent infection"), locally impaired ("sequestered infection") or fails altogether ("persistent" or "chronic infection"). Host and pathogen are thus capable of negotiating various modes of coexistence and the simultaneous presence of detectable virus titers and humoral or cellular immunity is not uncommon in many viral diseases (OLDSTONE 1998; AHMED and CHEN 1999). One of the distinctive advantages of the LCMV system lies in the potential to model these differing degrees of virus persistence as a function of virus isolate, route and dosage of infection, immune status of the host etc. (Fig. 1). Virus infection in utero or at birth represents an especially pronounced immunosuppressive event since it overwhelms the developing immune system by induction of at least partial tolerance to the virus. Keys to understanding the mechanism for control of such a profound infection are found in the recurrent and protracted infections and thus more subtle courses of immune failure observed in adults. Here, the gradual loss of immune control or the eventual clearance of virus allows for an evaluation of individual parameters critical for virus-host interactions. The protracted and recurrent types of acute infections also illustrate that although $CD8^+$ T cells are the cardinal effectors in virus control, other cells are intimately and at times irreducibly involved in the development of effective and lasting immunity.

3.1 The Role of $CD4^+$ T Cells in Acute LCMV Infection

The role of $CD4^+$ T cells in controlling acute LCMV infections illustrates the differential requirements for virus control as a function of specific virus-host interactions. Reports by several groups have demonstrated effective virus control after acute LCMV infection of CD4-depleted mice (MOSKOPHIDIS et al. 1987; LEIST et al. 1987; AHMED et al. 1988; KASAIAN et al. 1991), results that were subsequently corroborated in genetically CD4-deficient mice (RAHEMTULLA et al. 1991; TISHON et al. 1995; VON HERRATH et al. 1996). Although two of these studies documented reduced primary LCMV-specific $CD8^+$ CTL activity (LEIST et al. 1987; AHMED et al. 1988), the biological significance of this finding only became apparent when LCMV isolates were used that cause a protracted but eventually controlled infection in immunocompetent mice. In contrast, inoculation of even temporarily CD4-depleted mice with viscerotropic LCMV strains (including clone 13) caused virus persistence and loss of specific $CD8^+$ CTL memory (MATLOUBIAN et al. 1994). The effect of both virus dosage and strain on immune control was further demonstrated by the capacity of CD4-deficient mice to resolve a low dose infection with LCMV WE whereas inoculation with high dose LCMV WE or low dose LCMV docile caused persistent infection (BATTEGAY et al. 1994). Similarly, MHC class II-deficient mice, which have a pronounced reduction of $CD4^+$ T cells, showed delayed virus clearance after infection with LCMV Traub, impaired memory $CD8^+$ CTL function, eventual CTL exhaustion and virus recrudescence (CHRISTENSEN et al. 1994; THOMSEN et al. 1996). A correlation between MHC class II function and virus control was furthermore attested by the observation that reduced LCMV-specific CTL responses generated after infection of MHC class II-deficient mice with the less lymphotropic WE strain were partially restored in mice with impaired but not abolished assembly and transport of MHC class II heterodimers due to lack of the class II associated invariant chain (BATTEGAY et al. 1996). Nevertheless, even after low dose WE infection, virus eventually reappeared in MHC class II-deficient mice (PLANZ et al. 1997). Finally, careful examination of $CD8^+$ T cell memory generated after LCMV Armstrong infection of CD4-deficient mice demonstrated declining bulk CTL activity and CTL precursor frequencies during the memory phase of the immune response (TISHON et al. 1995; VON HERRATH et al. 1996). Impaired $CD8^+$ memory in CD4-deficient mice vaccinated with vaccinia virus expressing the LCMV nucleoprotein (NP) or glycoprotein (GP) was further shown to result in reduced protection from lethal intracerebral virus challenges (VON HERRATH et al. 1996).

More recently, an elegant study by Zajac and colleagues has analyzed the basis for establishing virus persistence in more detail by using a combination of functional (ELISPOT) and phenotypic (MHC:peptide tetramers) assays to enumerate virus-specific $CD8^+$ T cells (ZAJAC et al. 1998). Mice infected with the t1b or clone 13 LCMV isolates generated but subsequently lost $CD8^+$ T cells specific for the immunodominant $NP_{396-404}$ epitope. However, the mice eventually resolved infection due to $CD8^+$ T cells specific for the codominant GP_{33-41} epitope. Interestingly, only 10%–20% of these cells were functionally active. While infected CD4-deficient mice similarly deleted $NP_{396-404}$-specific $CD8^+$ T cells and retained GP_{33-41}-specific $CD8^+$ T cells at frequencies comparable to wild-type mice, none their GP_{33-41}-specific cells, although highly activated as judged by a surface marker profile of $CD69^{hi}$, $CD44^{hi}$ and $CD62^{lo}$ expression, displayed effector functions and the virus infection was never controlled (ZAJAC et al. 1998). However, progress in the past few years has established APCs as the crucial intermediaries.

The exact molecular basis for CD4:CD8 T cell cooperation remains unclear. However, progress in the past few years has established antigen-presenting cells (APCs) as the crucial intermediaries. Activation or "conditioning" of APCs by $CD4^+$ T cells through CD40:CD40L mediated interactions has been proposed as a mechanism to provide help for $CD8^+$ T cells (SCHOENBERGER et al. 1998; BENNETT et al. 1998; RIDGE et al. 1998). Although CD40L expression is not restricted to $CD4^+$ T cells, lack of CD40L produced a phenotype in the LCMV system reminiscent of $CD4^+$ T cell deficiency. While CD40L-deficient mice controlled an acute infection with LCMV Armstrong or WE (BORROW et al. 1996; OXENIUS et al. 1996), generation of virus-specific $CD8^+$ CTL was impaired resulting in reduced levels of memory T cells (BORROW et al. 1996, 1998). Furthermore, generation of virus-specific $CD4^+$ T cells was significantly impaired leading to decreased protection against rechallenge (WHITMIRE et al. 1999) and after infection with LCMV Traub, impaired T cell responses eventually resulted in recurrence of virus (THOMSEN et al. 1998; ANDREASEN et al. 2000). Other interactions of $CD4^+$ T cells and APCs involving members of the TNF and TNF receptor superfamily such as OX40:OX40L and TRANCE:TRANCE-R have been evaluated in the LCMV system (KOPF et al. 1999; BACHMANN et al. 1999). Although defects have been so far only observed in the generation of specific $CD4^+$ T cell immunity, future research will have to determine whether inhibition of these particular T cell:APC interactions also affect long-term $CD8^+$ T cell immunity.

3.2 The Role of B Cells in Acute LCMV Infection

Similar to $CD4^+$ T cells, the role of B cells and antibody in preventing the establishment of pervasive virus persistence following LCMV infection is a direct function of virus isolate, immune status of the animal and experimental protocol. Here, I will only briefly discuss immune responses in the absence of B cells and antibodies following acute LCMV infection. Neither lethal immunopathology nor primary or secondary $CD8^+$ T cell responses were considerably affected in early experiments using B cell-depleted mice (JOHNSON et al. 1978; CERNY et al. 1986; CERNY et al. 1988; LEHMANN-GRUBE 1987). When tested in B cell-deficient mice (KITAMURA et al. 1991), primary CTL activity and virus clearance kinetics after LCMV Armstrong infection were unaffected as was the establishment and maintenance of $CD8^+$ CTL memory (ASANO and AHMED 1996; HOMANN et al. 1998). Similar observations were made after low dose infection with LCMV WE (BRUNDLER et al. 1996). However, although B cell-deficient mice challenged with high dose LCMV WE generated primary $CD8^+$ T cell responses comparable to wild-type controls, high dose challenges could not be reliably controlled and resulted in virus persistence (BRUNDLER et al. 1996) and even low dose inoculations were subsequently shown to result in late virus recrudescence (PLANZ et al. 1997). Likewise, LCMV Traub infections were only temporarily controlled in the absence of B cells. By the time virus reappeared 3 months after infection, no secondary $CD8^+$ CTL activity was demonstrable (THOMSEN et al. 1996). Adding further complexity to these observations is the finding that $CD4^+$ T cell immunity is compromised in B cell-deficient mice (HOMANN et al. 1998; MACAULAY et al. 1998; VAN ESSEN et al. 2000; BAUMGARTH et al. 2000; LINTON et al. 2000). In the LCMV system this defect resulted in significantly reduced IFN-γ and interleukin-2 (IL-2) production by virus-specific memory T cells generated and maintained in the absence of B cells (HOMANN et al. 1998).

4 Persistent LCMV Infection and Tolerance

Naturally occurring persistent infection was first observed by Traub (TRAUB 1936), who noted the presence of infectious virus in most murine tissues without causing apparent harm. Persistently infected dams transmit the virus in utero and cause a disseminated infection of the fetus (including the fetal thymus, KING et al. 1992). Hotchin and Cintis subsequently demonstrated that such persistent infection can be modeled by inoculating newborn mice with LCMV within the first 24–48h of life (HOTCHIN and CINTIS 1958). As postulated early on, thymic expression of viral antigens should lead to clonal deletion of na LCMV-specific precursor T cells (COLE et al. 1973; CIHAK and LEHMAN-GRUBE 1978; THOMSEN et al. 1979). This hypothesis was formally proven by neonatal LCMV infection of TCR transgenic mice that specifically recognize an MHC class I-restricted LCMV epitope. Drastically reduced numbers of thymocytes and peripheral $CD8^+$ T cells directly demonstrated that virus-specific T cells were clonally deleted (PIRCHER et al. 1989). For a detailed discussion of the biology of persistent LCMV infection, see Chap. 7 of this volume.

The global extent of persistent virus infection has been illustrated by in situ hybridization on whole mouse sections with a LCMV S RNA segment-specific probe (Fig. 2) (SOUTHERN et al. 1986). More recently, a flow cytometry-based method has allowed the detailed phenotyping of individual infected cells of the immune system (Fig. 3) (SEVILLA et al. 2000; D. Homann and M.B.A. Oldstone, unpublished data) thus allowing for the characterization of persistent infection from single cell to whole body. The profound degree of infection makes the LCMV model system particularly suitable for the evaluation of efficacy and mechanisms of immunocytotherapy operative in different tissues.

5 Immunocytotherapy of Persistent Infection

5.1 Historical Background

Between 1962 and 1967, Volkert and colleagues provided a detailed analysis of immunocytotherapy for neonatally infected mice using the LCMV Traub strain

---

Fig. 2A–E. The capacity of immunocytotherapy to terminate a persistent viral infection: removal of infectious virus and viral nucleic acids. **A** Whole body sections of mice persistently infected with LCMV were prepared prior to as well as 15 and 120 days after adoptive immunocytotherapy. 30-µm sections were hybridized with a 545-bp ^{32}P-labelled cDNA probe specific for the glycoprotein gene of LCMV Armstrong as described (SOUTHERN et al. 1984; OLDSTONE et al. 1984). Note the failure of virus clearance following adoptive transfer of MHC-mismatched immune cells (*upper panels*) and virus clearance after transfer of MHC-matched immune cells (*lower panels*). Values indicate titers of infectious virus determined by plaque forming units (*PFU*) per ml of serum. **B,C** Immunohistochemistry for detection of viral nucleoprotein (NP) in the CNS was performed using the monoclonal antibody 1.1.3. to LCMV NP. Panel B demonstrates virus in neurons of a persistently infected mouse. Virus is no longer detectable in neurons 100 days after immunocytotherapy as shown in panel C. **D** Immunocytotherapy is associated with differential virus clearance kinetics from various tissues. Before and 60 or 120 days after adoptive transfer of 5×10^7 LCMV-immune lymphocytes obtained from spleen, infectious virus was determined in the indicated tissues by infectious center assay or plaque assay on Vero cell layers as described (OLDSTONE et al. 1984; TISHON et al. 1993). Note the delayed virus clearance from CNS and kidneys. **E** Clearance of viral nucleic acid sequences from tissues of persistently infected mice. At 137 days after adoptive transfer of LCMV-immune lymphocytes, RNA was extracted from treated mice as well as uninfected (*NIL*) and persistently infected (+) controls and hybridized with a ^{32}P-labeled cDNA-specific LCMV NP probe as described (TISHON et al. 1993). At this time point, viral nucleic acids were only found in the kidneys of one in three immunocytotherapeutically treated recipients

and infectious virus titers as their principal experimental readout. Their findings include the proof of principle for cure of a lifelong persistent LCMV infection by adoptive transfer of spleen and lymph node cells from LCMV-immune donors as

D Detection of infectious virus following immunocytotherapy by infectious center assay or plaque assay

Time after transfer	Infected cells per 10^6 PBL	Log PFU of virus/ml serum of gm tissue					
		Serum	Spleen	Liver	Lung	CNS	Kidney
Day 0	3.2×10^3	4.5	5.9	5.2	4.8	5.2	6.2
Day 60	<1	NIL	NIL	NIL	NIL	3.0	4.1
Day 120	<1	NIL	NIL	NIL	NIL	1.9	3.1

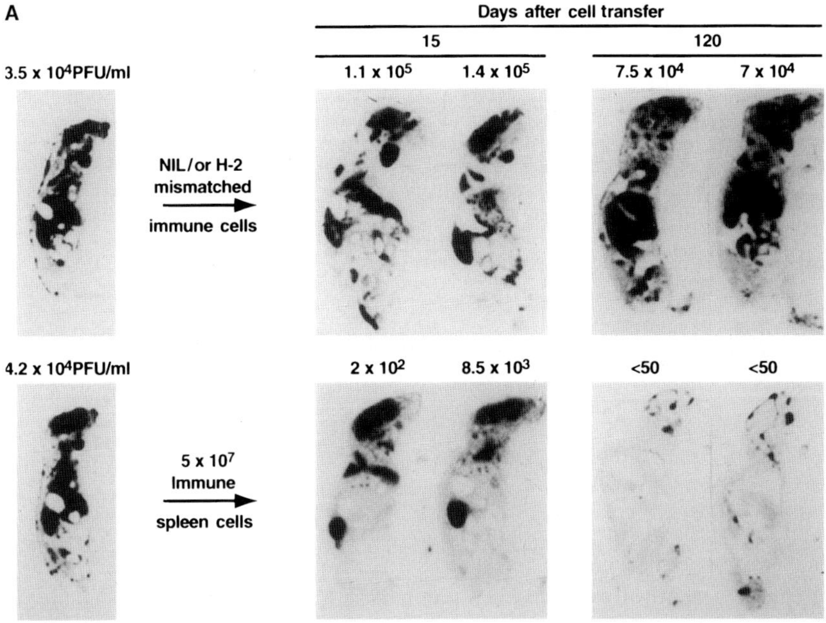

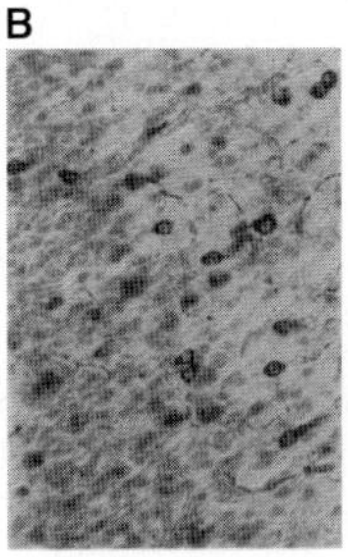

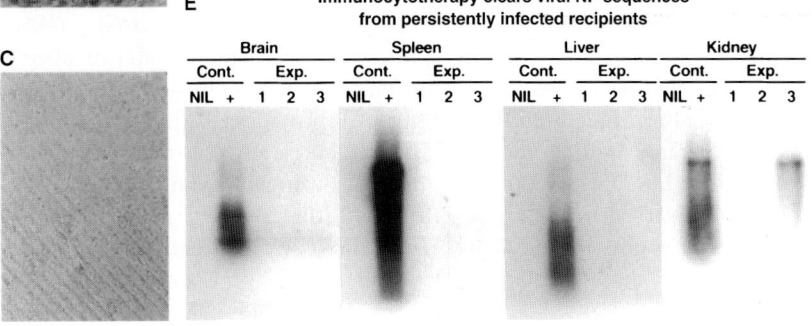

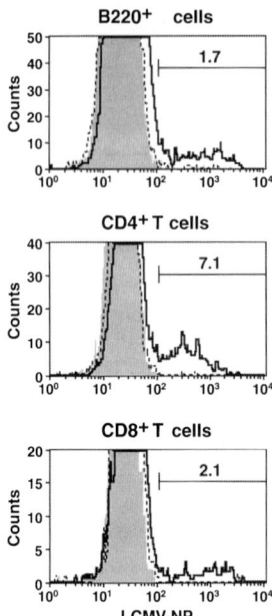

Fig. 3. Detection and enumeration of LCMV-infected lymphocytes by flow cytometry. Spleen cells obtained from uninfected (*gray filled histograms*), LCMV-immune (*broken line histogram*) and persistently infected mice (*solid line histogram*) were stained for B220, CD8 and CD4 followed by intracellular staining for LCMV nucleoprotein (*NP*) using the purified monoclonal 113 antibody specific for LCMV-NP and conjugated to Alexa 488. Numbers indicate the percentage of LCMV-NP positive cells among B cells (*upper panel*), $CD4^+$ T cells (*middle panel*) and $CD8^+$ T cells (*lower panel*). Background staining of uninfected and LCMV-immune controls was consistently <0.1%. Note the preferential infection of $CD4^+$ T cells. This observation is an extension of earlier findings from this laboratory obtained by infectious center assay, immunofluorescence and in situ hybridization (AHMED et al. 1987a; TISHON et al. 1988). The principal advantage of this novel method is the rapid (~4h) and accurate quantitation of virus infection among minor cell population such as rare APC subsets as well as the direct functional comparison of infected and uninfected cells (D. Homann and M.B.A Oldstone, unpublished data)

well as, on occasion, the fatal outcome of such a procedure (VOLKERT 1962a). Optimal vaccination schedule for donors (no multiple or booster vaccinations required), therapeutically effective cell numbers (5×10^7–2×10^8), independence of route of transfer (i.v. or i.p.) and sex matching as well as the superiority of memory ("late") as opposed to effector ("early") cells were also established early on (VOLKERT 1962b). Immunocytotherapy was shown to be effective in different mouse strains and the curative effect was long-lasting in spite of sequestered virus retained in kidneys (VOLKERT and LARSEN 1964). While transfusion with hyperimmune serum proved ineffective, successful virus control following transfer of LCMV-immune cells was shown to be MHC-dependent *avant la lettre* (VOLKERT 1962b; VOLKERT and LARSEN 1965b), i.e., a decade before the LCMV system was used to propose the concept of MHC-restricted CTL activity (ZINKERNAGEL and DOHERTY 1974). Interestingly, even newborn mice could be effectively immunized rather than tolerized with very low doses of virus (VOLKERT and LARSEN 1965a). Finally, it was demonstrated that, in principle, virus elimination can also be achieved with lymphocytes obtained from naive, i.e., not virus-immune donors. However, virus control required the transfer of up to 10^9 cells (in fact more than replacing the entire murine immune system) and was not reliably effective (VOLKERT 1962b; LARSEN and VOLKERT 1967). Further elaboration on "early" and "late" immune cells led to the proposal that effector cells generated after immunization of donors are for the most part terminally differentiated and to be distinguished from those mediating long-term immunity (VOLKERT et al. 1974; JOHNSON and COLE 1975).

5.2 Persistent LCMV Infection, Immunocytotherapy, and the CNS

Oldstone and colleagues demonstrated that removal of viral proteins and nucleic acids after transfer of MHC-matched but not -mismatched LCMV-immune lymphocytes occurred rapidly from most organs (e.g., liver, lung, spleen by day 15) while brain and kidney showed delayed kinetics (day 120 or later, Fig. 2) (OLDSTONE et al. 1986). This study, together with similar observations made by Ahmed and associates (AHMED et al. 1987b), emphasizes the importance of compartmentalization and sequestration of virus and its consequences for immunotherapeutical control. After adoptive transfer of immune cells, the CNS of recipients demonstrated histopathological evidence of gliosis but not, as opposed to other organs, lymphocytic infiltration or structural damage such as neuronal "drop-out" (OLDSTONE et al. 1986). Due to the lack of MHC class I expression by neurons, these cells are not susceptible to direct lysis by LCMV-specific $CD8^+$ CTL (JOLY et al. 1991). Instead, interferons, subsequently shown to be essential for resolving a persistent virus infection (TISHON et al. 1995; PLANZ et al. 1997), may purge virus in a "bystander" fashion. Very recent results, based on the observation of MHC class I induction on electrically silenced neurons (CORRIVEAU et al. 1998), have demonstrated that neurons treated with tetrodotoxin and IFN-γ and sensitized with the LCMV GP_{33-41} peptide can be killed in vitro by specific $CD8^+$ CTL in a Fas/FasL-dependent but perforin-independent pathway (MEDANA et al. 2000). However, details of this unique mechanism for virus control operative in the CNS in vivo remain unclear.

An intriguing possibility, although not experimentally validated in vivo, is the local self-limitation of virus infection. Hotchin has shown that in vitro LCMV infection of tissue culture cells can be cyclical and transient (HOTCHIN 1973, 1974), and de la Torre and coworkers have demonstrated that virus replication is restricted in terminally differentiated neurons in vitro (DE LA TORRE et al. 1993). By removing infectious virus in the periphery, immunocytotherapy may perhaps effectively deprive the CNS from the "reseeding" of virus derived from the periphery. CNS infection would thus cease with abortive replication of LCMV in the neurons.

Finally, functional but not overt anatomical neuronal damage incurred during lifelong exposure to persistent virus is demonstrable in the form of behavioral and cognitive alterations (HOTCHIN and SEEGAL 1977, 1978; GOLD et al. 1994), and is not alleviated after virus clearance following immunocytotherapy (BROT et al. 1997). Since infected neurons are purged from virus rather than destroyed and replaced, extended virus exposure may have introduced subtle but irreversible changes. A correlate for neurocognitive impairments has been demonstrated by altered GAP-43 expression in LCMV-infected neurons in vivo (DE LA TORRE et al. 1996). GAP-43 plays a central role in neuronal plasticity processes accompanying learning and memory. Without affecting synaptic density, GAP-43 expression was shown to be greatly decreased in the hippocampus, an area of heightened viral replication in the CNS of persistently infected mice. Subsequent in vitro studies suggested that LCMV interferes downstream of protein A and C kinases in the nerve growth factor signal transduction pathway (CAO et al. 1997). However, to be sure, the precise mechanism of virus clearance from the CNS still awaits experimental elucidation (TISHON et al. 1993) as does the molecular basis of functional neuronal alterations during persistent LCMV infection.

5.3 The Limits of T Cell Tolerance in Persistent Infection

Jamieson and Ahmed showed that after clearance of LCMV from neonatally infected mice by immunocytotherapy, donor CTL could persist for extended periods of time and eliminate virus after secondary transfer into persistently infected recipients (JAMIESON and AHMED 1989). Furthermore, the cured recipients could mount a host-derived virus-specific CTL response and resist a second LCMV challenge (JAMIESON and AHMED 1988). Interestingly, transferred donor T cells were capable of entering the thymus and clearing the virus from medullary and subsequently cortical regions (GOSSMANN et al. 1991; KING et al. 1992) at a time when replicating virus was still present in brain, kidney and testes (JAMIESON et al. 1991). In spite of sequestered virus replication, host T cell precursors derived from the bone marrow could effectively differentiate into LCMV-specific precursors in the absence of viral antigen in the thymus before exiting into the periphery to participate in the reconstitution of a naive T cell pool capable of developing into effector cells upon suitable stimulation. Thus, immunocytotherapy could successfully reverse specific tolerance to LCMV in the persistently infected hosts (JAMIESON and AHMED 1988; JAMIESON et al. 1991).

5.4 The Role of Virus-Specific $CD8^+$ T Cells and Clonal Diversity

What are the cellular requirements for successful immunocytotherapy? As in acute LCMV infection, virus-specific $CD8^+$ T cells are indispensable for the control of a persistent LCMV infection by adoptive immunotherapy. Depletion of $CD8^+$ T cells prior to adoptive transfer of LCMV-immune lymphocytes into persistently infected recipients completely abrogated the capacity for virus clearance (OLDSTONE et al. 1986; JAMIESON et al. 1987; reviewed OLDSTONE 1987). However, while necessary, specific $CD8^+$ T cells are not sufficient to terminate a persistent infection. Even though transfer of LCMV-specific $CD8^+$ T cell clones could significantly lower virus titers in an acute infection, infusion of 10^7 or more cloned CTL into persistently infected recipients proved fatal (BYRNE et al. 1986). The potential pitfalls of limited diversity were also illustrated by infection of transgenic mice expressing a TCR specific for a dominant LCMV epitope on their $CD8^+$ T cells. All mice mounted LCMV-specific $CD8^+$ CTL after infection with high dose WE but failed to clear virus due to the generation of viral escape mutants (PIRCHER et al. 1990). Furthermore, when naive or memory TCR transgenic $CD8^+$ T cells were transferred into recipients harboring high virus titers, they were rapidly deleted (MOSKOPHIDIS et al. 1993a,b). A recent reevaluation of these deletion kinetics has demonstrated longer survival rates for virus-specific $CD8^+$ T cells (OXENIUS et al. 1998) but it remains clearly evident that such monospecific $CD8^+$ T cell preparations are incapable of eliminating a persistent virus infection. This conclusion is supported by recent mathematical modeling which demonstrated an association between a broad CTL response, low virus load and immunological control of infection. Conversely, failure of virus control was linked to a short-lived CTL response of narrow specificity (WODARZ and NOWAK 2000).

5.5 The Role of Virus-Specific $CD4^+$ T Cells in Immunocytotherapy

In early experiments, depletion of $CD4^+$ T cells prior to adoptive transfer of immune lymphocytes into persistently infected recipients did not significantly affect the efficacy of virus clearance (OLDSTONE et al. 1986; JAMIESON et al. 1987). However, both groups cautioned that due to incomplete depletion techniques some $CD4^+$ T cells were in fact transferred. If $CD4^+$ T cells were thus to play a role in mediating virus clearance, their numerical contribution was expected to be rather small. The use of CD4-deficient LCMV-immune donors eventually demonstrated that $CD4^+$ T cells are in fact indispensable for the control of chronic LCMV infection (TISHON et al. 1995). These results were subsequently confirmed for chronic infections with both WE and Armstrong using advanced cell separation techniques (PLANZ et al. 1997; BERGER et al. 2000). Berger and colleagues demonstrated in addition that the $CD4^+$ T cells supplied in the transferred population had to be specific for LCMV and could not be substituted with $CD4^+$ T cells specific for Pichinde virus, a virus belonging to the arenavirus family but without known cross reactivity at the $CD4^+$ T cell level (Fig. 4). In vivo titration experiments showed that a minimum of 350,000 virus-specific $CD8^+$ and as few as 7,000 virus-specific $CD4^+$ T cells, i.e., specific T cells at a CD8:CD4 ratio of 50:1, were sufficient to mediate rapid and long lasting virus clearance after adoptive transfer into neonatally infected recipients (Fig. 3; BERGER et al. 2000). While any extrapolations to human viral diseases have to be made with caution, it was an intriguing finding that the reported cell numbers (calculated as cells/m² body surface area: $5 \times 10^7/m^2$ virus-specific $CD8^+$ and $1 \times 10^6/m^2$ $CD4^+$ T cells) lie in the range reported for a successful human trial with mixed EBV-specific T cell populations (2 doses of $2 \times 10^7/m^2$ polyclonal T cells at an average CD8:CD4 ratio of 3.7:1, ROONEY et al. 1998) and 20–40 times below the numbers of transferred $CD8^+$ CTL clones in some other trials.

While the requirement of specific $CD4^+$ T cells is by now well established, the mechanism(s) by which they support and maintain $CD8^+$ T cell activity remain poorly defined. A role for CD40L:CD40 interactions has been demonstrated for control of persistent LCMV infection (BORROW et al. 1996) and in vitro experiments suggest that potentiation of cytokine production (IFN-γ and IL-2) between virus-specific $CD8^+$ and $CD4^+$ T cells may be an important factor (BERGER et al. 2000).

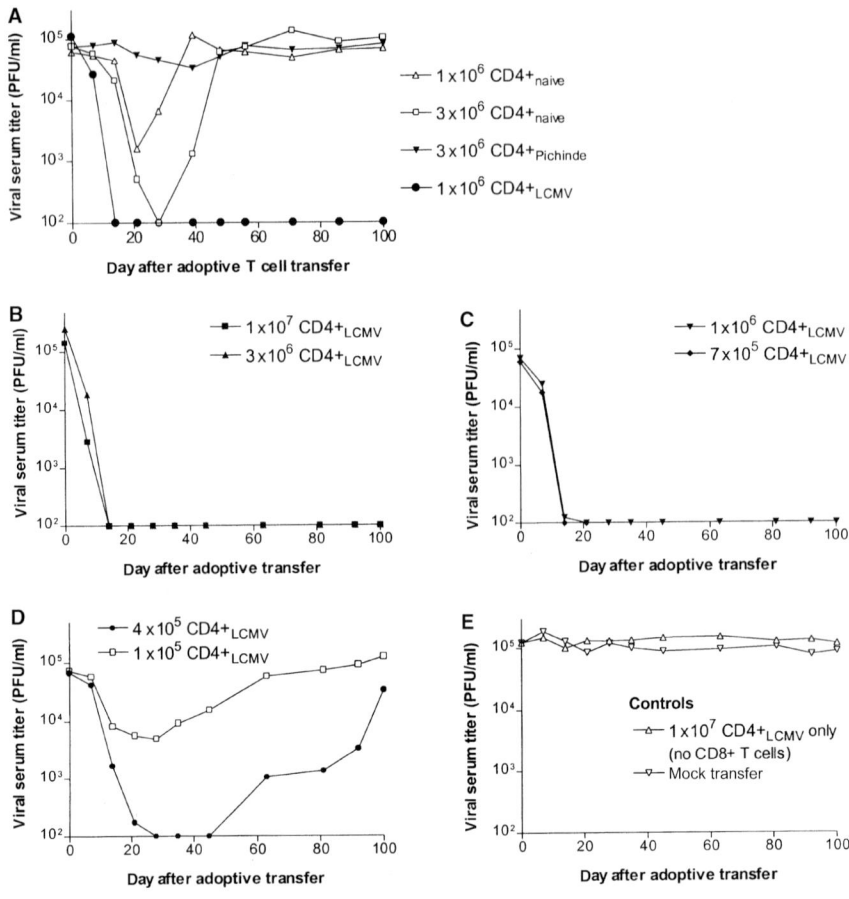

Fig. 4A–E. Qualitative and quantitative contribution of CD8$^+$ and CD4$^+$ T cells to the control of persistent LCMV infection by immunocytotherapy. Initial adoptive transfer experiments with LCMV-immune spleen cells harvested ~60 days after acute infection with 10^5 pfu LCMV Armstrong have shown that unfractioned populations containing at least 3.5×10^6 CD8$^+$ T cells are reproducibly effective in clearing the persistent virus infection (data not shown) **A** Persistently LCMV infected recipients were infused with CD4-depleted LCMV-immune splenocytes populations containing 3.5×10^6 CD8$^+$ T cells and purified LCMV-immune CD4$^+$ T cells (CD4$^+_{LCMV}$) Pichinde-immune CD4$^+$ T cells (CD4$^+_{Pichinde}$) or naive CD4$^+$ T cells (CD4$^+_{naive}$). Virus level in serum were monitored for 100 days following transfer. Only cotransfer of LCMV-primed CD4$^+$ T cells leads to effective and lasting virus clearance. **B–E** Cotransfer of 3.5×10^6 LCMV-primed CD8$^+$ T cells and graded numbers of CD4$^+_{LCMV}$ indicates that at least 7×10^5 CD4$^+$ T cells are required to clear the persistent virus infection. Analysis of LCMV-specific CD8$^+$ and CD4$^+$ frequencies in donor cell preparations by intracellular cytokine staining and ELISPOT demonstrated that ~1/10 CD8$^+$ and ~1/100 CD4$^+$ T cells are specific for LCMV. Thus, a minimum of 350,000 LCMV-specific CD8$^+$ and 7,000 CD4$^+$ T cells (i.e., at a CD8:CD4 ratio of 50:1) are sufficient to mediate virus clearance from persistently infected recipients. Virus clearance in various organs was confirmed at day 100 after adoptive transfer. (From BERGER et al. 2000)

Finally, when persistently infected recipients that had been transfused 60 days earlier with constant numbers of virus-specific CD8$^+$ and graded dilutions of CD4$^+$ T cells were evaluated, it was found that the percentage of functional

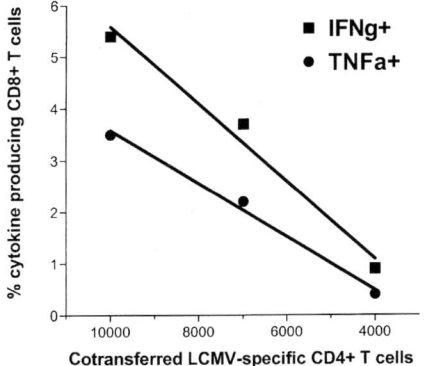

Fig. 5. Maintenance of virus-specific CD8$^+$ T cells by virus-specific CD4$^+$ T cells. Sixty days after cotransfer of 350,000 LCMV-specific CD8$^+$ and indicated numbers of virus-specific CD4$^+$ T cells, frequencies of IFN-γ- and TNF-α-producing virus-specific CD8$^+$ T cells were determined by intracellular cytokine staining (for experimental details see legend to Fig. 4). Results demonstrate that the maintenance of functional virus-specific CD8$^+$ T cells after transfer into a persistently infected recipient is a function of the amount of cotransferred virus-specific CD4$^+$ T cells. Note that cotransfer of 4,000 specific CD4$^+$ T cells, which does not promote virus clearance but leads to temporary virus titer reduction (Fig. 4D, *closed circles*), supports survival of some functional CD8$^+$ T cells. A similar correlation was observed for secondary CTL function and transferred CD4$^+$ T cells (data not shown). (From BERGER et al. 2000)

LCMV-specific CD8$^+$ T cells was directly proportional to the numbers of cotransferred virus-specific CD4$^+$ T cells (Fig. 5) (BERGER et al. 2000). Interestingly, virus-specific cytokine production and CTL activity could be demonstrated in some of the mice that failed to clear the persistent infection indicating that adoptively transferred LCMV-specific CD8$^+$ T cells may not be completely "exhausted" and deleted in the absence of sufficient CD4 help. The survival of functional CD8$^+$ CTL is significantly extended as compared to that of naive TCR transgenic CD8$^+$ T cells which were activated but functionally silenced within 10–15 days after adoptive transfer into persistently infected recipients (OXENIUS et al. 1998). The reason for these differences may be again an advantage of a clonally diverse vs. a monospecific population as well as the transfer of fully functional, committed memory cells vs naive cells that undergo activation, expansion, silencing and eventual deletion in an environment of undiminished high virus titers. Moreover, the cotransfer of 4,000 or less virus-specific CD4$^+$ in addition to CD8$^+$ T cells reported by Berger et al. resulted in a dose-dependent, transitory virus titer reduction and thus at least to a temporary advantage for the survival of CD8$^+$ T cells. Apparently, overwhelming virus titers pose a somewhat less hostile environment for CD4$^+$ than CD8$^+$ T cells since naive TCR transgenic CD4$^+$ T cells specific for the dominant MHC class II-restricted LCMV epitope and transferred into persistently infected recipients retained functionality for up to 6 weeks. Intriguingly, prior to but not after functional silencing, these CD4$^+$ T cells could mediate wasting disease and eventual death after transfer into immunodeficient, LCMV-infected mice (OXENIUS et al. 1998).

5.6 The Role of B Cells and Antibodies in Immunocytotherapy

Since T cells have been implicated early on as mediators of successful adoptive immunization, B cells have received considerably less attention in this experimental setting. Early studies suggested that LCMV-specific antibodies were incapable of controlling persistent infection and could only temporarily reduce virus titers (VOLKERT 1962b; VOLKERT and LARSEN 1965b). In fact, removal of B cells prior to adoptive transfer did not affect the kinetics or efficacy of virus clearance in the recipients (VOLKERT et al. 1975; JAMIESON et al. 1987; HOMANN et al. 1998). In contrast, a recent study has argued that B cells are crucial for virus control due to their capacity for generating neutralizing antibodies (PLANZ et al. 1997). Since neutralizing LCMV-specific antibodies appear relatively late after acute LCMV infection of donors, this study offered an explanation for the observation that "late" (i.e., memory) lymphocytes are more efficient than "early" (i.e., effector) lymphocytes in mediating virus clearance. Although it was subsequently shown that lymphocytes from LCMV-immune B cell-deficient donors failed to cure persistently infected recipients, reconstitution of immune T cells derived from B cell-deficient mice with immune B cells or immune serum equally failed to control virus (HOMANN et al. 1998). Rather than to the absence of B cells, therapeutical failure could be attributed to impaired T cell priming in a B cell-deficient microenvironment resulting in subtle T cell defects that came to the fore only under the pronounced demands put on donor cells in an immunocytotherapeutic setting. In particular, $CD4^+$ T cell function, which is critical for the control of persistent virus infection as discussed above, has been shown to be impaired in B cell-deficient mice in the LCMV model as well as in other systems (HOMANN et al. 1998; MACAULAY et al. 1998; VAN ESSEN et al. 2000; BAUMGARTH et al. 2000; LINTON et al. 2000).

5.7 Other Factors Contributing to Virus Control After Immunocytotherapy

Cytokines are powerful mediators of antiviral immunity (GUIDOTTI and CHISARI 1999; SLIFKA and WHITTON 2000) and LCMV-specific effector and memory $CD8^+$ and $CD4^+$ T cells are an important source for interferon gamma (IFN-γ) and tumor necrosis factor alpha (TNF-α) (MURALI-KRISHNA et al. 1998; BERGER et al. 2000). While the inability of perforin-deficient mice to control an LCMV infection (KAGI et al. 1994; WALSH et al. 1994) indicates that cytokines alone are not sufficient to clear the virus, their role in LCMV control is dependent on the nature of the infection. An acute infection with LCMV Armstrong could be resolved in the absence of IFN-γ or TNF-α as studies using antibodies to deplete these cytokines have shown (KLAVINSKIS et al. 1989; LEIST and ZINKERNAGEL 1990). These results were subsequently confirmed in IFN-γ-receptor- and IFN-γ-deficient mice (HUANG et al. 1993; TISHON et al. 1995). However, in mice that deleted LCMV-specific high avidity CD8$^+$ T cell precursors due to thymic expression of the LCMV nucleoprotein, virus control by peripheral low avidity T cells relied on mechanisms involving IFN-γ (VON HERRATH et al. 1997). In addition, IFN-γ-deficient mice infected i.v. with LCMV Traub were unable to control the virus and succumbed to an immunopathologically mediated wasting disease (NANSEN et al. 1999). Very recent results by the latter group of researchers suggest that LCMV Armstrong may also fail to be cleared in IFN-γ-deficient mice. Interestingly, the ensuing virus persistence did not lead to CD8$^+$ CTL exhaustion since memory CTL remained detectable at high levels for extended periods of time (BARTHOLDY et al. 2000). IFN-γ becomes unequivocally essential for immunocytotherapeutical control of a persistent infection as shown by use of IFN-γ-deficient donors (TISHON et al. 1995) or IFN-γ-receptor-deficient recipients (PLANZ et al. 1997). It should be noted that different organs appear to have differential susceptibility to the antiviral effects of IFN-γ. Guidotti and colleagues have shown that persistent LCMV can be purged from hepatocytes but not splenocytes by the activity of IFN-γ alone. IFN-γ is thus necessary but not sufficient to eradicate a persistent LCMV infection (GUIDOTTI et al. 1999). The study by Planz et al. also indicated a role for type I interferons in adoptive immunotherapy since persistently infected mice deficient for IFN-α/β-receptors could not be cured by immunotherapy (PLANZ et al. 1997). This finding is in agreement with earlier reports that have shown an increased susceptibility to acute LCMV infection in such mice (MULLER et al. 1994).

6 Other Applications of Immunocytotherapy

The use of dendritic cell (DC) transfers to induce immunity has only recently been explored in the LCMV system. A series of studies by Ludewig and coworkers has demonstrated that DC loaded with LCMV-derived peptides and adoptively transferred into a naive host can be used to induce protective T cell immunity (LUDEWIG et al. 1998a). Apparently, this form of immunization requires repeated restimulation, i.e., additional adoptive DC transfers, for optimal maintenance (LUDEWIG et al. 1999). Furthermore, it was shown that the LCMV-specific T cell response induced by repeated DC transfer can lead to autoimmunity in the LCMV-transgenic mouse model for type 1 diabetes (LUDEWIG et al. 1998b). This model was also employed to demonstrate the potential risks of anti-tumor DC therapy if targeted tumor antigens are shared by host cells. LCMV-specific T cells were generated by DC therapy to eradicate a tumor expressing LCMV antigens. However, cross-reaction with an LCMV transgene expressed by pancreatic beta-cells led to their destruction and clinically overt diabetes (LUDEWIG et al. 2000). Further studies with LCMV epitope-expressing tumors and DC therapy have also suggested that immune responses to multiple rather than single epitopes are important for effective anti-tumor immunity (RAWSON et al. 2000).

7 Conclusions

Exploring the mechanisms required for virus control of a persistent LCMV infection by immunocytotherapy is a study in extremes. LCMV infection in utero or of neonates leads to a chronic infection that affects virtually all tissues tested. The immune system itself is not spared (Fig. 3). Although not more than $\sim$2%–3% of immune cells carry viral materials in peripheral lymphoid organs, LCMV has a pronounced tropism for APCs (SEVILLA et al. 2000) and the degree of infection can exceed 50% in some APC subsets. Furthermore, infection rates may vary in different anatomical sites. For example $CD4^+$ T cells, the major lymphocyte population affected in persistently infected mice (AHMED et al. 1987a; TISHON et al. 1988), carry detectable viral antigen in $\sim$5% of peripheral but up to 20% of bone marrow $CD4^+$ T cell populations (D. Homann and M.B.A. Oldstone, unpublished data). Immunocytotherapy using LCMV-immune splenocytes and/or lymph node preparations are remarkably effective and efficient in removing the vast majority of the virus without causing pronounced collateral damage. The LCMV system thus represents an ideal model to test hypotheses, analyze mechanisms and establish guide lines for the immunocytotherapy of persistent virus infections in humans.

As discussed in this chapter, dissection of the mechanisms contributing to viral clearance has to take into account a wide variety of specific virus-host interactions. The emerging theme from the studies discussed here is an imperative on diversity. The immune system has evolved to comprise "multiple arms" of defense which we have come to differentiate conceptually into innate vs adaptive, cellular vs. humoral responses etc. The chances of curing virus persistence are greatly increased if multiple arms of the immune system are engaged. Virus-specific $CD8^+$ T cells are potent effectors that can efficiently destroy or purge virus-infected cells and thus control cell to cell spread of infection. Neutralizing antibodies on the other hand have the capacity to rapidly neutralize virus without the necessity of antigen-presentation, cell proliferation or activation. And virus-specific $CD4^+$ T cells are essential for both of these effector functions to unfold their full potential. Remarkably, relatively few specific $CD4^+$ T cells appear to be sufficient (and necessary) to support efficient and effective virus clearance. The diversity on the cellular level, i.e., $CD8^+$, $CD4^+$ and B cells/antibodies, is echoed at the molecular level. TCR and BCR polyclonality instead of monoclonality and an array of effector functions instead of monofunctionality appear to be the critical ingredients for successful immunocytotherapy. Thus, the diversity of transferred immune lymphocyte populations necessitated by MHC polymorphism in an outbred population such as humans finds its correlate in the composition of the immune cells to be transferred: diverse, multifunctional, polyspecific.

References

Ahmed R, King CC, Oldstone MB (1987a) Virus-lymphocyte interaction: T cells of the helper subset are infected with lymphocytic choriomeningitis virus during persistent infection in vivo. J Virol 61:1571–1576

Ahmed R, Butler LD, Bhatti L (1988) T4+ T helper cell function in vivo: differential requirement for induction of antiviral cytotoxic T-cell and antibodyresponses. J Virol 62:2102–2106

Ahmed R, Jamieson BD, Porter DD (1987b) Immune therapy of a persistent and disseminated viral infection. J Virol 61:3920–3929

Ahmed R, Chen ISY (1999) Persistent viral infections. John Whiley & Sons Ltd., West Sussex, England

Andreasen SO, Christensen JE, Marker O, Thomsen AR (2000) Role of CD40 ligand and CD28 in induction and maintenance of antiviral CD8+ effector T cell responses. J Immunol 164:3689–3697

Asano MS, Ahmed R (1996) CD8 T cell memory in B cell-deficient mice. J Exp Med 183:2165–2174

Bachmann MF, Wong BR, Josien R, Steinman RM, Oxenius A, Choi Y (1999) TRANCE, a tumor necrosis factor family member critical for CD40 ligand-independent T helper cell activation. J Exp Med 189:1025–1031

Battegay M, Moskophidis D, Rahemtulla A, Hengartner H, Mak TW, Zinkernagel RM (1994) Enhanced establishment of a virus carrier state in adult CD4+ T-cell-deficient mice. J Virol 68:4700–4704

Battegay M, Bachmann MF, Burhkart C, Viville S, Benoist C, Mathis D, Hengartner H, Zinkernagel RM (1996) Antiviral immune responses of mice lacking MHC class II or its associated invariant chain. Cell Immunol 167:115–121

Bartholdy C, Christensen JP, Wodarz D, Thomsen AR (2000) Persistent virus infection despite chronic cytotoxic T-lymphocyte activation in gamma interferon-deficient mice infected with lymphocytic choriomeningitis virus. J Virol 74:10304–10311

Baumgarth N, Jager GC, Herman OC, Herzenberg LA (2000) CD4+ T cells derived from B cell-deficient mice inhibit the establishment of peripheral B cell pools. Proc Natl Acad Sci USA 97:4766–4771

Bennett SR, Carbone FR, Karamalis F, Flavell RA, Miller JF, Heath WR (1998) Help for cytotoxic-T-cell responses is mediated by CD40 signalling. Nature 393:478–480

Berger DP, Homann D, Oldstone MB (2000) Defining parameters for successful immunocytotherapy of persistent viral infection. Virology 266:257–263

Borrow P, Evans CF, Oldstone MB (1995) Virus-induced immunosuppression: immune system-mediated destruction of virus-infected dendritic cells results in generalized immune suppression. J Virol 69:1059–1070

Borrow P, Tishon A, Lee S, Xu J, Grewal IS, Oldstone MB, Flavell RA (1996) CD40L-deficient mice show deficits in antiviral immunity and have an impaired memory CD8+ CTL response. J Exp Med 183:2129–2142

Borrow P, Oldstone MB (1997) Lymphocytic choriomeningitis virus. In: Nathanson, et al. (eds) Viral Pathogenesis. Lippincott-Raven Publishers, Philadelphia, pp 593–627

Borrow P, Tough DF, Eto D, Tishon A, Grewal IS, Sprent J, Flavell RA, Oldstone MB (1998) CD40 ligand-mediated interactions are involved in the generation of memory CD8(+) cytotoxic T lymphocytes (CTL) but are not required for the maintenance of CTL memory following virus infection. J Virol 72:7440–7749

Brot MD, Rall GF, Oldstone MB, Koob GF, Gold LH (1997) Deficits in discriminated learning remain despite clearance of long-term persistent viral infection in mice. J Neurovirol 3:265–273

Brundler MA, Aichele P, Bachmann M, Kitamura D, Rajewsky K, Zinkernagel RM (1996) Immunity to viruses in B cell-deficient mice: influence of antibodies on virus persistence and on T cell memory. Eur J Immunol 26:2257–2262

Buchmeier MJ, Welsh RM, Dutko FJ, Oldstone MB (1980) The virology and immunobiology of lymphocytic choriomeningitis virus infection. Adv Immunol 30:275–331

Byrne JA, Oldstone MB (1986) Biology of cloned cytotoxic T lymphocytes specific for lymphocytic choriomeningitis virus. VI. Migration and activity in vivo in acute and persistent infection. J Immunol 136:698–704

Cao W, Oldstone MB, De La Torre JC (1997) Viral persistent infection affects both transcriptional and posttranscriptional regulation of neuron-specific molecule GAP43. Virology 230:147–154

Cerny A, Huegin AW, Sutter S, Bazin H, Hengartner HH, Zinkernagel RM (1986) Immunity to lymphocytic choriomeningitis virus in B cell-depleted mice: evidence for B cell and antibody-independent protection by memory T cells. Eur J Immunol 16:913–917

Cerny A, Sutter S, Bazin H, Hengartner H, Zinkernagel RM (1988) Clearance of lymphocytic choriomeningitis virus in antibody- and B-cell-deprived mice. J Virol 62:1803–1807

Christensen JP, Marker O, Thomsen AR (1994) The role of CD4$^+$ T cells in cell-mediated immunity to LCMV: studies in MHC class I and class II deficient mice. Scand J Immunol 40:373–382

Cihak J, Lehmann-Grube F (1978) Immunological tolerance to lymphocytic choriomeningitis virus in neonatally infected virus carrier mice: evidence supporting a clonal inactivation mechanism. Immunology 34:265–275

Ciurea A, Klenerman P, Hunziker L, Horvath E, Odermatt B, Ochsenbein AF, Hengartner H, Zinkernagel RM (1999) Persistence of lymphocytic choriomeningitis virus at very low levels in immune mice. Proc Natl Acad Sci USA 96:11964–11969

Cole GA, Prendergast RA, Henney CS (1973) In vitro correlates of LCM virus-induced immune response. In: Lehmann-Grube F (ed) Lymphocytic choriomeningitis virus and other arena viruses. Springer, Berlin Heidelberg New York, pp 61–71

Corriveau RA, Huh GS, Shatz CJ (1998) Regulation of class I MHC gene expression in the developing and mature CNS by neural activity. Neuron 21:505–520

de la Torre JC, Rall G, Oldstone C, Sanna PP, Borrow P, Oldstone MB (1993) Replication of lymphocytic choriomeningitis virus is restricted in terminally differentiated neurons. J Virol 67:7350–7359

de la Torre JC, Mallory M, Brot M, Gold L, Koob G, Oldstone MB, Masliah E (1996) Viral persistence in neurons alters synaptic plasticity and cognitive functions without destruction of brain cells. Virology 220:508–515

Doherty PC, Christensen JP (2000) Accessing complexity: the dynamics of virus-specific T cell responses. Annu Rev Immunol 18:561–592

Gold LH, Brot MD, Polis I, Schroeder R, Tishon A, de la Torre JC, Oldstone MB, Koob GF (1994) Behavioral effects of persistent lymphocytic choriomeningitis virus infection in mice. Behav Neural Biol 62:100–109

Gossmann J, Lohler J, Lehmann-Grube F (1991) Entry of antivirally active T lymphocytes into the thymus of virus-infected mice. J Immunol 146:293–297

Guidotti LG, Chisari FV (1999) Cytokine-induced viral purging–role in viral pathogenesis. Curr Opin Microbiol 2:388–391

Guidotti LG, Borrow P, Brown A, McClary H, Koch R, Chisari FV (1999) Noncytopathic clearance of lymphocytic choriomeningitis virus from the hepatocyte. J Exp Med 189:1555–1564

Homann D, Tishon A, Berger DP, Weigle WO, von Herrath MG, Oldstone MB (1998) Evidence for an underlying CD4 helper and CD8 T-cell defect in B-cell-deficient mice: failure to clear persistent virus infection after adoptive immunotherapy with virus-specific memory cells from μMT/μMT mice. J Virol 72:9208–9216

Hotchin JE, Sintis M (1958) Lymphocytic choriomeningits infection of mice as a model for the study of latent virus infection. Can J Microbiol 4:149–163

Hotchin J (1973) Transient virus infection: spontaneous recovery mechanism of lymphocytic choriomeningitis virus-infected cells. Nat New Biol 241:270–272

Hotchin J (1974) The role of transient infection in arenavirus persistence. Prog Med Virol 18:81–93

Hotchin J, Seegal R (1977) Virus-induced behavioral alteration of mice. Science 196:671–674

Hotchin J, Seegal R (1978) Alterations in behavior resulting from persistent lymphocytic choriomeningitis virus infection. Birth Defects Orig Artic Ser 14:171–178

Huang S, Hendriks W, Althage A, Hemmi S, Bluethmann H, Kamijo R, Vilcek J, Zinkernagel RM, Aguet M (1993) Immune response in mice that lack the interferon-gamma receptor. Science 259:1742–1745

Jamieson BD, Ahmed R (1988) T-cell tolerance: exposure to virus in utero does not cause a permanent deletion of specific T cells. Proc Natl Acad Sci USA 85:2265–2268

Jamieson BD, Ahmed R (1989) T cell memory. Long-term persistence of virus-specific cytotoxic T cells. J Exp Med 169:1993–2005

Jamieson BD, Butler LD, Ahmed R (1987) Effective clearance of a persistent viral infection requires cooperation between virus-specific Lyt2+ T cells and nonspecific bone marrow-derived cells. J Virol 61:3930–3937

Jamieson BD, Somasundaram T, Ahmed R (1991) Abrogation of tolerance to a chronic viral infection. J Immunol 147:3521–3529

Johnson ED, Cole GA (1975) Functional heterogeneity of lymphocytic choriomeningitis virus-specfic T lymphocytes. I. Identification of effector and memory subsets. J Exp Med 141:866–881

Johnson ED, Monjan AA, Morse HC 3rd (1978) Lack of B-cell participation in acute lymphocyte choriomeningitis disease of the central nervous system. Cell Immunol 36:143–150

Joly E, Mucke L, Oldstone MB (1991) Viral persistence in neurons explained by lack of major histocompatibility class I expression. Science 253:1283–1285

Kagi D, Ledermann B, Burki K, Seiler P, Odermatt B, Olsen KJ, Podack ER, Zinkernagel RM, Hengartner H (1994) Cytotoxicity mediated by T cells and natural killer cells is greatly impaired in perforin-deficient mice. Nature 369:31–37

Kasaian MT, Leite-Morris KA, Biron CA (1991) The role of CD4+ cells in sustaining lymphocyte proliferation during lymphocytic choriomeningitis virus infection. J Immunol 146:1955–1963

Klavinskis LS, Geckeler R, Oldstone MB (1989) Cytotoxic T lymphocyte control of acute lymphocytic choriomeningitis virus infection: interferon gamma, but not tumour necrosis factor alpha, displays antiviral activity in vivo. J Gen Virol 70:3317–3325

King CC, Jamieson BD, Reddy K, Bali N, Concepcion RJ, Ahmed R (1992) Viral infection of the thymus. J Virol 66:3155–3160

Kitamura D, Roes J, Kuhn R, Rajewsky K (1991) A B cell-deficient mouse by targeted disruption of the membrane exon of the immunoglobulin mu chain gene. Nature 350:423–426

Kopf M, Ruedl C, Schmitz N, Gallimore A, Lefrang K, Ecabert B, Odermatt B, Bachmann MF (1999) OX40-deficient mice are defective in Th cell proliferation but are competent in generating B cell and CTL responses after virus infection. Immunity 11:699–708

Larsen JH, Volkert M (1967) Studies on immunological tolerance to LCM virus 7 Adoptive immunization of virus carrier mice by grafts of normal syngeneic lymphoid cells. Acta Path et Microbiol Scandinav 70:95–106

Lehmann-Grube F (1987) Mechanism of recovery from acute virus infection. In: Bauer H, Klenk HD, Scholtissek C (eds) Modern trends in virology. Springer, Berlin Heidelberg New York, pp 49–64

Leist TP, Cobbold SP, Waldmann H, Aguet M, Zinkernagel RM (1987) Functional analysis of T lymphocyte subsets in antiviral host defense. J Immunol 138:2278–2281

Leist TP, Ruedi E, Zinkernagel RM (1988) Virus-triggered immune suppression in mice caused by virus-specific cytotoxic T cells. J Exp Med 167:1749–1754

Leist TP, Zinkernagel RM (1990) Treatment with anti-tumor necrosis factor alpha does not influence the immune pathological response against lymphocytic choriomeningitis virus. Cytokine 2:29–34

Linton PJ, Harbertson J, Bradley LM (2000) A critical role for B cells in the development of memory CD4 cells. J Immunol 165:5558–5565

Ludewig B, Ehl S, Karrer U, Odermatt B, Hengartner H, Zinkernagel RM (1998a) Dendritic cells efficiently induce protective antiviral immunity. J Virol 72:3812–3818

Ludewig B, Odermatt B, Landmann S, Hengartner H, Zinkernagel RM (1998b) Dendritic cells induce autoimmune diabetes and maintain disease via de novo formation of local tissue. J Exp Med 188:1493–1501

Ludewig B, Oehen S, Barchiesi F, Schwendener RA, Hengartner H, Zinkernagel RM (1999) Protective antiviral cytotoxic T cell memory is most efficiently maintained by restimulation via dendritic cells. J Immunol 163:1839–1844

Ludewig B, Ochsenbein AF, Odermatt B, Paulin D, Hengartner H, Zinkernagel RM (2000) Immunotherapy with dendritic cells directed against tumor antigens shared with normal host cells results in severe autoimmune disease. J Exp Med 191:795–804

Macaulay AE, DeKruyff RH, Umetsu DT (1998) Antigen-primed T cells from B cell-deficient JHD mice fail to provide B cell help. J Immunol 160:1694–1700

Matloubian M, Concepcion RJ, Ahmed R (1994) CD4+ T cells are required to sustain CD8+ cytotoxic T-cell responses during chronic viral infection. J Virol 68:8056–8063

McHeyzer-Williams MG, Ahmed R (1999) B cell memory and the long-lived plasma cell. Curr Opin Immunol 11:172–179

Medana IM, Gallimore A, Oxenius A, Martinic MMA, Wekerle H, Neumann H (2000) MHC class I-restricted killing of neurons by virus-specific CD8+ T lymphocytes is effected through the Fas/FasL, but not the perforin pathway. Eur J Immunol 30:3623–3633

Moskophidis D, Cobbold SP, Waldmann H, Lehmann-Grube F (1987) Mechanism of recovery from acute virus infection: treatment of lymphocytic choriomeningitis virus-infected mice with monoclonal antibodies reveals that Lyt-2+ T lymphocytes mediate clearance of virus and regulate the antiviral antibody response. J Virol 61:1867–1874

Moskophidis D, Laine E, Zinkernagel RM (1993a) Peripheral clonal deletion of antiviral memory CD8+ T cells. Eur J Immunol 23:3306–3311

Moskophidis D, Lechner F, Pircher H, Zinkernagel RM (1993b) Virus persistence in acutely infected immunocompetent mice by exhaustion of antiviral cytotoxic effector T cells. Nature 362:758–761

Muller U, Steinhoff U, Reis LF, Hemmi S, Pavlovic J, Zinkernagel RM, Aguet M (1994) Functional role of type I and type II interferons in antiviral defense. Science 264:1918–1921

Murali-Krishna K, Altman JD, Suresh M, Sourdive DJ, Zajac AJ, Miller JD, Slansky J, Ahmed R (1998) Counting antigen-specific CD8 T cells: a reevaluation of bystander activation during viral infection. Immunity 8:177–187

Nansen A, Jensen T, Christensen JP, Andreasen SO, Ropke C, Marker O, Thomsen AR (1999) Compromised virus control and augmented perforin-mediated immunopathology in IFN-gamma-deficient mice infected with lymphocytic choriomeningitis virus. J Immunol 163:6114–6122

Oldstone MB, Dixon FJ (1967) Lymphocytic choriomeningitis: production of antibody by "tolerant" infected mice. Science 158:1193–1195

Oldstone MB, Blount P, Southern PJ, Lampert PW (1986) Cytoimmunotherapy for persistent virus infection reveals a unique clearance pattern from the central nervous system. Nature 321:239–243

Oldstone MB (1987) Immunotherapy for virus infection. Curr Top Microbiol Immunol 134:211–229

Oldstone MB (1998) Viral persistence: mechanisms and consequences. Curr Opin Microbiol 1:436–441

Oxenius A, Campbell KA, Maliszewski CR, Kishimoto T, Kikutani H, Hengartner H, Zinkernagel RM, Bachmann MF (1996) CD40-CD40 ligand interactions are critical in T-B cooperation but not for other anti-viral CD4+ T cell functions. J Exp Med 183:2209–2218

Oxenius A, Zinkernagel RM, Hengartner H (1998) Comparison of activation versus induction of unresponsiveness of virus-specific CD4+ and CD8+ T cells upon acuteversus persistent viral infection. Immunity 9:449–457

Pircher H, Burki K, Lang R, Hengartner H, Zinkernagel RM (1989) Tolerance induction in double specific T-cell receptor transgenic mice varies with antigen. Nature 342:559–561

Pircher H, Moskophidis D, Rohrer U, Burki K, Hengartner H, Zinkernagel RM (1990) Viral escape by selection of cytotoxic T cell-resistant virus variants in vivo. Nature 346:629–633

Planz O, Ehl S, Furrer E, Horvath E, Brundler MA, Hengartner H, Zinkernagel RM (1997) A critical role for neutralizing-antibody-producing B cells, CD4(+) T cells, and interferons in persistent and acute infections of mice with lymphocytic choriomeningitis virus: implications for adoptive immunotherapy of virus carriers. Proc Natl Acad Sci USA 94:6874–6879

Rawson P, Hermans IF, Huck SP, Roberts JM, Pircher H, Ronchese F (2000) Immunotherapy with dendritic cells and tumor major histocompatibility complex class I-derived peptides requires a high density of antigen on tumor cells. Cancer Res 60:4493–4498

Rahemtulla A, Fung-Leung WP, Schilham MW, Kundig TM, Sambhara SR, Narendran A, Arabian A, Wakeham A, Paige CJ, Zinkernagel RM, et al (1991) Normal development and function of CD8+ cells but markedly decreased helper cell activity in mice lacking CD4. Nature 353:180–184

Ridge JP, Di Rosa F, Matzinger P (1998) A conditioned dendritic cell can be a temporal bridge between a CD4+ T-helper and a T-killer cell. Nature 393:474–478

Rooney CM, Smith CA, Ng CY, Loftin SK, Sixbey JW, Gan Y, Srivastava DK, Bowman LC, Krance RA, Brenner MK, Heslop HE (1998) Infusion of cytotoxic T cells for the prevention and treatment of Epstein-Barr virus-induced lymphoma in allogeneic transplant recipients. Blood 92:1549–1555

Schoenberger SP, Toes RE, van der Voort EI, Offringa R, Melief CJ (1998) T-cell help for cytotoxic T lymphocytes is mediated by CD40-CD40L interactions. Nature 393:480–483

Sevilla N, Kunz S, Holz A, Lewicki H, Homann D, Yamada H, Campbell KP, de La Torre JC, Oldstone MB (2000) Immunosuppression and resultant viral persistence by specific viral targeting of dendritic cells. J Exp Med 192:1249–1260

Slifka MK, Whitton JL (2000) Antigen-specific regulation of T cell-mediated cytokine production. Immunity 12:451–457

Smelt CS, Borrow P, Kunz S, Cao W, Tishon A, Lewicki H, Campbell KP, Oldstone MBA (2001) Differences in affinity of binding of Lymphocytic Choriomeningitis virus strains to cellular receptor α-dystroglycan correlate with viral tropism and disease kinetics. J Virol 75:448–457

Southern PJ, Blount P, Oldstone MB (1984) Analysis of persistent virus infections by in situ hybridization to whole-mouse sections. Nature 312:555–558

Tishon A, Southern PJ, Oldstone MB (1988) Virus-lymphocyte interactions. II. Expression of viral sequences during the course of persistent lymphocytic choriomeningitis virus infection and their localization to the L3T4 lymphocyte subset. J Immunol 140:1280–1284

Tishon A, Eddleston M, de la Torre JC, Oldstone MB (1993) Cytotoxic T lymphocytes cleanse viral gene products from individually infected neurons and lymphocytes in mice persistently infected with lymphocytic choriomeningitis virus. Virology 197:463–467

Tishon A, Lewicki H, Rall G, Von Herrath M, Oldstone MB (1995) An essential role for type 1 interferon-gamma in terminating persistent viral infection. Virology 212:244–250

Thomsen AR, Volkert M, Marker O (1979) The timing of the immune response in relation to virus growth determines the outcome of the LCM infection. Acta Pathol Microbiol Scand [C] 87C:47–54

Thomsen AR, Johansen J, Marker O, Christensen JP (1996) Exhaustion of CTL memory and recrudescence of viremia in lymphocytic choriomeningitis virus-infected MHC class II-deficient mice and B cell-deficient mice. J Immunol 157:3074–3080

Thomsen AR, Nansen A, Christensen JP, Andreasen SO, Marker O (1998) CD40 ligand is pivotal to efficient control of virus replication in mice infected with lymphocytic Choriomeningitis virus. J Immunol 161:4583–4590

Traub E (1936) Persistence of Lymphocytic choriomeningits vieus in immune animals and its relation to immunity. J Exp Med 63:847–861

van Essen D, Dullforce P, Gray D (2000) Role of B cells in maintaining helper T-cell memory. Phil Trans R Soc Lond B Biol Sci 355:351–355

Volkert M (1962a) Studies on immunological tolerance to LCM virus. Treatment of virus carrier mice by adoptive immunization. Acta Path et Microbiol Scandinav 56:305–310

Volkert M (1962b) Studies on immunological tolerance to LCM virus 2. A preliminary report on adoptive immunization of virus carrier mice. Acta Path et Microbiol Scandinav 57:465–487

Volkert M, Larsen JH (1964) Studies on immunological tolerance to LCM virus 3. Duration and maximal effect of adoptive immunization of virus carrier mice. Acta Path et Microbiol Scandinav 60:577–587

Volkert M, Larsen JH, Pfau CJ (1964) Studies on immunological tolerance to LCM virus 4. The question of immunity in adoptively immunized virus carriers. Acta Path et Microbiol Scandinav 61:268–282

Volkert M, Larsen JH (1965a) Studies on immunological tolerance to LCM virus 5. The induction of tolerance to the virus. Acta Path et Microbiol Scandinav 63:161–171

Volkert M, Larsen JH (1965b) Studies on immunological tolerance to LCM virus 6. Immunity conferred on tolerant mice by immune serum and by grafts of homologous lymphoid cells. Acta Path et Microbiol Scandinav 63:172–180

Volkert M, Lundstedt C (1968) The provocation of latent lymphocytic choriomeningitis virus infections in mice by treatment with antilymphocytic serum. J Exp Med 127:327–339

Volkert M, Marker O, Bro-Jorgensen K (1974) Two populations of T lymphocytes immune to the lymphocytic choriomeningitis virus. J Exp Med 139:1329–1343

Volkert M, Bro-Jorgensen K, Marker O, Rubin B, Trier L (1975) The activity of T and B lymphocytes in immunity and tolerance to the lymphocytic choriomeningitis virus in mice. Immunology 29:455–464

von Herrath MG, Yokoyama M, Dockter J, Oldstone MB, Whitton JL (1996) CD4-deficient mice have reduced levels of memory cytotoxic T lymphocytes after immunization and show diminished resistance to subsequent virus challenge. J Virol 70:1072–1079

von Herrath MG, Coon B, Oldstone MB (1997) Low-affinity cytotoxic T-lymphocytes require IFN-gamma to clear an acute viral infection.Virology 229:349–359

von Herrath MG, Berger DP, Homann D, Tishon T, Sette A, Oldstone MB (2000) Vaccination to treat persistent viral infection. Virology 268:411–419

Walsh CM, Matloubian M, Liu CC, Ueda R, Kurahara CG, Christensen JL, Huang MT, Young JD, Ahmed R, Clark WR (1994) Immune function in mice lacking the perforin gene. Proc Natl Acad Sci USA 91:10854–10858

Whitmire JK, Flavell RA, Grewal IS, Larsen CP, Pearson TC, Ahmed R (1999) CD40-CD40 ligand costimulation is required for generating antiviral CD4 T cell responses but is dispensable for CD8 Tcell responses. J Immunol 163:3194–3201

Whitton JL, Oldstone MBA (2001) The immune response to viruses. In: Knipe D, Howley P, Griffin D, Lamb R, Martin M, Straus S (eds) Fields Virology (4th edn.) Lippincott Williams & Wilkins, Philadelphia, pp 285–320

Wodarz D, Nowak MA (2000) CD8 memory, immunodominance, and antigenic escape. Eur J Immunol 30:2704–2712

Zajac AJ, Blattman JN, Murali-Krishna K, Sourdive DJ, Suresh M, Altman JD, Ahmed R (1998) Viral immune evasion due to persistence of activated T cells without effector function. J Exp Med 188:2205–2213

Zinkernagel RM, Doherty PC (1974) Restriction of in vitro T cell-mediated cytotoxicity in lymphocytic choriomeningitis within a syngeneic or semiallogeneic system. Nature 248:701–702

Mechanisms of Humoral Immunity Explored Through Studies of LCMV Infection

M.K. Slifka

1	Introduction	67
2	LCMV Model System	68
3	Long-Term Humoral Immunity Following Viral Infection	68
4	Anatomical Sites of Antiviral Antibody Production	69
5	Mechanisms of Long-Term Humoral Immunity	70
5.1	Antigen-Dependent Models for Maintaining Antibody Production	70
5.2	Antigen-Independent Model for Maintaining Antibody Production: Long-Lived Plasma Cells	74
6	Conclusions	78
References		78

1 Introduction

Prolonged synthesis of antigen-specific antibody is one of the cardinal features of successful vaccination and one of the key parameters used to demonstrate immunological memory. For years, long-term antibody production was attributed solely to the continuous stimulation and differentiation of memory B cells into antibody-secreting plasma cells. However, several studies have now revealed that long-lived plasma cells, most of which reside in the bone marrow, represent a previously overlooked mechanism for sustaining long-term humoral immunity. In this regard, analysis of the antiviral antibody response that occurs following acute infection with lymphocytic choriomeningitis virus (LCMV) has been instrumental in developing our current understanding of the mechanisms underlying long-term antibody synthesis.

OHSU Vaccine and Gene Therapy Institute, 505 NW 185th Avenue, Beaverton, OR 97006, USA

2 LCMV Model System

LCMV is a common murine pathogen that provides an excellent model for studying the interactions between a virus and the immune system of its natural host (Oldstone and Dixon 1967; Buchmeier et al. 1980; Lehmann-Grube et al. 1983; Oldstone et al. 1985; Ahmed and Gray 1996). Infection of adult mice with LCMV-Armstrong results in an acute infection that is cleared within 1 to 2 weeks in a perforin-mediated (Walsh et al. 1994; Kagi et al. 1994), $CD8^+$ T cell-dependent manner (Zinkernagel and Doherty 1979; Buchmeier et al. 1980; Byrne and Oldstone 1984; Ahmed et al. 1984; Lehmann-Grube et al. 1985; Whitton et al. 1993; Welsh and McNally 1999). B cells and/or neutralizing antibodies are not required for viral clearance (Cerny et al. 1988; Asano and Ahmed 1996; Brundler et al. 1996) but may play a substantial role in decreasing viral load (Cerny et al. 1988; Wright and Buchmeier 1991; Battegay et al. 1993; Seiler et al. 1998a,b) and act synergistically with antiviral T cells to prevent reinfection (Thomsen and Marker 1988; Baldridge et al. 1997). Recovery from acute LCMV infection invariably leads to long-term T cell and B cell memory (Buchmeier et al. 1980; Lau et al. 1994; Slifka et al. 1998; Ochsenbein et al. 2000) and persistence of antiviral serum antibody for the life span of immunocompetent mice (Kimmig and Lehmann-Grube 1979; Slifka et al. 1995, 1998; Ochsenbein et al. 2000).

3 Long-Term Humoral Immunity Following Viral Infection

How long can antigen-specific antibody responses be maintained in humans? Following systemic viral infection, it is not uncommon for antiviral antibody responses to be maintained for decades, even in the absence of re-exposure to the viral pathogen. Of more than ten different viral infections that lead to documented antiviral antibody responses lasting 10 years or more (Slifka and Ahmed 1996b), one of the most impressive examples is an elegant study by W.A. Sawyer demonstrating life-long humoral immunity against yellow fever virus (Sawyer 1931). Protection experiments were performed in which immune serum (containing neutralizing virus-specific antibody) from previously infected human donors was transferred into Rhesus monkeys prior to challenge with a lethal dose of virus. These studies demonstrated that most of the yellow fever virus-immune donors maintained high levels of neutralizing antibody in their serum that could easily protect primates from a lethal viral infection – some without even mounting a fever or showing any signs of disease. This observation becomes even more striking when one considers the duration and magnitude of the protective antibody responses maintained by the donors – many had not been re-exposed to virus or virus-infected mosquitoes for up to 75 years after their initial childhood infection.

4 Anatomical Sites of Antiviral Antibody Production

Despite extensive documentation of long-term antiviral antibody responses, there is still much to be learned about the mechanisms and anatomical location of antiviral antibody synthesis in vivo. We began to examine these issues by determining the anatomical sites of virus-specific antibody production in mice following acute LCMV infection (SLIFKA et al. 1995). For these studies, we used the ELISPOT assay to distinguish LCMV-specific plasma cells from memory B cells. Plasma cells spontaneously secrete high levels of antibody [up to 10,000 molecules per second (HELMREICH et al. 1961; HIBI and DOSCH 1986)] and can be easily detected by in vitro incubation on the surface of antigen-coated nitrocellulose-bottomed plates for as little as 4–5h (SLIFKA et al. 1995; SLIFKA and AHMED 1996a). In contrast, memory B cells do not spontaneously secrete antibody and require more than 24h of antigenic stimulation before they proliferate and differentiate into antibody secreting cells in vitro (SLIFKA and AHMED 1996a). Using the ELISPOT technique, we quantitated the number of virus-specific plasma cells in the spleen and bone marrow of mice infected with LCMV. The spleen was found to be the initial site of antibody production following acute viral infection. The number of virus-specific antibody secreting cells in the spleen peaked by around 8 days post-infection before declining by more than 95% over the course of the following 2–4 weeks, a finding similar to a previous report (MOSKOPHIDIS and LEHMANN-GRUBE 1984). The loss of splenic plasma cells, however, did not correlate with serum antibody titers, which are maintained at nearly steady-state levels for the life of the mouse (KIMMIG and LEHMANN-GRUBE 1979; SLIFKA et al. 1995, 1998; OCHSENBEIN et al. 2000). Quantitation of total plasma cell numbers revealed that the bone marrow soon became the major anatomical site of long-term antibody production and thus provided the mechanism to explain how serum antibody levels were so steadfastly maintained following the substantial loss of virus-specific plasma cells in the spleen.

Plasma cell accumulation in the bone marrow is not unique to the LCMV model system. Indeed, several acute viral infections (BACHMAN et al. 1994; HYLAND et al. 1994) and booster vaccination with non-replicating antigens (HILL 1976; BENNER et al. 1981; DILOSA et al. 1991; SMITH et al. 1997; TAKAHASHI et al. 1998; McHEYZER-WILLIAMS et al. 2000) have been found to induce a similar migration/ accumulation of antigen-specific plasma cells to the bone marrow compartment. Interestingly, we found that during the peak of the antibody response in the spleen (8 days post-infection), LCMV-specific plasma cell numbers were below detection in the bone marrow. This is somewhat surprising since actively cytolytic virus-specific $CD8^+$ T cells could be readily isolated from the bone marrow at this time point (SLIFKA et al. 1997). Later, when virus-specific plasma cells in the spleen declined, there was a corresponding increase in the number of virus-specific plasma cells found in the bone marrow. One explanation for the somewhat puzzling initial delay in the accumulation of antigen-specific plasma cells in the bone marrow may be that selection for high-affinity plasma cells must occur prior to or during the induction of the bone marrow-derived antibody response. This hypothesis is supported by studies showing that, based on the distribution of somatic mutations and secretion of high-affinity antibody, the bone marrow compartment is selectively enriched for high-affinity plasma cells (SMITH et al. 1997; TAKAHASHI et al. 1998). Utilization of bone marrow as the anatomical reservoir of antigen-specific plasma cells may be advantageous for the host in that this reduces the immunological burden of long-term antibody production by peripheral lymphoid tissues such as the spleen and lymph nodes. Indeed, a recent study has shown that the spleen has a finite capacity for sustaining plasma cell numbers following vaccination (between 20–100 plasma cells/mm^2) and that there is a rapid loss of surplus plasma cells when this number is exceeded (SZE et al. 2000). Therefore, by localizing high-affinity plasma cells in the bone marrow, the peripheral lymphoid organs are not held at maximum storage capacity and instead are able to recede to homeostasis, ready to mount new immune responses upon later encounter with stimulatory antigens.

5 Mechanisms of Long-Term Humoral Immunity

The persistence of virus-specific antibody production is a well-recognized but poorly understood aspect of immunological memory (SLIFKA and AHMED 1996b; AHMED and GRAY 1996). Because antibody molecules have a serum half-life that ranges from a few days to at most 2–3 weeks (FAHEY and SELL 1965; TALBOT and BUCHMEIER 1987; VIEIRA and RAJEWSKY 1988; SLIFKA et al. 1998), continuous production of immunoglobulin is required if high levels of circulating antibody are to be maintained. Most researchers would agree that antibody production is maintained by the combined efforts of two specialized groups of B cells: plasma cells and memory B cells. However, the relative contribution of each of these two highly specialized groups of antigen-specific cells to the long-term maintenance of humoral immunity remains a subject of intense debate (SLIFKA et al. 1998; OCHSENBEIN et al. 2000). As stated above, plasma cells are the main form of antibody-producing cells whereas memory B cells do not secrete antibody per se, but instead become stimulated to proliferate and differentiate into plasma cells following antigen contact. Several conventional models of long-term humoral immunity will be discussed in the following sections and compared to the experimental results obtained through studies of the LCMV model system in which long-lived plasma cells have emerged as an important component of persistent antiviral antibody production.

5.1 Antigen-Dependent Models for Maintaining Antibody Production

There are many potential mechanisms involved with the long-term maintenance of humoral immunity (Fig. 1). Antigen-dependent models for sustaining antiviral

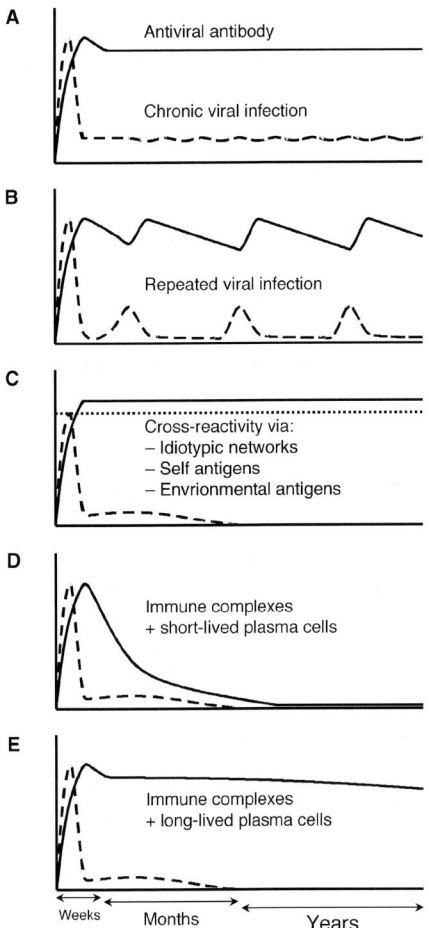

Fig. 1A–E. Proposed mechanisms of long-term antiviral antibody production. There are several potential mechanisms that may be involved in the maintenance of persistent humoral immunity and some of the most commonly described models are shown here. In each panel, the *solid line* represents antiviral antibody production and the *dashed lines* represent relative antigen levels. **A** Low-grade chronic viral infection provides a simple model in which virus-specific B cells are continuously stimulated to become plasma cells due to prolonged exposure to replicating viral antigens. **B** Repeated exposure to the same virus can boost antibody titers by restimulating memory B cells. The primary infection would result in the highest antigen load whereas each subsequent infection would be greatly reduced due to the pre-existing T cell and B cell memory of the immune host. **C** Following acute viral infection and the subsequent decline in viral antigen, idiotypic networks or cross-reactivity to either self or environmental antigens (*dotted line*) might result in continuous stimulation of virus-specific memory B cells. **D** The most widely accepted theory explaining long-term antibody production is that memory B cells are continuously stimulated by antigen trapped in the form of immune complexes. Following resolution of the acute viral infection, viral antigen levels would drop rapidly for the first few weeks. Thereafter, low levels of viral antigen would persist in antigen-antibody complexes. These immune complexes would decline slowly until the antigen reservoir is eventually exhausted. If plasma cells were short-lived, then antibody titers would correspondingly decrease following the loss of stimulatory antigen. **E** Prolonged maintenance of antibody production by long-lived plasma cells has been demonstrated in a variety of systems and could function simultaneously with stimulation of memory B cells by other mechanisms such as immune complexes. The main difference, however, is that once the antigen reservoir was depleted, serum antibody titers would be sustained for an extended period of time as a function of the survival rate of the antigen-independent plasma cell populations in the spleen and bone marrow compartments

antibody production include; chronic viral infection, repeated exposure to acute viral infections, idiotypic networks, cross-reactivity to self or environmental antigens, or antigen persistence in the form of immune complexes deposited on the surface of dendritic cells. The common factor among each of these models is the assumption that plasma cells are short-lived and that continuous stimulation of memory B cells is required for maintaining plasma cell numbers and long-term antibody production.

Several viruses are known to cause either latent or persistent infections that would be expected to present viral antigens to the immune system over extended periods of time. Although the amount of viral antigen expressed would differ greatly depending on the nature of the chronic infection, Fig. 1A depicts a generalized example in which the initial viral infection would initiate host immune responses including the production of antiviral antibodies. Later, the infection might become either latent or persist at varying levels in which case either intermittent or continuous re-stimulation of memory B cells would occur. Mice that are persistently infected with LCMV provide an experimental model for studying antiviral antibody responses induced by chronic exposure to high levels of viral antigen (OLDSTONE and DIXON 1967, 1969; BUCHMEIER et al. 1980). Under these conditions, circulating antiviral antibodies bind their respective antigens and are deposited in the renal glomeruli, arteries, and choroid plexus, resulting in immune complex disease that is most severe in mice strains that express the highest level of antiviral antibodies (OLDSTONE and DIXON 1970). Virus-specific antibody (including some virus-neutralizing antibody) can be detected for 4 months to at least 1 year in the presence of persistent viral replication (OLDSTONE et al. 1980; THOMSEN et al. 1985) and demonstrates how a balance between free antibody, viral antigen, and immune complex formation/deposition can exist in vivo.

In addition to viruses known to cause chronic infection, some viruses such as measles and rubella virus, which are typically classified as acute infections, may on rare occasions persist for several years after the initial infection (SCHNEIDER-SCHAULIES et al. 1995; CHANTLER et al. 1981). However, not all viruses cause latent or persistent infections, making it unlikely that persistent infection is the sole mechanism underlying the decades of antiviral antibody production observed in various epidemiological studies (PANUM 1847; PAUL et al. 1951; UKKON et al. 1984; SLIFKA and AHMED 1996b).

If low-grade viral persistence does not occur, then multiple re-exposure to the same virus might serve as a similar mechanism for sustaining antiviral antibody production (Fig. 1B). The primary infection would most likely lead to the highest viral load and induce the initial antiviral antibody response. If the host were re-exposed to the same virus, as might be the case for several common respiratory infections, viral replication would be severely reduced and infectious virus would be more rapidly cleared due to the pre-existing immunity of the host. Nevertheless, re-infection would likely be sufficient to "boost" memory B cell responses and restore potentially declining antibody levels. In this model, one might expect short-lived antiviral antibody responses to be common unless the host lived in endemic areas in which intermittent re-exposure to virus occurred on a relatively frequent basis. However, several viral infections, including vaccinia and yellow fever virus, lead to long-term humoral immunity in the absence of known persistence or re-exposure to virus (SLIFKA and AHMED 1996b). This raises serious doubts about whether re-exposure to infectious virus is an absolute requirement for maintaining long-term antibody responses. Moreover, other examples of long-term humoral immunity achieved by booster vaccination with non-replicating antigens (diphtheria or tetanus toxoid) result in protective levels of antigen-specific antibody production for more than 20 years in the absence of re-exposure or further vaccination (GOTTLIEB et al. 1964; KJELDSEN et al. 1985; COHEN et al. 1994). Similar to replicating antigens, the use of alum as an adjuvant results in the slow release of antigen to stimulate memory B cells over an extended period of time. Although antigen release is prolonged, elegant serological and histological studies have demonstrated that alum/antigen complexes are only immunogenic in vivo for a few weeks after vaccination before they become fully encapsulated in a granuloma-type structure that is no longer actively recognized by the immune system (HOLT 1950).

Idiotypic networks and cross-reactivity to either self or environmental antigens are also potential mechanisms for aiding in the persistence of humoral immunity, but their relevance to antiviral antibody production has yet to be identified. In each of these cases, the original antigen does not directly drive antibody production, but instead the antibody response relies on the stimulatory capacity of similar antigens that mimic the initial B cell epitopes of the virus (Fig. 1C). Idiotypic networks, for instance, are examples in which an antibody molecule is cross-reactive with another antibody molecule that is mistakenly recognized as the initiating antigen. Theoretically, this leads to a complex series of antibodies and anti-idiotypic antibodies that cross-react and stimulate each other in a self-perpetuating cascade (JERNE 1984). Most models of humoral immunity to viruses do not rely on idiotypic networks or cross-reactivity to self or environmental antigens because these mechanisms leave us with several unanswered questions. For example, how would affinity maturation of antiviral antibodies occur if the antibody responses were maintained by cross-reactivity to unrelated antigens? If antibody production was maintained by exposure to environmental antigens, then where would the $CD4^+$ T cell help for T cell-dependent antibody responses come from? What is the likelihood that an environmental antigen would contain both cross-reactive B cell epitopes and the correspondingly immunogenic cross-reactive T cell epitopes? Furthermore, most anti-self reactive B cells are deleted (CORNALL et al. 1995) or change their receptors to a non-self specificity by somatic mutation or receptor editing, thus reducing (although not necessarily eliminating) the capacity for autoreactive antibody responses to occur (NEMAZEE 2000). Despite these unanswered questions, it is not possible to formally rule out the potential for virus-specific antibody responses to be altered by mechanisms such as idiotypic networks or cross-reactivity to self or environmental antigens. However, if these mechanisms of continuous B cell stimulation are responsible for sustaining humoral immunity then it is difficult to explain why affinity maturation ceases within a month or two following vaccination with inert antigens and is resumed only after booster vaccination (MACLENNAN et al. 1997).

The most widely accepted antigen-dependent theory explaining long-term antibody production is that memory B cells are continuously stimulated into antibody production by exposure to specific antigen in the form of immune complexes deposited on the surface of follicular dendritic cells (FDC). This model is depicted in Fig. 1D, in which viral antigen would peak during the early stages of the acute infection, stimulate high levels of antiviral antibody production, and then drop rapidly as infectious virus is cleared from the host. According to this model, viral antigen would then persist at a low level in the form of immune complexes and stimulate memory B cells to differentiate into short-lived plasma cells for several weeks, months, or possibly years. Later, as the level of antigen-antibody complexes declined, the antibody response would be expected to show a corresponding decrease due to the lower amounts of antigen available for memory B cell stimulation and differentiation into antibody-secreting plasma cells. The immune complex hypothesis stems from early studies which clearly demonstrated that antigen can be trapped and maintained on the surface of FDC with an extended half-life ($T_{1/2}$ = ~8 weeks) (TEW and MANDEL 1979). The authors estimated that about 300,000 molecules of antigen could be bound by a single FDC. With a half-life of 8 weeks, these levels of trapped antigen would disappear within about 3 years, presumably due to mechanisms such as protein degradation or consumption by memory B cells during activation and acquisition of T cell help (SLIFKA and AHMED 1996b). Unlike other models of humoral immunity that are based on cross-reactivity to self or environmental antigens, persistence of specific antigen in the form of immune complexes is compatible mechanistically with the observations of affinity maturation during the early stages of antigen-specific antibody responses. Indeed, most models of affinity maturation are based on the competition of germinal center B cells for specific antigen trapped by antibody on the surface of FDC (TARLINTON 1998; MCHEYZER-WILLIAMS and AHMED 1999; WABL et al. 1999; KELSOE 1999). However, based on the experimentally derived half-life of immune complexes, it is unlikely that protein antigens can be maintained at an immunogenic level for a time span measured in decades. Moreover, several studies have found that dendritic cells have a rapid turnover rate indicative of a half-life of a few days to, at most, a few weeks (SALOMON et al. 1998; KAMATH et al. 2000; RUEDL et al. 2000). How protein antigens are maintained for months or (theoretically) years on the surface of short-lived cells with a rapid turnover rate poses an important question worthy of further investigation.

5.2 Antigen-Independent Model for Maintaining Antibody Production: Long-Lived Plasma Cells

Plasma cells spontaneously secrete enormous amounts of antibody directly ex vivo in the absence of any form of antigenic stimulation. In addition, they typically down-regulate both membrane-bound immunoglobulin and MHC Class II expression (SLIFKA et al. 1998), indicating that antigen persistence (or the lack thereof) would have little impact on the longevity of these terminally differentiated cells. Several studies indicated that plasma cells were short-lived, with a half-life of only a few days (COOPER 1961; SCHOOLEY 1961; MAKELA and NOSSAL 1962; LEVY et al. 1987) to at most a few weeks (Ho et al. 1986). To explain the immunological paradox of how short-lived plasma cells could sustain long-lived antibody responses, most models of humoral immunity have been based on the hypothesis that continuous stimulation of memory B cells is required to maintain the short-lived plasma cell pool. However, interpretation of the early studies on plasma cell longevity is complicated because plasma cell numbers were only monitored during the first 2 weeks following vaccination, when B cells are rapidly dividing and being deleted during the course of selection and affinity maturation. Also, plasma cells were only monitored in the spleen and/or lymph nodes and subsequent studies have now shown that plasma cell numbers in these lymphoid organs drop precipitously within a few weeks after vaccination as the bone marrow becomes the major site of long-term antibody production (HILL 1976; BENNER et al. 1981; DILOSA et al. 1991; SMITH et al. 1997; TAKAHASHI et al. 1998; MCHEYZER-WILLIAMS et al. 2000; BACHMAN et al. 1994; HYLAND et al. 1994; SLIFKA et al. 1995). To better understand the role of plasma cells in sustaining long-term humoral immunity, we re-examined this issue by determining plasma cell longevity following acute LCMV infection.

LCMV-Armstrong infection of adult immunocompetent mice results in virus-specific memory B cells and antiviral antibody responses that are maintained for the life of the host (SLIFKA et al. 1998; OCHSENBEIN et al. 2000). This provides a useful model for determining the relative contributions of plasma cells and memory B cells to the maintenance of long-term antibody production. To examine the life span of virus-specific plasma cells, we took advantage of the differences in the radiosensitivity of memory B cells and plasma cells. Memory B cells, like most lymphocytes and other dividing cells, are highly sensitive to ionizing radiation and are efficiently depleted by whole body irradiation at doses above 500 RAD (MAKINODAN et al. 1962; ANDERSON and LEFKOVITS 1980; OKUDAIRA and ISHIZAKA 1981). Plasma cells, on the other hand, are terminally differentiated, non-dividing cells that are resistant to up to 10,000 RAD (MAKINODAN et al. 1967). Plasma cell longevity was measured by irradiating LCMV immune mice and then following the rate at which virus-specific antibody levels declined in the serum and the rate at which virus-specific plasma cells were lost from the spleen and bone marrow compartments. As expected, total body irradiation (600–800 RAD) resulted in the loss of more than 95% of peripheral lymphocytes. Following this high dose of irradiation, virus-specific memory B cell numbers dropped below the limits of detection within 24h as determined by both limiting dilution analysis (SLIFKA and AHMED 1996a) and adoptive transfer/challenge experiments. One of the main concerns with using virus-immune animals for these studies was the possibility that infectious virus or viral antigen trapped in immune complexes would persist after memory B cell depletion and thus stimulate new memory B cells to differentiate into plasma cells

(OCHSENBEIN et al. 2000). For these reasons, we waited between 60 to 80 days postinfection before irradiating LCMV-immune mice. At these late time points, LCMV-Armstrong can not be detected by plaque assay, histology, or RT-PCR analysis (LAU et al. 1994; SLIFKA et al. 1995). Following irradiation, the mice were reconstituted with allotypic B cells (e.g., $IgH^b \rightarrow IgH^a$) in order to distinguish host B cells from donor B cells. If infectious virus or high levels of viral antigen remained after irradiation, then the "sentinel" B cells would be expected to mount antiviral antibody responses. Although the LCMV immune mice became fully reconstituted with donor B cells, the LCMV-specific antibody response was sustained only by host plasma cells; donor B cell-derived antiviral antibody ($IgG2a^b$) was not detected. These results demonstrated that we could monitor plasma cell longevity in vivo in an environment devoid of memory B cells and lacking enough antigen to stimulate naive B cells into antibody production.

In striking contrast to the rapid loss of memory B cells, virus-specific serum antibody levels remained high for more than 500 days post-irradiation and followed a slope of decay similar to that shown in Fig. 1E. Antiviral antibody production was not maintained at steady-state levels, but instead declined slowly as a function of the number of remaining plasma cells. Importantly, virus-specific plasma cells were identified more than 1 year after memory B cell depletion – not bad for an animal with a maximum life span of only 2–3 years. Virus-specific plasma cell survival rates were also determined in different anatomical compartments using the ELISPOT technique. About 90% of virus-specific plasma cells were found in the bone marrow and their estimated half-life was 94 days whereas virus-specific plasma cells in the spleen (accounting for ~10% of the total number of antigen-specific plasma cells) had an estimated half-life of 172 days. Together, these numbers corresponded well with the results obtained from measuring serum antibody levels, suggesting that not only are plasma cells long-lived, but they appear to maintain their extremely high rates of antibody secretion for extended periods of time.

To confirm the existence of long-lived plasma cells by an alternative method and to reduce the concerns over the role of persisting antigen in peripheral tissues, we performed adoptive transfer experiments in which bone marrow containing virus-specific plasma cells was injected into naive recipients. The cells were treated with mitomycin C prior to transfer to ensure that any contaminating memory B cells in the bone marrow (OCHSENBEIN et al. 2000) would be unable to participate in the antibody response observed in the recipient hosts. Mitomycin C is an antiproliferative agent that permanently abrogates cell division due to its covalent interactions with DNA (TOMASZ et al. 1974) and completely blocks memory B cell differentiation into plasma cells in vitro (SLIFKA and AHMED 1996a). Following adoptive transfer of LCMV-specific plasma cells, we observed prolonged virus-specific antibody responses that declined slowly over the course of several months. Consistent with our findings in irradiated (i.e., memory T cell and memory B cell-depleted) mice, these results indicated that the transferred plasma cells were long-lived with an average half-life of more than 70 days. The antigen-specific antibody response observed in the recipient mice was not due to the transfer of virus or viral antigen because if the bone marrow cells were lysed by sonication prior to transfer, no antibody was detected in the serum of the recipient hosts. These results extended our previous studies by demonstrating that plasma cells were not only long-lived, but were capable of survival in the absence of specific antigen, a result later verified by an independent study (MANZ et al. 1998).

The experiments on plasma cell longevity in the LCMV model system are further supported by convincing proof of long-lived plasma cells identified by metabolically labeling cells in vivo with either ^{3}H-thymidine or bromodeoxyuridine (BrdU). Following vaccination, ^{3}H-labeled plasma cells in the draining lymph nodes of rats follow a biphasic rate of decline (MILLER 1964). Radiolabeled plasma cells disappeared rapidly from the draining nodes for the first several weeks. However, further analysis indicated that plasma cell decline was biphasic in that the rate of plasma cell loss became much slower at later time points and a number of ^{3}H-labeled plasma cells could still be identified in the draining lymph nodes as late as 6 months post-vaccination. Previous studies (COOPER 1961; SCHOOLEY 1961; MAKELA and NOSSAL 1962; NOSSAL and MAKELA 1962) had monitored plasma cell numbers only during the first 1–2 weeks post-vaccination and were probably unaware of the long-lived plasma cells that remained after the local immune response had diminished and homeostasis was re-established. A similar biphasic plasma cell response was also observed following acute LCMV infection (SLIFKA et al. 1998). Virus-specific plasma cell numbers in the spleen at early time points (8–30 days post-infection) declined with an apparent half-life of 5.7 days whereas splenic plasma cell longevity at later time points (60–300 days post-infection) averaged a half-life of 172 days. The important message obtained by comparing these two models, is that plasma cells induced at early stages of an immune response are likely to be short-lived whereas at later time points, i.e., after affinity maturation has occurred, long-lived plasma cells emerge and function to maintain antigen-specific antibody levels.

Metabolic labeling of ovalbumin-specific plasma cells with BrdU provides independent confirmation of extended plasma cell longevity (MANZ et al. 1997). This study demonstrated that more than 60% of the long-lived plasma cells identified at 120 days post-booster vaccination were labeled during the first 19 days after antigen exposure, indicating that they had a non-dividing life span of greater than 90 days. Second, and perhaps most surprisingly, the authors noted that nearly all of the plasma cells were produced within the first 2 months post-vaccination with very few additional plasma cells developing after this relatively short period of time. This would suggest that antigen-dependent proliferation and differentiation of memory B cells may play a role during early stages of the immune response, but appears to be of little consequence to the long-term maintenance of plasma cell numbers in the bone marrow. Although a specific plasma cell half-life was not calculated by either of these studies (MILLER 1964; MANZ et al. 1997), the existence of long-lived plasma cells was nevertheless clearly documented. These studies are consistent with the identification of long-lived plasma cells following acute LCMV infection and together provide irrefutable proof of plasma cells with an extended life span.

6 Conclusions

Following acute LCMV infection, virus-specific serum antibody titers are maintained for the life of the murine host. To better understand the nature of such long-term antibody production, the anatomical location of virus-specific plasma cells and their relative life span in vivo were determined. The spleen and lymph nodes are the initial sites of antibody synthesis, but after resolution of the acute viral infection and affinity maturation, plasma cell numbers in these organs drop precipitously as the bone marrow becomes the predominate reservoir of virus-specific plasma cells. Why plasma cells are targeted to the bone marrow is still a mystery but this phenomenon may provide a selective advantage for the host by reducing the immunological burden of peripheral lymphoid organs that have a finite capacity for sustaining plasma cell numbers. Conventional models of long-term antibody production have been based on the dogma that plasma cells have a short life span and that continuous antigenic stimulation of memory B cells is required to repopulate short-lived plasma cell populations. However, this notion has been drawn into question by studies in the LCMV model system in which serum antibody production can be maintained for more than a year by long-lived virus-specific plasma cells in the absence of a functional memory B cell pool. The existence of long-lived plasma cells has been confirmed independently by both adoptive transfer experiments as well as by studies of plasma cell turnover rates following in vivo labeling of cells with either ^{3}H-thymidine or BrdU. Together, these studies highlight the important role of long-lived plasma cells in the maintenance of long-term humoral immunity and open a new field of research as we try to understand the underlying mechanisms of plasma cell longevity.

References

Ahmed R, Gray D (1996) Immunological memory and protective immunity: understanding their relation. Science 272:54–60

Ahmed R, Salmi A, Butler LD, Chiller JM, Oldstone MBA (1984) Selection of genetic variants of lymphocytic choriomeningitis virus in spleens of persistently infected mice. Role in suppression of cytotoxic T lymphocyte response and viral persistence. J Exp Med 160:521–540

Anderson RE, Lefkovits I (1980) Effects of irradiation on the in vitro immune response. Expl Cell Biol 48:255–278

Asano MS, Ahmed R (1996) CD8 T cell memory in B cell-deficient mice. J Exp Med 183:2165–2174

Bachman MF, Kundig TM, Odermatt B, Hengartner H, Zinkernagel RM (1994) Free recirculation of memory B cells versus antigen-dependent differentiation to antibody forming cells. J Immunol 153:3386–3397

Baldridge JR, McGraw TS, Paoletti A, Buchmeier MJ (1997) Antibody prevents the establishment of persistent arenavirus infection in synergy with endogenous T cells. J Virol 71:755–758

Battegay M, Kyburz D, Hengartner H, Zinkernagel RM (1993) Enhancement of disease by neutralizing antiviral antibodies in the absence of primed antiviral cytotoxic T cells. Eur J Immunol 23:3236–3241

Benner R, Hijmans W, Haaijman JJ (1981) The bone marrow: the major source of serum immunoglobulins, but still a neglected site of antibody formation. Clin Exp Immunol 46:1–8

Brundler MA, Aichele P, Bachmann M, Kitamura D, Rajewsky K, Zinkernagel RM (1996) Immunity to viruses in B cell-deficient mice: influence of antibodies on virus persistence and on T cell memory. Eur J Immunol 26:2257–2262

Buchmeier MJ, Welsh RM, Dutko FJ, Oldstone MBA (1980) The virology and immunobiology of lymphocytic choriomeningitis virus infection. Adv Immunol 30:275–331

Byrne JA, Oldstone MBA (1984) Biology of cloned cytotoxic T lymphocytes specific for lymphocytic choriomeningitis virus: clearance of virus in vivo. J Virol 51:682–686

Cerny A, Sutter S, Bazin H, Hengartner H, Zinkernagel RM (1988) Clearance of lymphocytic chorio-meningitis virus in antibody- and B-cell- deprived mice. J Virol 62:1803–1807

Chantler JK, Ford DK, Tingle AJ (1981) Rubella-associated arthritis: rescue of rubella virus from peripheral blood lymphocytes two years postvaccination. Infect Immun 32:1274–1280

Cohen D, Green MS, Katzenelson E, Slepson R, Bercovier H, Wiener M (1994) Long-term persistence of anti-diptheria toxin antibodies among adults in Israel. Eur J Epidemiol 10:267–270

Cooper EH (1961) Production of lymphocytes and plasma cells in the rat following immunization with human serum albumin. Immunology 4:219–231

Cornall RJ, Goodnow CC, Cyster JG (1995) The regulation of self-reactive B cells. Curr Opin Immunol 7:804–811

Dilosa RM, Maeda K, Masuda A, Szakal AK, Tew JG (1991) Germinal center B cells and antibody production in the bone marrow. J Immunol 146:4071–4077

Fahey JL, Sell S (1965) The immunoglobulins of mice: The metabolic (catabolic) properties of five immunoglobulin classes. J Exp Med 122:41–58

Gottlieb S, McLaughlin FX, Levine L, Latham WC, Edsall G (1964) Long-term immunity to tetanus – a statistical evaluation and its clinical implications. Am J Pub Health 54:961–971

Helmreich E, Kern M, Eisen HN (1961) The secretion of antibody by isolated lymph node cells. J Biol Chem 236:464–473

Hibi T, Dosch HM (1986) Limiting dilution analysis of the B cell compartment in human bone marrow. Eur J Immunol 16:139–145

Hill SW (1976) Distribution of plaque-forming cells in the mouse for a protein antigen. Evidence for highly active parathymic lymph nodes following intraperitoneal injection of hen lysozyme. Immunology 30:895–906

Ho F, Lortan JE, MacLennan I, Khan M (1986) Distinct short-lived and long-lived antibody-producing cell populations. Eur J Immunol 16:1297–1301

Holt LB (1950) Developments in diphtheria prophylaxis. William Heinemann, Ltd. London.

Hyland L, Sangster M, Sealy R, Coleclough C (1994) Respiratory virus infection of mice provokes a permanent humoral immune response. J Virol 68:6083–6086

Jerne NK (1984) Idiotypic networks and other preconceived ideas. Immunol Rev 79:5–24

Kagi D, Ledermann B, Burki K, Seiler P, Odermatt B, Olsen KJ, Podack ER, Zinkernagel RM, Hengartner H (1994) Cytotoxicity mediated by T cells and natural killer cells is greatly impaired in perforin-deficient mice. Nature 369:31–37

Kamath AT, Pooley J, O'Keeffe MA, Vremec D, Zhan Y, Lew AM, D'Amico A, Wu L, Tough DF, Shortman K (2000) The Development, Maturation, and Turnover Rate of Mouse Spleen Dendritic Cell Populations. J Immunol 165:6762–6770

Kelsoe G (1999) V(D)J hypermutation and receptor revision: coloring outside the lines. Curr Opin Immunol 11:70–75

Kimmig B, Lehmann-Grube F (1979) The immune response of the mouse to lymphocytic choriomeningitis virus. I. Circulating antibodies. J Gen Virol 45:703–710

Kjeldsen K, Heron I, Simonsen O (1985) Immunity against diphtheria 25–30 years after primary vaccination in childhood. Lancet 1:900–902

Lau LL, Jamieson BD, Somasundaram T, Ahmed R (1994) Cytotoxic T-cell memory without antigen. Nature 369:648–652

Lehmann-Grube F, Assmann U, Loliger C, Moskophidis D, Lohler J (1985) Mechanism of recovery from acute virus infection. I. Role of T lymphocytes in the clearance of lymphocytic choriomeningitis virus from spleens of mice. J Immunol 134:608–615

Lehmann-Grube F, Peralta M, Bruns M, Lohler J (1983) Persistent infection of mice with the lymphocytic choriomeningitis virus. In: (Fraenkel-Conrat H, Wagner RR (eds) Comprehensive virology. Plenum Publishing, New York, pp 43–103

Levy M, Vieira P, Coutinho A, Freitas A (1987) The majority of "natural" immunoglobulin-secreting cells are short-lived and the progeny of cycling lymphocytes. Eur J Immunol 17:849–854

Maclennan ICM, Casamayor-Palleja M, Toellner KM, Gulbranson-Judge A, Gordon J (1997) Memory B-cell clones and the diversity of their members. Semin Immunol 9:229–234

Makela O, Nossal GJV (1962) Autoradiographic studies on the immune response. J Exp Med 115: 231–245

Makinodan T, Kastenbaum MA, Peterson WJ (1962) Radiosensitivity of spleen cells from normal and pre-immunized mice and its significance to intact animals. J Immunol 88:31–37

Makinodan T, Nettesheim P, Morita T, Chadwick CJ (1967) Synthesis of antibody by spleen cells after exposure to kiloroentgen doses of ionizing radiation. J Cell Physiol 69:355–366

Manz RA, Lohning M, Cassese G, Thiel A, Radbruch A (1998) Survival of long-lived plasma cells is independent of antigen. Int Immunol 10:1703–1711

Manz RA, Thiel A, Radbruch A (1997) Lifetime of plasma cells in the bone marrow. Nature 388:133–134

McHeyzer-Williams LJ, Cool M, McHeyzer-Williams MG (2000) Antigen-specific B cell memory. Expression and replenishment of a novel b220(−) memory b cell compartment. J Exp Med 191:1149–1166

McHeyzer-Williams MG, Ahmed R (1999) B cell memory and the long-lived plasma cell. Curr Opin Immunol 11:172–179

Miller JJ (1964) An autoradiographic study of plasma cell and lymphocyte survival in rat popliteal lymph nodes. J Immunol 92:673–681

Moskophidis D, Lehmann-Grube F (1984) The immune response of the mouse to lymphocytic choriomeningitis virus. IV. Enumeration of antibody-producing cells in spleens during acute and persistent infection. J Immunol 133:3366–3370

Nemazee D (2000) Role of B cell antigen receptor in regulation of V(D)J recombination and cell survival. Immunol Res 21:259–263

Nossal GJV, Makela O (1962) Autoradiographic studies on the immune response. I. The kinetics of plasma cell proliferation. J Exp Med 115:209–230

Ochsenbein AF, Pinschewer DD, Sierro S, Horvath E, Hengartner H, Zinkernagel RM (2000) Protective long-term antibody memory by antigen-driven and T help-dependent differentiation of long-lived memory B cells to short-lived plasma cells independent of secondary lymphoid organs. Proc Natl Acad Sci USA 97:13263–13268

Okudaira H, Ishizaka K (1981) Reaginic antibody formation in the mouse. XI. Participation of long-lived antibody-forming cells in persistent antibody formation. Cell Immunol 58:188–201

Oldstone MB, Buchmeier MJ, Doyle MV, Tishon A (1980) Virus-induced immune complex disease: specific anti-viral antibody and C1q binding material in the circulation during persistent lymphocytic choriomeningitis virus infection. J Immunol 124:831–838

Oldstone MB, Dixon FJ (1970) Persistent lymphocytic choriomeningitis viral infection. 3. Virus-antiviral antibody complexes and associated chronic disease following transplacental infection. J Immunol 105:829–837

Oldstone MBA, Ahmed R, Byrne J, Buchmeier MJ, Riviere Y, Southern P (1985) Virus and immune responses: lymphocytic choriomeningitis virus as a prototype model of viral pathogenesis. Brit Med Bull 41:70–74

Oldstone MBA, Dixon FJ (1967) Lymphocytic choriomeningitis: production of antibody by "tolerant" infected mice. Science 158:1193–1195

Oldstone MBA, Dixon FJ (1969) Pathogenesis of chronic disease associated with persistent lymphocytic choriomeningitis viral infection. I. Relationship of antibody production to disease in neonatally infected mice. J Exp Med 129:483–505

Panum PL (1847) Beobachtungen uber das Maserncontagium. Virch Arch 1:492

Paul JR, Riordan JT, Melnick JL (1951) Antibodies to three different antigenic types of poliomyelitis virus in sera from North Alaskan Eskimos. Am J Hyg 54:275–285

Ruedl C, Koebel P, Bachmann M, Hess M, Karjalainen K (2000) Anatomical origin of dendritic cells determines their life span in peripheral lymph nodes. J Immunol 165:4910–4916

Salomon B, Cohen JL, Masurier C, Klatzmann D (1998) Three populations of mouse lymph node dendritic cells with different origins and dynamics. J Immunol 160:708–717

Sawyer WA (1931) Persistence of yellow fever immunity. J Prev Med 5:413–428

Schneider-Schaulies J, Dunster LM, Schneider-Schaulies S, ter Meulin V (1995) Pathogenitic aspects of measles virus infections. Vet Microbiol 44:113–125

Schooley JC (1961) Autoradiographic observations of plasma cell formation. J Immunol 86:331–337

Seiler P, Brundler MA, Zimmermann C, Weibel D, Bruns M, Hengartner H, Zinkernagel RM (1998a) Induction of protective cytotoxic T cell responses in the presence of high titers of virus-neutralizing antibodies:implications for passive and active immunization. J Exp Med 187:649–654

Seiler P, Kalinke U, Rulicke T, Bucher EM, Bose C, Zinkernagel RM, Hengartner H (1998b) Enhanced virus clearance by early inducible lymphocytic choriomeningitis virus-neutralizing antibodies in immunoglobulin- transgenic mice. J Virol 72:2253–2258
Slifka MK, Ahmed R (1996a) Limiting dilution analysis of virus-specific memory B cells by an ELISPOT assay. J Immunol Methods 199:37–46
Slifka MK, Ahmed R (1996b) Long-term humoral immunity against viruses: revisiting the issue of plasma cell longevity. Trends Microbiol 4:394–400
Slifka MK, Antia R, Whitmire JK, Ahmed R (1998) Humoral immunity due to long-lived plasma cells. Immunity 8:363–372
Slifka MK, Matloubian M, Ahmed R (1995) Bone marrow is a major site of long-term antibody production after acute viral infection. J Virol 69:1895–1902
Slifka MK, Whitmire JK, Ahmed R (1997) Bone marrow contains virus-specific cytotoxic T lymphocytes. Blood 90:2103–2108
Smith KGC, Light A, Nossal GJV, Tarlinton DM (1997) The extent of affinity maturation differs between the memory and antibody-forming cell compartments in the primary immune response. EMBO 16:2996–3006
Sze DM, Toellner KM, de Vinuesa CG, Taylor DR, MacLennan IC (2000) Intrinsic constraint on plasmablast growth and extrinsic limits of plasma cell survival. J Exp Med 192:813–822
Takahashi Y, Dutta PR, Cerasoli DM, Kelsoe G (1998) In situ studies of the primary immune response to (4-hydroxy-3- nitrophenyl)acetyl. V. Affinity maturation develops in two stages of clonal selection. J Exp Med 187:885–895
Talbot PJ, Buchmeier MJ (1987) Catabolism of homologous murine monoclonal hybridoma IgG antibodies in mice. Immunology 60:485–489
Tarlinton D (1998) Germinal centers:form and function. Curr Opin Immunol 10:245–251
Tew JG, Mandel TE (1979) Prolonged antigen half-life in the lymphoid follicles of specifically immunized mice. Immunology 37:69–76
Thomsen AR, Marker O (1988) The complementary roles of cellular and humoral immunity in resistance to re-infection with LCM virus. Immunology 65:9–15
Thomsen AR, Volkert M, Marker O (1985) Different isotype profiles of virus-specific antibodies in acute and persistent lymphocytic choriomeningitis virus infection in mice. Immunology 55:213–223
Tomasz M, Mercado CM, Olson J, Chatterjie ON (1974) The mode of interaction of mitomycin C with deoxyribonucleic acid and other polynucleotides in vitro. Biochemistry 13:4878–4887
Ukkon P, Hovi T, von Bonsdorff C-H, Saikku P, Penttinen K (1984) Age-specific prevalence of complement-fixing antibodies to sixteen viral antigens: a computer analysis of 58,500 patients covering a period of eight years. J Med Virol 13:131–148
Vieira P, Rajewsky K (1988) The half-lives of serum immunoglobulins in adult mice. Eur J Immunol 18:313–316
Wabl M, Cascalho M, Steinberg C (1999) Hypermutation in antibody affinity maturation. Curr Opin Immunol 11:186–189
Walsh CM, Matloubian M, Liu CC, Ueda R, Kurahara CG, Christensen JL, Huang MT, Young JD, Ahmed R, Clark WR (1994) Immune function in mice lacking the perforin gene. Proc Natl Acad Sci USA 91:10854–10858
Welsh RM, McNally JM (1999) Immune deficiency, immune silencing, and clonal exhaustion of T cell responses during viral infections. Curr Opin Microbiol 2:382–387
Whitton JL, Sheng N, Oldstone MBA, McKee TA (1993) A "string-of-beads" vaccine, comprising linked minigenes, confers protection from lethal-dose virus challenge. J Virol 67:348–352
Wright KE, Buchmeier MJ (1991) Antiviral antibodies attenuate T-cell-mediated immunopathology following acute lymphocytic choriomeningitis virus infection. J Virol 65:3001–3006
Zinkernagel RM, Doherty PC (1979) MHC-restricted cytotoxic T cells:studies on the biological role of polymorphic major transplantation antigens determining T-cell restriction-specificity, function, and responsiveness. Adv Immunol 27:51–177

Biology and Pathogenesis of Lymphocytic Choriomeningitis Virus Infection

M.B.A. Oldstone

1	Introduction	83
2	Balance Between Acute and Persistent Infection	86
3	Persistent Infection	92
3.1	Establishment with LCMV: General	92
3.2	Virus-Antibody Immune Complex Formation and Disease in Persistent Viral Infection	97
4	Persistent Virus Infection Alters the Function of Differentiated Cells Leading to Disease	100
4.1	General Concept and In Vitro Findings	100
4.2	Growth Hormone Deficiency Syndrome	102
4.3	Alterations in Behavior and Learning Associated with Persistent LCMV Infection	108
5	Conclusions	109
	References	111

1 Introduction

The strength of the lymphocytic choriomeningitis virus (LCMV) model rests on the following five foundations. First, the virus in vivo in its natural host, the mouse, or in vitro in cultured cells is non-cytolytic. This quality allows clear separation of effects caused by the virus from those caused by the host immune system. Consequently, the host cell control of viral infection as opposed to how virus interacts with cells to distort their functions without killing them can be decoded. Second, reactions to LCMV infection can encompass a widely diverse range of immune responses (Fig. 1). Usually when immunocompetent adult mice are injected with LCMV, they generate a marked immune response to eliminate the infectious agent. Although their innate responses include the production of interferon (IFN), macrophages and natural killer (NK) cells (MULLER et al. 1994; see reviews BUCHMEIER et al. 1980; BORROW and OLDSTONE 1997; see chapter by Biron et al., this volume), it is the adoptive immune response – primarily the virus-specific $CD8^+$ CTL response – that is responsible for virus clearance. This protective response proceeds

---

Division of Virology, Department of Neuropharmacology, The Scripps Research Institute, 10550 N. Torrey Pines Road, La Jolla, CA 92037, USA

Age of host	Route	Generation CTL	Phenotype	Immunologic reason
Virus → Adult	IP, SC, IV	↑CTL	Clear viral infection, immunity	Positively selected LCMV antigen specific CD8+ CTL in the periphery are clonally expanded and lyse virus infected cells. If virus infects leptomeningeal or choroid plexus cells in the brain CTL induces death.
	IC	↑CTL	Leptomeningitis, death	
Virus → Newborn	IP, IV, IC	↓CTL	Persistent infection	High affinity LCMV antigen specific T cells are negatively selected out by the thymus. Low affinity antigen specific cells reside in the periphery.
Immuno-suppressive virus → Adult	IV	↓CTL	Persistent infection	Immunosuppressive viruses bind preferentially and at high affinity to α-DG, a cellular receptor for arenaviruses. Among immune cells, α-DG is preferentially expressed on interdigitating dendritic cells (DEC 205+, CD11c+).
Persistently infected adult mouse	+ Transfer immune memory T cells (CD8 + CD4)		Clear viral infection	Mice with low number and low affinity LCMV antigen specific CD8+ and CD4+ T cells are reconstituted with high affinity LCMV antigen specific CD8+ and CD4+ T cells. Virus cleared by CTL directed lysis and by action of IFN-γ and TNF-α cytokines.

Fig. 1. The several disparate scenarios following different routes of infection in different aged mice challenged with LCMV and the role of cytotoxic T lymphocytes (*CTL*)

mostly by lysing virally infected cells through a perforin-mediated pathway. The sensitivity and specificity of this viral antigen-specific $CD8^+$ and the viral antigen-specific $CD4^+$ T cell response have allowed experimentation that revealed how these cells function in acute infection and in the maintenance of immunologic memory or its loss during persistent infection. Quantitative analysis of such antigen-specific T cell responses during specific stages of expansion, burst size, retraction and maintenance was also achieved (see chapters by; McNally and Welsh and Homann, this volume). Third, LCMV easily establishes persistent infection in vivo via in utero or congenital vertical transmission, in newborn mice, or in immunocompetent adult mice inoculated with immunosuppressive viral variants. The basic mechanism(s) by which virus persists and the resultant immunosuppression that maintains this long-term infection are amenable to experimental analysis. For example, LCMV infection by all these modalities, except for immunosuppressive viral variants in adult mice, results in an aborted or inefficient anti-LCMV T cell response even though the same animals' immune responses to other viruses or humoral responses to particulate or soluble foreign antigens are not impaired (TISHON et al. 1993a). In contrast, a generalized suppression immune response occurs in normal adult mice following high-dose (2×10^6 PFU or higher i.v.) LCMV inoculation (TISHON et al. 1993a). Although all these models accommodated persistent infection, the mechanisms by which it occurs differ. That induced in utero, congenitally or in newborn mice stems from the infection of thymic cells and the specific removal (negative selection) of lymphocytes with potential responsiveness to LCMV (reviewed in BORROW and OLDSTONE 1997). In contrast, the persistence and generalized immunosuppression following infection of adult mice with immunosuppressive variants is caused by selective LCMV infection of

DEC205$^+$ and CD11c$^+$ interdigitating dendritic cells (SEVILLA et al. 2000; SMELT et al. 2001), the major antigen-presenting cells of the host (Figs. 1, 2). Regardless of the mechanism of initiating infection, continuous expression of viral genes in differentiated cells like neurons, endocrine cells, or immune system cells, can alter homeostasis and cause disease (OLDSTONE 1989, 1993). This phenomenon can also be studied in cultures of specialized cells that make differentiation products such as GH, neurotransmitters, cytokines, or immunoglobulins (Ig). As with the in vivo scenario, infection in vitro affects more than 95% of cells within a few days, and the infection lasts during continuous passages. The fourth basic asset of LCMV is a simple one in that this virus contains four genes, two each placed on two distinct

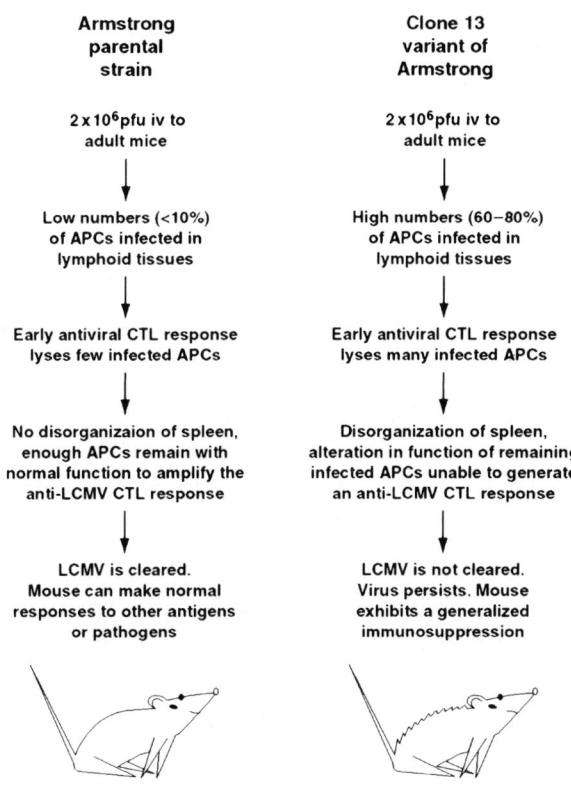

Viral tropism for antigen presenting cells is of critical importance. APCs are DEC 205$^+$ and CD11c$^+$ interdigitating dendritic cells and among immune cells preferentially express the cellular receptor for LCMV, α-DG on their surface. Because of a single acid mutation in the viral glycoprotein, clone 13 binds at 2.5 logs higher affinity to α-DG than does the Armstrong strain.

Fig. 2. Biologic differences between wild-type LCMV ARM, which generates a robust CTL response and clears the virus infection, and its variant Clone 13, which fails to generate a sufficient primary day 7 CTL response to clear the viral infection resulting in a persistent viral infection. The mechanism by which this occurs is shown in Fig. 5 (also see SEVILLA et al. 2000; SMELT et al. 2001)

pieces of genomic RNA. Thus, one can make genetic reassortants between LCMV strains or variants, each with differing phenotypes to map the viral gene(s) involved. Further, (see chapter by Lee and de la Torre, Vol. I) the ability to use reverse genetics to complement the impressive biologic observations from experiments with LCMV infection offers a unique opportunity to dissect and understand basic building blocks of viral pathogenesis. Fifth, and most germane, events initially uncovered by the study of LCMV in its natural rodent hosts have proved applicable to infections caused by a wide spectrum of DNA and RNA viruses and with cell-mediated immune control of bacteria, parasites and tumors in humans.

The foregoing precepts reflect the beauty of LCMV as a model of infection and its value to the field not only of viral but also of microbial pathogenesis. Table 1 in the Introduction to this volume and the contribution by Zinkernagel (this volume) list the remarkable number of key concepts in immunology and virology that were first defined in studies using this virus. These concepts include basic immunologic phenomena such as the major histocompatibility complex (MHC) restriction of T cell recognition (ZINKERNAGEL and DOHERTY 1974), that tissue injury is mediated by antiviral immune responses associated with acute and persistent viral infections, and that the non-cytopathic LCMV can cause disease by interfering with the differentiated functions of infected cells.

2 Balance Between Acute and Persistent Infection

The understanding of virus-induced immune response disease and the related immunopathology has its roots in the studies reported by ROWE (1954). He showed that suppression of immune responses changed the ordinarily lethal, acute infection with LCMV to a persistent infection in susceptible mice. The work of ROWE and later of other investigators who used neonatal thymectomy, genetically athymic mice, irradiation, antilymphoid drugs, or antithymocyte sera, etc. (reviewed in BORROW and OLDSTONE 1997), extended the concept of immune response-mediated injury during viral infection. Thereafter, the role of lymphocytes in mediating virus-induced immunologic injury was delineated. First, experiments performed inde-

Table 1. Immune factors that control acute and persistent LCMV infection

	Control of:	
	Acute viral infection	Persistent viral infection
CD^8+ T cells	+++	+++
Perforin	+++	+++
CD^4+ T cells	No	+++
IFN-γ	No	+++
B cells	No	No
L-selectin	No	+++
CD40 ligand	No	+++

pendently by LUNDSTEDT (1969) and OLDSTONE et al. (1969) showed that lymphocytes obtained 6–9 days after an acute LCMV infection killed LCMV-infected targets in vitro. Next, such lymphocytes were found to be of thymic origin and to bear Thy1.2 markers on their surfaces (COLE et al. 1972; MARKER and VOLKERT 1973). GILDEN et al. (1972a,b) showed that mice surviving an ordinarily lethal dose of LCMV when immunosuppressed during acute LCMV disease died when reconstituted with syngeneic, immune T lymphocytes. ZINKERNAGEL and DOHERTY (1974) defined the need for two signals between the cytolytic T lymphocyte and its infected target, namely, virus specificity and H-2 restriction, so that the killing of LCMV-infected cells could proceed. Armed with these data, others (MIMS and BLANDEN 1972; ZINKERNAGEL and WELSH 1976) used virus-specific splenic lymphocytes to reconstitute immunosuppressed, infected H-2-matched mice and thereby lowered their viral titers.

In a series of mice whose genes encoding various T cell components, cytokine genes or effector molecules, like perforin, were knocked out or neutralized by adding antibodies to the products of these genes (YOUNG et al. 1989; WALSH et al. 1994; KAGI et al. 1995; TISHON et al. 1995; KLAVINSKIS et al. 1989a; LEIST and ZINKERNAGEL 1990), investigators showed that $CD8^+$ T cells and perforin were the cardinal players in the clearance of acute viral infection and that neither $CD4^+$ T cells, IFN-γ, nor tumor necrosis factor (TNF)-α were required (Table 1).

In perforin-deficient mice, $CD8^+$ T cells became activated with LCMV infection, but the virus was not eliminated. By contrast, mice carrying the spontaneous mutations *gld* (nonfunctional Fas ligand) or *lpr* (inactive Fas) controlled acute LCMV infection with normal kinetics, indicating that perforin-dependent cytotoxicity is crucially involved in the clearance of acute LCMV infection but that no measurable involvement of Fas-dependent pathways occurs (KAGI et al. 1995; WALSH et al. 1994). Further, the fact that perforin-deficient mice were unable to clear LCMV indicates that cytokine production alone by $CD8^+$ cytotoxic T lymphocytes (CTL) or $CD4^+$ T cells is likely to be insufficient to mediate the clearance of an acute viral infection. In agreement with these findings are results showing resolution of the acute LCMV infection in mice treated with antibodies to IFN-γ or TNF-α, and IFN-γ-deficient (knockout) mice are able to clear an acute infection caused by this virus (TISHON et al. 1995; KLAVINSKIS et al. 1989a; LEIST and ZINKERNAGEL 1990). However, antiviral cytokines can modulate the virus-immune response balance, and thus play a role in combating acute LCMV infection (GUIDOTTI et al. 1999). Recently, the use of tetramer- (MURALI-KRISHNA et al. 1998) and intracellular cytokine-staining techniques (BUTZ and BEVAN 1998; MURALI-KRISHNA et al. 1998) combined with the identification of MHC class I D^b-restricted peptides and the use of CTL clones (OLDSTONE et al. 1988; KLAVINSKIS et al. 1989a; YANAGI et al. 1990; GAIRIN et al. 1995; HUDRISIER et al. 1996; BAENZIGER et al. 1986; BYRNE et al. 1984; ANDERSON et al. 1985) and class II I^b-(VARGA and WELSH 1998; OXENIUS et al. 1998) has allowed fine mapping of the total anti-LCMV antigen-specific $CD8^+$ and $CD4^+$ T cell response (Fig. 3; BERGER et al. 2000; HOMANN et al. 2001).

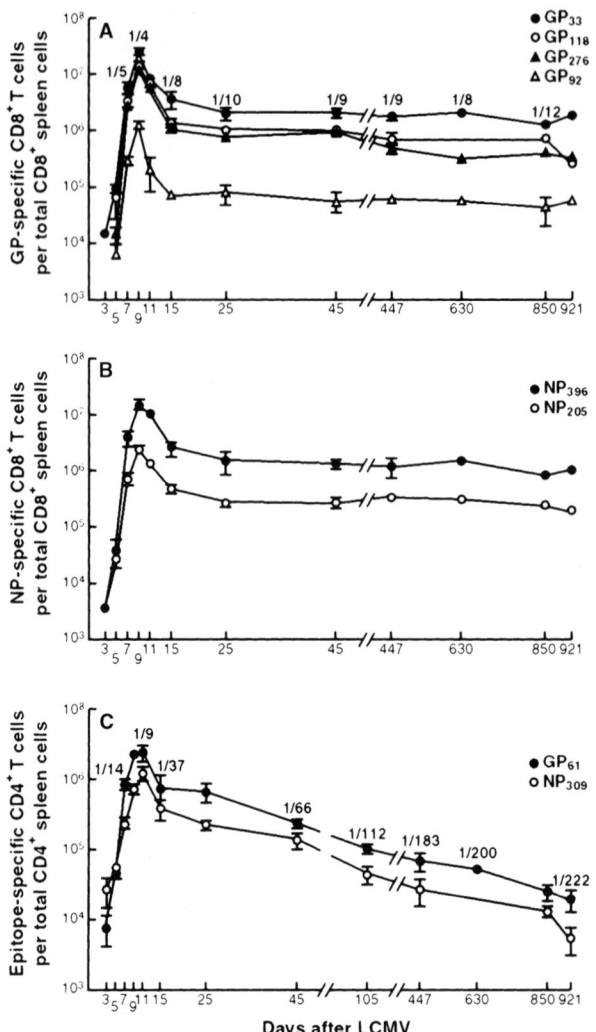

Fig. 3. Generation of LCMV-specific $CD8^+$ CTL and $CD4^+$ T helper cells during the course of an acute LCMV infection and the differences between maintenance of $CD8^+$ and $CD4^+$ antigen-specific memory (see the chapter by Homann, this volume; HOMANN et al. 2001)

Both class I MHC-restricted (largely $CD8^+$) and class II MHC-restricted (generally $CD4^+$) T cell responses are elicited after a virus infection (Fig. 3). "Professional" antigen-presenting cells, primarily dendritic cells (5–10,000× greater in activity than macrophages), and mainly also macrophages, process peptides made from viral proteins for presentation in association with class I and II molecules. MHC class II molecules induce $CD4^+$ T cell responses, whereas MHC class I induce primarily $CD8^+$ T cell responses. In most instances, peptides presented in association with class I are derived exclusively from proteins synthesized within the cell. As viruses replicate intracellularly, peptide fragments of the viral proteins become associated and presented with class I MHC molecules. This sequence targets $CD8^+$ T cell activation and subsequent interaction with virus-infected cells. Identification of the epitopes of LCMV reactive with $CD8^+$ T cells has confirmed that non-structural internal virion proteins are frequently recognized in addition to virion surface proteins. Antiviral T cells enjoy several advantages from the recognition of non-structural, internal viral proteins: (1) the number of potential viral epitopes is increased; (2) these proteins tend to be more conserved among viruses than are virion surface proteins. Thus, T cell responses can cross-react between strains thereby providing protection against related viruses; and (3) targeting of T cells to the first non-structural viral proteins available in an infected cell promotes destruction of the infected cell before production of progeny virions is complete; consequently, the factory manufacturing infectious virus is removed.

The recent development of new techniques for quantitating antigen-specific T cells has allowed the reassessment of the magnitude and kinetics of both virus-specific $CD4^+$ and, in particular, $CD8^+$ T cell responses initially in LCMV infected mice (MURALI-KRISHNA et al. 1998; BUTZ and BEVAN 1998) and subsequently in virus-infected humans. Antigen-specific T cells can now be identified by staining with fluorescently labeled tetrameric complexes of MHC molecules folded around a specific peptide or, alternatively, on the basis of their ability to produce cytokines in response to antigen stimulation. These assays quantitated single cytokine-producing cells, either by ELISPOT or by intracellular staining of cells from which cytokine secretion was blocked using brefeldin A. The use of these techniques has shown that antigen-specific $CD8^+$ T cells can expand to far higher numbers than previously appreciated during the acute phase of a virus infection, e.g., more than 50% of the $CD8^+$ T cells in the spleen of mice and on occasion 80% were found to be virus-specific at the peak of the acute response to LCMV (MURALI-KRISHNA et al. 1998; BUTZ and BEVAN 1998; HOMANN et al. 2001).

LCMV can escape recognition by virus-specific $CD8^+$ T cells by acquiring mutations that prevent binding of viral peptide epitopes to MHC molecules or prevent recognition by the T cell receptor (TCR) (AEBISCHER et al. 1991; LEWICKI et al. 1995a,b; PIRCHER et al. 1990). Several rules govern viral peptide-MHC interactions (reviewed by WHITTON and OLDSTONE 2001; BJORKMAN et al. 1987; FALK et al. 1991; ROTZSCHKE et al. 1990; GARCIA et al. 1996; RAMMENSEE et al. 1995; GAIRIN et al. 1995; YOUNG et al. 1994; MATSUMURA et al. 1992; HUDRISIER et al. 1996). The bound peptide sequence is linear, resulting from proteolytic fragmentation of a viral protein synthesized within the cell. Studies of endogenously processed viral peptides indicate that they vary in length from 8–11 amino acids (aa) and display MHC allele-specific motifs. Mutational and crystallographic studies of MHC molecules complexed with viral peptide show that the molecule's flexible conformation allows these 8–11aa to bind within the MHC groove once their anchoring residues are fixed (BJORKMAN et al. 1987; RAMMENSEE et al. 1995; FALK et al. 1991; MATSUMURA et al. 1992). Analysis of residues flanking the anchoring residue(s) indicates the critical importance of minor pockets of MHC-binding clefts in the peptide selection process, leading to the concept that these structural factors are likely to be responsible for the preferential choice of specific peptides observed in interactions between viral components and MHC molecules (HUDRISIER et al. 1996). Mutation of residues in a viral epitope that is important for MHC binding can ablate peptide MHC binding altogether, or reduce the affinity of a peptide-MHC interaction so that the peptide-MHC complex has an extremely short half-life and is unlikely to trigger T cell activation. Alternatively, mutations at residues involved in MHC binding can cause peptides to bind to MHC in a distorted conformation, so that the TCR contact surface is altered (BERTOLETTI et al. 1994a; reviewed in KOUP 1994 and OLDSTONE 1991, 1997). TCR recognition of peptide-MHC complexes can also be affected by changes in the TCR contact residues of an epitope. Altered peptide-MHC complexes may fail to interact with a particular TCR altogether, or they may be recognized but the T cell may receive a reduced or even a different type of signal when it recognizes this complex. The responding T cell may thus be only partly activated, e.g., to proliferate but not perform effector functions, or even anergized. Presentation of certain mutant peptides to T cells can thereby inhibit the response to the complex not just by competing with it for binding to MHC, but by negatively signaling the responding T cell population, a phenomenon known as T cell antagonism (BERTOLETTI et al. 1994b; KLENERMAN et al. 1994).

Studies with LCMV have shown that, under CTL selection pressure in vitro, it is possible to select viral variants bearing CTL escape-conferring mutations in all three of the most immunodominant epitopes recognized in $H-2^b$ mice (LEWICKI et al. 1995a). However, as long as at least one of these three immunodominant epitopes has not mutated, the virus is cleared efficiently and effectively with the usual kinetics in vivo (LEWICKI et al. 1995b). Importantly, when all three immunodominant epitopes mutate in the same virus, the host can still mount lower affinity CTL responses against subdominant epitopes in the virus and control infection, although in this case more time is required to achieve viral clearance. Thus, under CTL pressure, escape variants can arise (PEWE et al. 1998; GOULDER et al. 1997; MOSKOPHIDIS and ZINKERNAGEL 1995; PHILLIPS et al. 1991), but the host has other options and can use epitopes lower in the hierarchy when immunodominant epitopes mutate. The inference drawn is that, unless a particular antiviral response is restricted to one epitope (e.g., in LCMV TCR transgenic mice when the majority of the host T cells are directed against a single viral epitope) (PIRCHER et al. 1990), mutations in dominant CTL epitopes are unlikely to make a major contribution to viral persistence in vivo. This view, derived from work with LCMV, is reinforced by observations of selected CTL escape variants in human immunodeficiency virus (HIV-1) infection (BORROW et al. 1997). Hence, results with LCMV suggest that during natural infections, the selection of CTL escape viral variants is not likely to constitute the sole mechanism for achieving viral persistence. Exceptions are the presence of a uniquely mono-restricted (focused) $CD8^+$ T cell response (PIRCHER et al. 1990) or a unique MHC/HLA haplotype (DE CAMPOS-LIMA et al. 1993). The MHC diversity and breadth of the TCR repertoire in most individuals allows for recognition of viruses even when one or more of the most dominant epitopes have mutated.

Naturally occurring persistent LCMV infection was first recorded by TRAUB (1936), who observed retention of infectious virus in tissues and circulation throughout the lifespan of mice. Two decades later HOTCHIN and CINTIS (1958) described a model whereby newborn mice, inoculated with LCMV within the first 24h of life, survived an ordinarily lethal intracerebral dose of virus (the same dose or one several logs less was lethal to adult mice) and showed continuous infection in sera and tissues over the natural course of life. This model provided a clinical and biological picture similar to the naturally occurring disease described by Traub. Several investigators have demonstrated that such persistently infected mice are relatively deficient in virus-specific immune lymphocytes, but are able to mount high titers of antiviral antibodies (reviewed in BUCHMEIER et al. 1980; BORROW and OLDSTONE 1997). Recently, use of a LCMV CTL vaccine showed that low affinity CTL can be generated in persistently infected mice (VON HERRATH et al. 2000).

Figure 1 shows elements that affect the balance between acute and persistent LCMV infections, and Table 1 lists immune factors that control the course of both acute and persistent infection. Inoculation of immunocompetent adult mice with LCMV by the intracerebral route usually leads to acute leptomeningitis, choroiditis, and inflammation of the ventricle with death occurring within 6–8 days. Cumulative evidence indicates that CTL are the major effector cells in this reaction (DOHERTY and ALLAN 1986; reviewed in BORROW and OLDSTONE 1997; JOLY et al. 1991; GAIRIN et al. 1991). When virus is introduced into adult immunocompetent animals by the peripheral route, animals may survive or die dependent on the balance struck favoring either immunity or immunopathologic injury. When newborn mice, less than 24h old, are inoculated with LCMV by any route, including intracerebral, virus persists in tissues and sera throughout the animals' lives. However, reconstitution of these newborns with LCMV-immune lymphocytes leads to clearance of infectious virus and viral nucleic acids (OLDSTONE et al. 1986; TISHON et al. 1993b; BERGER et al. 2000). When adult immunocompetent mice are either inoculated with an ordinarily lethal dose of virus coupled with an immunosuppressive agent, i.e., irradiation, thymectomy, antilymphocyte serum, cytoxan, or inoculated with an immunosuppressive virus variant (see Fig. 1 and 2), a persistent infection follows. In a persistent infection, established through inoculation of newborn animals, or when virus is transmitted vertically mother-to-fetus (Fig. 4), immunosuppression is restricted to the LCMV T cell response (OLDSTONE et al. 1973). That is, T cell responses are generated to other viruses as well as antibody responses to soluble and particulate antigens or viruses including LCMV. In contrast, when persistent infection is generated by inoculation of immunosuppressive variants (Fig. 2 and 4), a generalized suppression of the immune system occurs (TISHON and OLDSTONE 1991).

The requirements for clearance of a persistent LCMV infection differ from those of an acute infection (Table 1). Whereas virus-specific $CD8^+$ T cells alone control acute LCMV infections, long-term control of persisting virus requires $CD4^+$ T cell help to sustain the $CD8^+$ T cell response (reviewed in BORROW and OLDSTONE 1997; ALTHAGE et al. 1992; JAMIESON et al. 1991; MATLOUBIAN et al. 1994; BATTEGAY et al. 1994; AHMED et al. 1987; OLDSTONE et al. 1986; TISHON

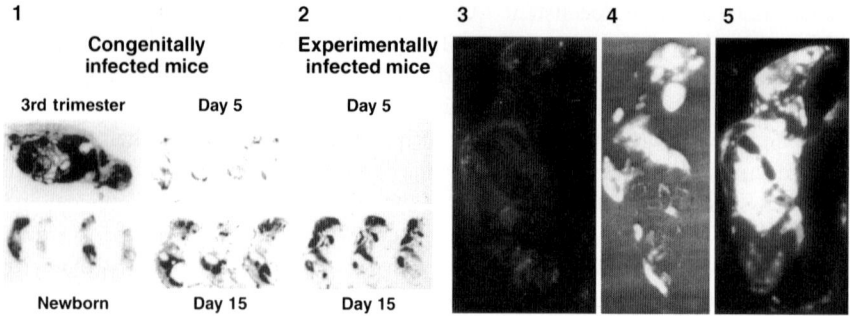

Fig. 4. Transmission of LCMV by the congenital route (*panel 1*) or after experimental inoculation of infectious virus (*panel 2* and *4*): inoculation of newborn mice; *panel 5*, inoculation of adult mice with immunosuppressive variant Clone 13) (*panel 3*, probing uninfected mouse) as observed in whole animal sections and with a ^{32}P riboprobe to LCMV NP. Note infected fetus in the mouse infected by vertical (congenital) transmission. For details of the technology for using whole mouse sections and probes to detect and follow the expression of LCMV or other viral sequences see BLOUNT et al. (1986) and LIPKIN et al. (1989)

et al. 1993b). The precise number of $CD8^+$ T cells and $CD4^+$ T cells required to clear a persistent LCMV infection was quantitated recently (BERGER et al. 2000). In adoptive transfers, the minimum number of cells needed to purge the virus and cure persistent infection is 3×10^6 LCMV-specific $CD8^+$ T cells and 0.7×10^6 LCMV-specific $CD4^+$ T helper cells, a ratio of 50 antigen-specific $CD8^+$ T cells to 1 antigen-specific $CD4^+$ T cell. In addition, although neither IFN-γ nor TNF-α is required for clearance of an acute LCMV infection, both are essential to eliminate persistent LCMV infection (Table 1) (TISHON et al. 1995; A. Tishon and M.B.A. Oldstone, unpublished data).

3 Persistent Infection

3.1 Establishment with LCMV: General

Persistent infection with LCMV can be established by negative selection when this virus invades the thymus. The result is the deletion of potential T lymphocytes of high to moderate affinity that would usually respond to LCMV. Low affinity LCMV-specific cells pass to the periphery and can be activated to respond to LCMV infection or vaccination (reviewed in BORROW and OLDSTONE 1997; VON HERRATH et al. 1994, 2000; BERGER et al. 2000). In addition to this well appreciated mechanism of central tolerance, LCMV can cause peripheral tolerance by infecting interdigitating $DEC205^+$ and $CD11c^+$ dendritic cells (SEVILLA et al. 2000; SMELT et al. 2001).

Initiation of immunosuppression in immunocompetent adult mice by LCMV variants was initially discovered over a decade ago (AHMED et al. 1984b; AHMED and OLDSTONE 1988). Cloning and sequencing of the immunosuppressive variant LCMV Clone 13 revealed five nucleotide changes from the parental LCMV ARM which altered a single aa in the viral GP and single aa in the viral polymerase (SALVATO et al. 1988). Subsequent findings demonstrated the importance of the GP mutation for the selection of LCMV immunosuppressive variants in the spleen, its influence on the numbers of cells infected in lymphoid tissues and the fact that it has no or limited effect on infection of neurons (AHMED et al. 1991; BORROW et al. 1993; EVANS et al. 1994; MATLOUBIAN et al. 1990, 1994; VILLARETE et al. 1994; DOCKTER et al. 1996). Once inside the cell, a role for the viral polymerase in enhanced replication/transcription that affected viral yield per cell was reported (AHMED et al. 1988).

However, to enter the cell the virus first had to bind to its receptor. Alpha-dystroglycan (α-DG) is the receptor for several arenaviruses, including LCMV (CAO et al. 1998; see also Kunz et al., in Vol. I). DG plays a fundamental role in cell assembly and the organization of basement membranes (HENRY and CAMPBELL 1999). Several viral strains and variants, including LCMV Clone 13 (Figs. 2, 5), bind to α-DG at 2–3 logs higher affinity than other strains [including Armstrong (ARM); Figs. 2, 5] and variants of LCMV (SEVILLA et al. 2000; SMELT et al. 2001). This conclusion came from binding studies using α-DG immobilized on membranes or soluble α-DG to competitively block virus binding to cells expressing α-DG (Fig. 5). Further, viral reassortants made between Clone 13 and ARM and between Traub and ARM mapped attachment and infection of the virion to genes encoded in the S RNA (SMELT et al. 2001) (Fig. 5, panel 2). The S RNA contains the gene that encodes LCMV glycoprotein (GP) and nucleoprotein (NP). Since ARM and Clone 13 have the same NP aa sequence, but differ at a single aa position in their GP, i.e., ARM has phenylalanine (F) at aa 260, but Cl 13 has leucine (L) at that position, the GP1 mutation is implicated in the binding of LCMV to α-DG. Of over 35 cloned LCMV variants studied (SEVILLE et al. 2000), those having a small aliphatic aa at GP1 aa 260 [isoleucine (I) or L] or a serine (S) at GP1 153 bound at high affinity to α-DG. In contrast, viruses having bulky aliphatic aa F at GP1 aa 260 or aa 153 bound at low affinity to α-DG. Most important, strains that bound at high affinity to α-DG, when injected into adult immunocompetent mice, caused immunosuppression (failed to generate anti-LCMV CTL at day 7 post-inoculation) and were associated with persistent infection ($10^{3.5}$–10^5 PFU LCMV in sera) over the next several weeks (SEVILLA et al. 2000). In contrast, viruses that bound at low affinity to α-DG were able to generate a robust CTL response by day 7 and cleared the viruses by day 15. Other studies (ODERMATT et al. 1991; BORROW et al. 1995) showed that the immunosuppressive LCMV strains disorganized splenic architecture and preferentially bound to cells in the white pulp (BORROW et al. 1995; SEVILLA et al. 2000; SMELT et al. 2001) (Fig. 5, panel 3), whereas the non-immunosuppressive strains bound to red pulp. Again, reassortants between immunosuppressive and non-immunosuppressive strains mapped this tropism of the virus for the spleen's white pulp to genes encoded in the S RNA (SMELT et al. 2001).

Analysis of cells of the immune system for expression of α-DG revealed its preferential expression by professional antigen-presenting interdigitating dendritic

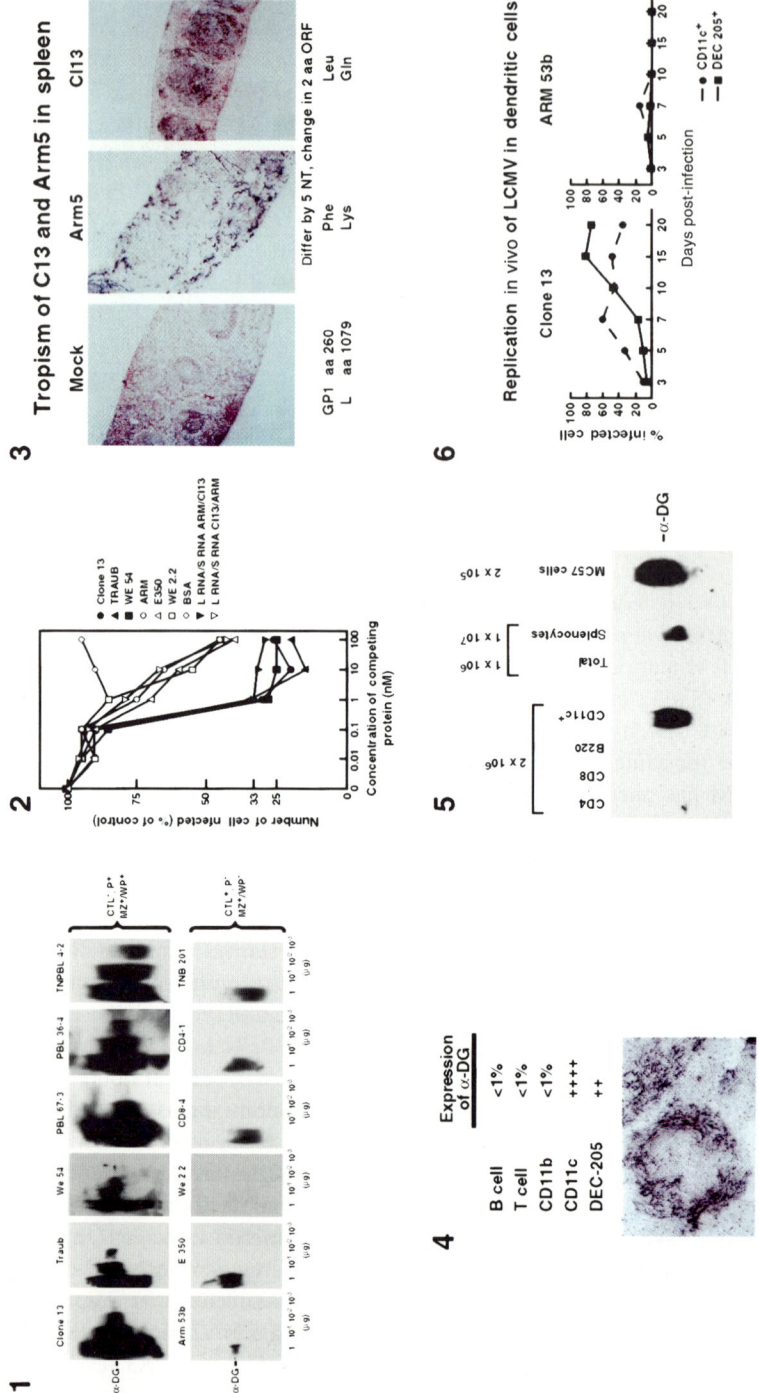

Fig. 5. Virus–dendritic cell interactions. Quasispecies selection of variants that bind at high affinity to the arenavirus cellular receptor α-DG leads to generalized immunosuppression (CAO et al. 1998; SEVILLA et al. 2000; SMELT et al. 2001). *Panel 1*, multiple strains of LCMV and variants generated during a persistent viral infection and isolated from a variety of immune cells bind at high affinity to α-DG immobilized on membranes (*top row*) segregating these strains from those (*bottom row*) that bind at low affinity to α-DG. *Panel 2*, division of viruses into two groups. Those that require 1–8nM of soluble α-DG to block virus attachment (high-affinity binders) and infection (*darkened symbols*) and those requiring >400nM (low-affinity binders) of soluble α-DG (*open circles*). *Panel 2* also shows that genes encoded by S RNA of LCMV (NP or GP) of a high affinity binder are responsible for binding to α-DG. Since the NP sequence of high affinity (Clone 13) and low affinity (ARM 53b) viruses is identical, binding to α-DG maps to the GP of these two strains, which differs by one aa. *Panel 3* shows that the high affinity Clone 13 α-DG binder preferentially localized to the white pulp of spleen, but the low affinity (ARM 53b) binder is tropic for red pulp. Note that the difference in GP aa residues is a single aa with Clone 13 having a leucine in position 260, as opposed to ARM 53b with a phenylalanine at that site. Study of over 50 virus strains and variants indicates the uniformity of the findings in *panels 1–3* and the binding of those viruses at high affinity to α-DG, which causes generalized immunosuppression and persistent infection. In contrast, the low α-DG binders generate a $CD8^+$ CTL response that clears the acute virus infection. *Panels 4–6* implicate infection of $CD11c^+$ and $DEC205^+$ dendritic cells in the binding to α-DG and generalized immunosuppression. The initial observation of organ-specific (spleen) selection of viral variants (AHMED and OLDSTONE 1988; AHMED et al. 1984b) preceded the discovery of α-DG as the receptor for LCMV (CAO et al. 1998) and the discovery of selected tropism for white and red pulp (BORROW et al. 1995; SEVILLA et al. 2000; SMELT et al. 2001). *Panel 4*, the use of selected monoclonal antibodies and a fluorescence activated cell sorter (FACS) shows that expression of α-DG is preferential on $CD11c^+$ and $DEC205^+$ dendritic cells but negligible on B and T cells and CD11b cells. *Panel 4* also displays virus nucleic acid sequences of a high affinity binding virus located to the white pulp of the spleen. *Panel 5*, quantitative studies of α-DG isolated from cells of the immune system again show preferential localization to dendritic cells ($CD11c^+$). *Panel 6*, replication of LCMV Clone 13 in $CD11c^+$ and $DEC205^+$ cells (up to 80% of such cells are infected) is shown. By comparison <12% of such cells are infected by LCMV ARM, a low α-DG binder

cells (IDC) (SEVILLE et al. 2000) (Fig. 5, panels 4–6). Study of the kinetics of IDC infection revealed that over 60% with as much as 80% of $CD11c^+$, $DEC205^+$ cells were infected with the immunosuppressive variant Clone 13, whereas less than 10% of these cells became infected with a non-immunosuppressive strain of the virus. Biochemical studies suggested the involvement of B7.1 coregulating factor in infected $DEC205^+$ and $CD11c^+$ cells in vivo (N. Sevilla et al., unpublished data), and in vitro analysis of a murine dendritic cell line infected with high affinity Clone 13 but not with low affinity LCMV ARM revealed participation of IL-12 transcription (A. Holz et al., unpublished data) as functional correlates for maintaining the generalized immunosuppression.

Thus, in conclusion, among cells of the immune system, $CD11c^+$ and $DEC205^+$ splenic dendritic cells primarily express the cellular receptor α-DG for LCMV. By selection, strains and variants of LCMV that bind α-DG with high affinity are associated with virus replication in the white pulp, show preferential replication in a majority of $CD11c^+$ and $DEC205^+$ cells, cause immunosuppression and establish persistent infection. Recent studies by Smelt et al. (unpublished data) show that LCMV strains entering $CD11c^+$ and $DEC205^+$ cells in the marginal zone are carried into the T cell-dependent area akin to the Trojan horse of Homer's *The Iliad*. In contrast, LCMV strains and variants that bind with low affinity to α-DG replicate mainly in the red pulp but only minimally in $CD11c^+$

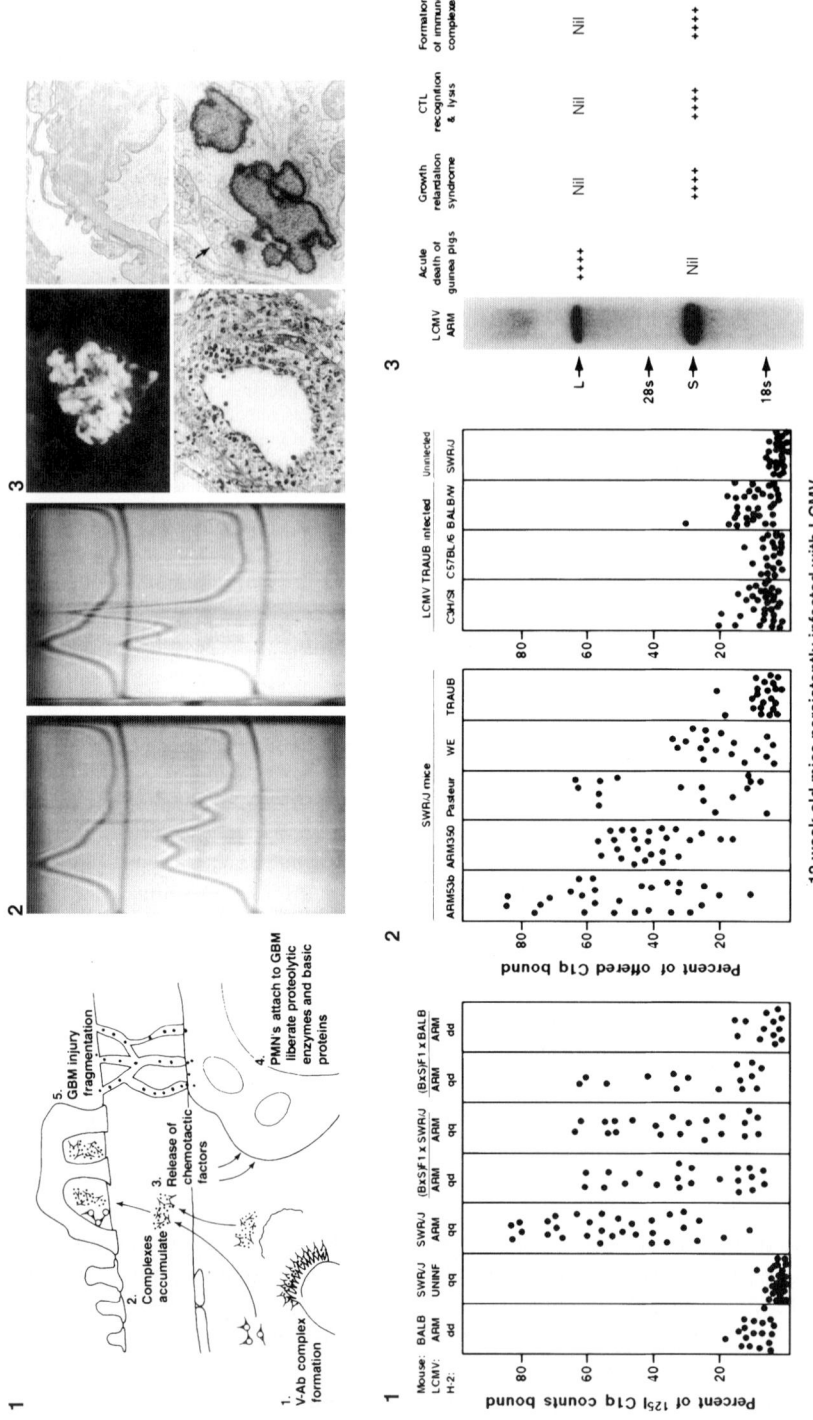

Fig. 6. Virus-antiviral (*V-Ab*) immune complexes. Generation, disease consequences, host MHC gene control and role of viral genes. *Panel 1 (top)*, cartoon illustrating how V-Ab immune complex form and deposit in the renal glomeruli. *Panel 2 (top)*, *upper portion* shows ultracentrifugation profile of sera from normal control mice; the *lower part* shows sera from persistently infected mice containing circulating complexes. These V-Ab immune complexes are infectious as proven when removal of Ig by immunoprecipitation with an antibody to Ig markedly reduces (2 logs or more) the viral titer in the sera, yet similar precipitation with an antibody to albumin does not lower the viral titer. *Panel 3 (top; clockwise beginning at 12 o'clock)* profiles tissue deposition of immune complexes in the renal glomerulus detected by immunofluorescence, detected by electron microscopy, immune complexes in the choroid plexus being engulfed by macrophages and immune complexes in an artery causing arteritis (OLDSTONE and DIXON 1969, 1972; LAMPERT and OLDSTONE 1974; OLDSTONE 1976). *Panel 1* and *panel 2 (bottom)* show circulating immune complexes detected in sera using C1q. In *panel 1*, after persistent infection with LCMV ARM strain BALB mice (H-2^d) generate low to negligible amounts of C1q binding immune complexes, whereas SWR/J mice (H-2^q) generate high levels of C1q binding complexes. F1 hybrids between these two strains have high levels of complexes, but mice generated by crossing these F1 animals to each parent have high levels of C1q circulating complexes only when one H-2^q gene is present. Low to negligible amounts require both H-2^d alleles. Use of recombinant inbred mice maps the host genetic control to MHC I^A gene(s) (OLDSTONE et al. 1983). LCMV titers in sera of H-2^d and H-2^q and hybrid mice are equivalent. *Panel 2 (bottom)* shows that SWR/J generate an abundance of V-Ab immune complexes when persistently infected with LCMV ARM 53b, E-350 or Pasteur strains; however infection with LCMV Traub or WE strains fails to generate significant levels of V-Ab immune complexes (TISHON et al. 1991). Reassortants between LCMV ARM 53b and LCMV Traub mapped the formation of V-Ab immune complexes to viral genes encoded on the S RNA (the GP and NP). *Panel 3 (bottom)* shows other examples in which reassortants mapped LCMV pathogenic genes. The acute death of guinea pigs infected with LCMV WE mimicked aspects of Lassa fever viral infections and mapped to the L RNA that encodes the Z and polymerase (*L*) genes (RIVIERE et al. 1985b). Both the generation and function of CTL and inception of GH disease mapped to genes encoded on the S RNA of LCMV (RIVIERE et al. 1985a; AHMED et al. 1984a; OLDSTONE et al. 1985)

and DEC205$^+$ cells and generate a robust anti-LCMV CTL response that clears the virus infection. Differences in binding affinities can be mapped to mutations in the viral GP1 ligand that binds to α-DG and in several instances to a single aa change. Thus, receptor/virus interaction on dendritic cells in vivo can be an essential step in the initiation of virus-induced immunosuppression and viral persistence (Fig. 2).

3.2 Virus-Antibody Immune Complex Formation and Disease in Persistent Viral Infection

In most virus infections, antibody to virus interacts with virus or viral antigens in the circulation, resulting in the formation of V-Ab complexes. V-Ab complexes themselves are potent pathogenic agents and, once deposited in tissues, induce a phlogogenic response. V-Ab immune complex disease occurs frequently in chronic viral infections and is a common pathogenic mechanism of animal and human nephritides and arteritides. Persistent LCMV infection has been a paradigm in which the occurrence, basis and genetic control of immune complex disease have been dissected (reviewed in OLDSTONE 1975; OLDSTONE et al. 1983; TISHON and OLDSTONE 1991) (Fig. 6).

Evidence for V-Ab immune complex disease lies in (1) demonstrating circulating V-Ab complexes, and (2) showing localization of virus, host Ig and complement at the sites of tissue injury. Several techniques used to show circulating V-Ab immune complexes (Fig. 6) are C1q precipitation, complement utilization, electron microscopy, analytical ultracentrifugation, monoclonal rheumatoid factor precipitation, platelet agglutination, precipitation of V-Ab complex with antibody directed toward the antibody or complement bound with the virus (anti-Ig or anti-C3 precipitation) and use of cultured cells that contain receptors for bound C3.

That virus travels in the circulation complexed with host Ig was demonstrated by precipitating Ig from the sera (using an anti-mouse IgG) of mice chronically infected with LCMV and showing significant reduction of the infectivity titer (reviewed in OLDSTONE 1976). These results with LCMV have been duplicated in mice chronically infected with lactic dehydrogenase virus and leukemia virus, mink infected with Aleutian disease virus, horses infected with equine infectious anemia virus, and humans infected with hepatitis B antigen (HB Ag), Epstein-Barr virus, cytomegalovirus (reviewed in OLDSTONE 1975) and HIV. In addition, in the sera of chronically infected mice, LCMV circulates complexed not only with Ig but also with C3, since immunologically specific precipitation of C3 as well as Ig removes two or more logs of infectivity.

Deposition of circulating V-Ab complexes in tissues has been confirmed by identification of viral antigens, host Ig and complement in a granular pattern along renal basement membranes. Immunofluorescence and electron microscopy of such tissues containing immune complexes yields the characteristic patterns pictured in Fig. 6. Identification and quantitation of the specific antiviral antibodies present in the deposited complex are accomplished by: (1) elution of the glomerular-bound Ig by either low ionic, high ionic or low pH buffers to dissociate V-Ab bonds; (2) recovery of the eluted Ig and its quantitation by immunoprecipitation; and (3) quantitation of the Ig after absorption with various virus, tissue and cellular antigens, again by any of several immunoprecipitation assays. Serum Ig is immunochemically isolated and assayed in a similar way. The ratio of antiviral antibody to total Ig in the tissue eluate over the antiviral antibody to total Ig in the serum depicts the concentration of antiviral antibody localized in the tissue. Although such elution studies have limitations – most notably loss of eluted antibody via recombination with eluted antigens, incomplete elution or denaturation of eluted antibody – they nevertheless provide the only direct quantitation of antibodies present in the injured tissues.

The major sites of V-Ab immune complex deposits are on the basement membranes of the renal glomerulus, endothelial walls, medium and small arteries and in the choroid plexus (Fig. 6) (OLDSTONE 1975). V-Ab immune complexes have also been found in other tissue, i.e., heart, lung, joints, skin.

Once complexes are formed and circulate, they are either phagocytosed and removed by cells of the reticuloendothelial system (RES) or deposited in tissues. Large complexes appear to be preferentially removed by the RES over small complexes. Also, some cells of the RES have receptors that recognize different classes of Ig and complement and show different binding affinities. Changes in structure or recognition units of Ig seem important in phagocytosis, since alteration by reduction and alkylation of antibody (before making the soluble complex) decreases the complex's rate of removal from the circulation. RES activity undergoes depression after prolonged exposure to circulating immune complexes. The resulting decreased efficiency in RES function precedes the onset of immune complex deposition and subsequent development of proteinuria and nephritis. The phagocytic properties of mesangial cells of the kidney may then protect against such deposition in the glomerular basement membrane until mesangial cell overload eventually allows deposition on the glomerular basement membrane.

During the active process of immune complex deposition, increased vascular permeability occurs and is associated with the release of vasoactive agents. With this increase in vascular permeability, large complexes, usually 19s or larger, deposit along filtering membranes.

Evidence that immunopathologic consequences of LCMV-Ab immune complexes cause disease is supported by several observations. First, if mice infected at birth with LCMV are nursed by LCMV-immune foster mothers, they have a more rapid and severe onset of immune complex glomerulonephritis and arteritis as well as a shorter lifespan than conventionally reared carrier mice (OLDSTONE and DIXON 1972). Maternal antibody is found complexed to LCMV antigens in the glomeruli. Second, adoptive transfer of anti-LCMV antibody into persistently infected mice or the parabiosis of an immune syngeneic mouse to a persistently infected mouse results in enhancement and severe manifestations of chronic LCMV disease (OLDSTONE and DIXON 1969). Third, induction of LCMV infection-associated immune complex disease depends on both the mouse strain and infecting LCMV isolate, and disease severity correlates with the level of V-Ab complexes produced (Fig. 6).

Disease-susceptible inbred mouse strains are those that make high levels of antiviral antibody after neonatal infection with LCMV. For example, SWR/J mice persistently infected with LCMV ARM (high responders) make 50-fold more LCMV NP- and GP-specific antibody and have sevenfold higher levels of circulating complement-binding immune complexes than persistently infected BALB/WEHI mice (low responders) (Fig. 6). The SWR/J strain shows heavier deposits of virus-antibody complexes in their tissues, although both mouse strains carry the same load of infectious virus (OLDSTONE et al. 1983; TISHON et al. 1991). Breeding studies between high responder mice, low responder mice, their hybrid offspring, backcrosses of the hybrids to both high and low responder parents, and use of selected recombinant inbred mouse strains map the formation of C1q binding V-Ab complexes to genes located in the Ir region of MHC H-2 loci (Fig. 6). Thus, MHC class II genes controlling the strength of antibody responses played a prominent role in the V-Ab immune complex disease of mice despite similar levels of infectious virus in high and low responders.

Other studies showed the importance of LCMV strain in causing V-Ab immune complex formation and disease during persistent infection (TISHON et al. 1991) (Fig. 6). Although SWR/J mice persistently infected with LCMV ARM or E-350 contain very high levels of circulating and trapped immune complexes, and LCMV WE or Pasteur produce somewhat less but still substantial levels, LCMV Traub elicits a much lower antiviral antibody response. Thus, SWR/J mice persistently infected with the Traub strain of LCMV have low to negligible levels of circulating immune complexes, with minimal immune complex deposition in tissues, even though their viral load is similar to that in animals infected with LCMV ARM (Fig. 6). Therefore, both host and viral determinants influence susceptibility to the development of immune complex disease, acting by modulating the level of antiviral antibody production, immune complex formation and resultant immunopathologic injury.

4 Persistent Virus Infection Alters the Function of Differentiated Cells Leading to Disease

4.1 General Concept and In Vitro Findings

Peter Medawar amply summarized viruses when he described them as "a piece of bad news wrapped in a protein coat" (MEDAWAR and MEDAWAR 1983). Viruses were first separated from other microbial agents by using the Pasteur-Chamberland filter. Viruses known or suspected to cause disease have been characterized, first, by the clinical picture and histological profile and, second, by virologic, immunologic, or molecular biologic assays. These first procedures, were initially applied by IVANOVSKI (1899) and BEIJERINCK (1899) to infection of tobacco plants with tobacco mosaic virus and by LOEFFLER and FROSCH (1898) to infection of cattle with foot and mouth virus. This and the second approach continue to be used today. Thus, the sine qua non for suspecting viral involvement has been a histopathologic picture of cellular necrosis, usually, but not always, including inflammatory infiltrates. Indeed, the histological picture frequently suggests the virus involved, e.g., "cytomegalo-like" cells during infection with cytomegalovirus, injury of anterior horn neurons in poliomyelitis, and the appearance of Negri or Lyssa bodies within certain neurons of the central nervous system in rabies. In other instances, such as encephalitis or myocarditis, the cellular necrosis and associated perivascular cuffing are not specific enough to suggest a single etiologic agent – any of several agents could be the cause. The decisive factors in the course of virally induced injury are initially the tropism of that virus for specific cells and then either the viral lytic ability per se or the interaction between the components of the host immune response and the infected cell or virion. In either case, viral infections are often classified in contemporary medical books according to the cells injured and the inflammation or transformation observed. As indicated, the type of disease is dependent on the specific cell infected (e.g., neuron, lymphocyte, or myocardial cell), and the severity reflects the number of host cells infected and the type injured. For example, the death of neurons is likely to be more devastating than death of fibroblasts.

A possibility of very recent vintage is that a virus might cause disease by altering the function of a cell without destroying that cell (reviewed in OLDSTONE

1984a,b, 1985, 1989, 1993). For example, while studying the effects of persistent LCMV infection on differentiated neuroblastoma cells, we noted abnormalities in the synthesis and degradation of acetylcholine caused by a decrease in the production of the appropriate acetylase or esterase (OLDSTONE et al. 1977) (Fig. 7, panels 1 and 2). Nevertheless, these cells were normal in morphology, growth rates, cloning efficiencies, and levels of total RNA, DNA, protein, and vital enzyme

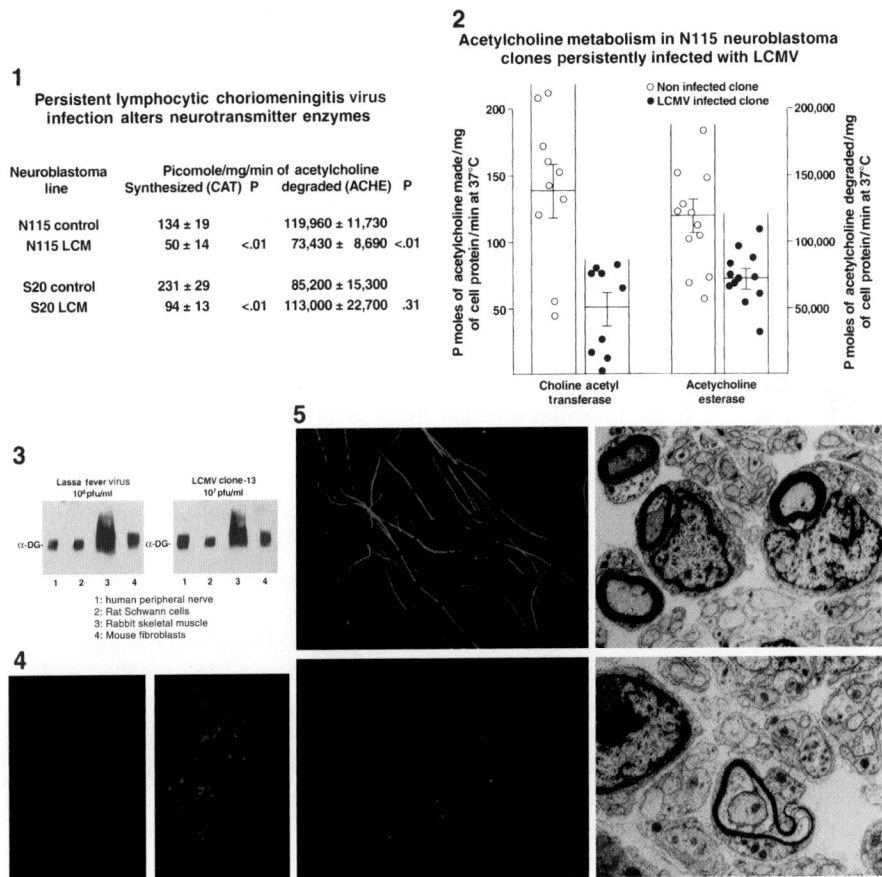

Fig. 7. Virus alters differentiated functions of neuronal cells in vitro (OLDSTONE et al. 1977; RAMBUKKANA and OLDSTONE 2001). *Panels 1* and *2* show diminished synthesis of the acetylase and the esterase of acetylcholine in neuroblastoma cells on the population (*panel 1*) and clonal cell level (*panel 2*). Panel 3 shows several cell lines, including Schwann cells, expressing high concentrations of the LCMV receptor α-DG; Lassa fever virus and LCMV Clone 13 display similar high affinity binding to α-DG. Low affinity α-DG binding LCMV ARM cannot infect Schwann cell cultures, high affinity binding LCMV Clone 13 (*panel 4*) does infect (*green fluorescence* >95% of cells express LCMV NP) but not kill these cells [host strain (*blue*) no cell necrosis], instead causing a persistent infection. *Panel 5* (*upper panel*) control cells (dorsal root ganglia and neuronal cells) or LCMV ARM infected cells in culture allow normal axon formation with myelination as shown by the myelin basic protein (*green fluorescence*), and by electron microscopy. In contrast (*lower panel*), Clone 13 infection prevents myelin genesis yet does not lyse these cells and minimal myelin formation is detected by EM

synthesis. Similarly, Holtzer and his associates (HOLTZER et al. 1975, 1982) showed that Rous sarcoma virus (Prague strain ts mutant), grown at temperatures that were permissive for infection of differentiated chick chondroblasts, myotubes, or melanoblasts, altered the unique functions of these specialized cells. Thus, infected chondroblasts failed to make sulfated proteoglycans; myotubes failed to synthesize heavy and light chains of myosin, and melanoblasts did not produce melanin. At non-permissive temperatures, manufacture of these differentiated products was normal. By changing the temperature of these infected cells, Holtzer and colleagues observed either the selected deficiencies of differentiated products or a return to normal functions.

Subsequently, viruses have been shown to alter immunologic functions in a similar way. CASALI et al. (1984) showed that infection with measles virus or influenza virus of peripheral blood lymphocytes aborted their expected specialized functions, including the manufacture of Ig and the capacity to act as cytotoxic effectors. Similar results were shown by SCHRIER and OLDSTONE (1986) during human cytomegalovirus (HCMV) infection which rendered human T cells unable to lyse HCMV targets. Hence, these RNA (measles, influenza) and DNA (HCMV) viruses altered the differentiated functions of lymphocytes without lysing or destroying them.

Recently, with an in vitro system comprising dorsal root ganglia cultured with Schwann cells, LCMV infection aborted axon myelination (RAMBUKKANA et al. 2001) (Fig. 7, panels 3–5). These cultured neurons developed myelin sheaths as documented by the expression of myelin basic protein on axons and by electron microscopy of myelin formation. Because these cells have heavy concentrations of α-DG (RAMBUKKANA et al. 2001; S. Kunz and M.B.A. Oldstone, unpublished observations), their infection by either Clone 13, a high affinity LCMV binder to α-DG, or LCMV ARM, a low affinity binder, was studied. Clone 13 infected >95% of the cells, while LCMV ARM infected <10%. Despite universal LCMV Clone 13 infection, the cells showed no apoptosis or physical injury. Yet, as dramatically shown in Fig. 7, myelin failed to form in LCMV Clone 13-infected but not uninfected cultures or cultures infected with LCMV ARM. Clearly, the virus can alter the differentiated or luxury function of cells without disturbing their vital functions. These findings have been extended to other cells/organs infected with LCMV and with other RNA and DNA viruses both in vitro and in vivo (Table 2). The two best-studied in vivo systems, the growth hormone (GH) deficiency syndrome and alterations in behavior and learning, are presented below.

4.2 Growth Hormone Deficiency Syndrome

C3H/St mice infected neonatally with the ARM strain of LCMV exhibit a GH deficiency syndrome (Fig. 8) manifested as growth retardation and severe hypoglycemia (OLDSTONE et al. 1982, 1984). Decreases in body weight and length become apparent when the mice are 5–7 days old, and by 15 days the weight of infected animals is approximately 50% that of uninfected control animals. Such

Table 2. Viruses cause disease by altering a cell's differentiation function and unbalancing homeostasis

Pathology in the absence of cell lysis

Growth retardation and hypoglycemia
- 50%   ↓   Growth hormone synthesis
- 5x   ↓   Growth hormone mRNA (steady state)
- 16x   ↓   Growth hormone mRNA (initiation of transcription)
- 5–10x ↓   Growth hormone transactivator GHF-1

Hypothyroidism
- 30%   ↓   T3, T4
- 5x   ↓   Thyroglobulin mRNA (steady state)

Chemical diabetes
- ↑ Blood glucose
- Normal or low pancreatic insulin

Neuronal dysfunction
- ↓ Gap 43
- ↓ Somatostatin mRNA
- ↓ GABA
- ↓ Cholinergic mRNA and protein
- [Normal GAD (GABA) mRNA
- Normal MuBr 8 (SNAP mRNA)
- Normal amyloid B protein mRNA
- Normal actin mRNA]
- ↓ Integrative function
- ↓ Ability to form myelin

Obesity
- ↓ Norepinephrine, ↓ Dopamine

Deformed whiskers

Cardiomyopathy

[Faulty antigen presentation by dendritic cells
  T lymphocyte suppression]
- Failure to clear virus
- Viral persistence

persistently infected mice also become severely hypoglycemic, a result of GH deficiency, which is likely to be the reason that most of them die. The mechanisms underlying this growth retardation have been analyzed in great detail, and this is probably the best understood example of a disease resulting from direct viral interference with the differentiated function of cells in the absence of cytolysis or inflammation.

The retardation of growth and severe hypoglycemia are dependent on mouse strain and virus isolate (OLDSTONE et al. 1984; BUREAU et al. 2001; TISHON and OLDSTONE 1990). Whereas C3H/St and certain other mouse strains, including CBA/N mice display the GH deficiency syndrome when infected neonatally with LCMV ARM, still other strains (e.g., BALB/WEHI and SWR/J mice) are disease resistant (TISHON and OLDSTONE 1990). Disease susceptibility is not MHC linked and involves multiple genes. Similarly, whereas the ARM and E-350 strains of LCMV

104 M.B.A. Oldstone

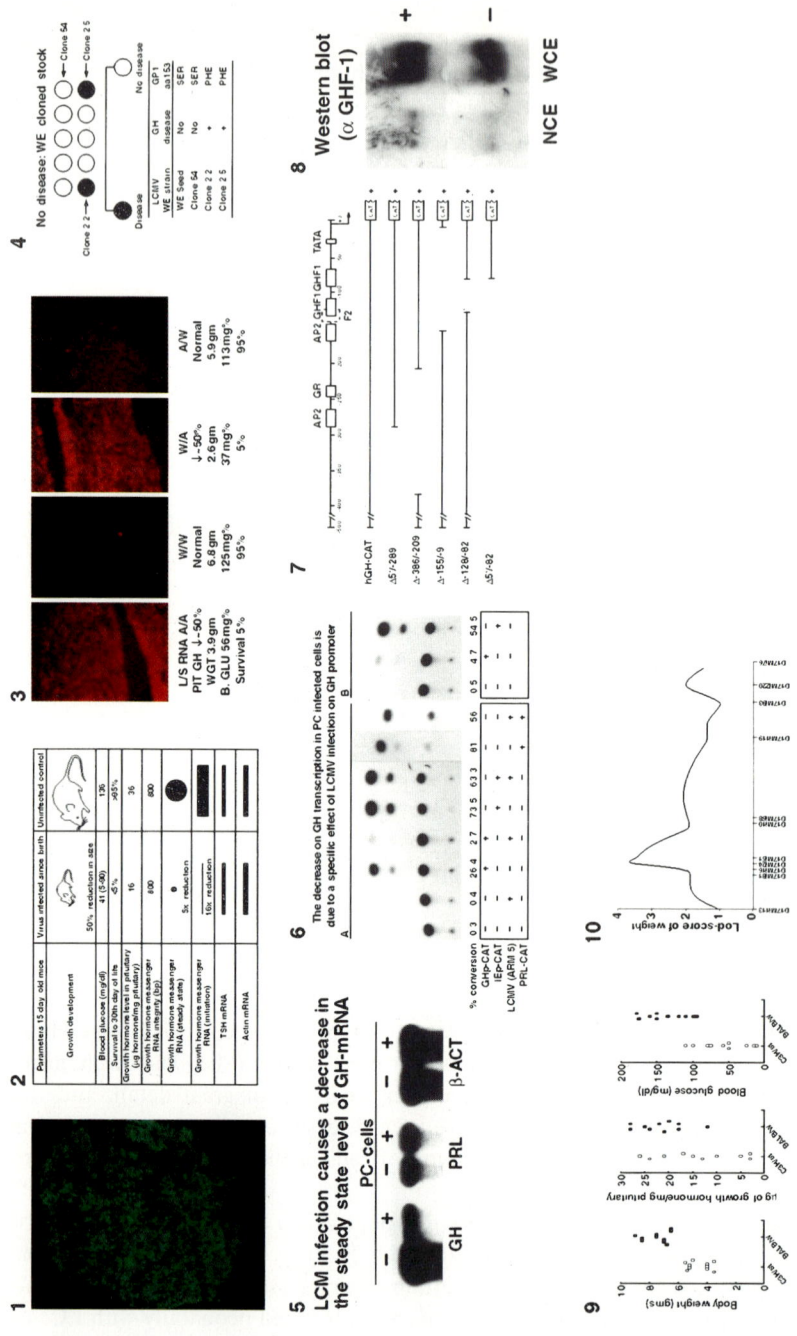

11

Curing of PC (Pi) by ribavarin restores PC WT phenotype

A) Characterization of PC (Pi) after ribavirin treatment

	Virus titer (PFU/ml)	Infect center (%)	Viral antigen (NP)	PCR Viral NP GP GAPDH
PC	-	-	-	- - -
PC (Pi)	5x10⁴	60	-	- - -
PC (Pi) RBV	-	-	-	- - -

B) PC (Pi) RBV display PC WT phenotype

PC (Pi)RBV (B)
PC (Pi)RBV (A)
PC WT

— NP

— GH

12

- **Two viral genes are involved in GH disease**
  1. GH Disease maps to S RNA of LCMV
  2. Viral GP: receptor binding/entry
  3. Viral NP: binding to GHFI transcription factor

- **Host susceptibility gene(s) map to locus on chromosome 17 by site H-2D/D17 Mit 24/D17 Mit 51**

Fig. 8. Persistent virus infection alters the differentiation (luxury) function of a cell without lysing it, but disturbs homeostasis and causes systemic disease: mapping of the viral and host genes involved. *Panel 1* illustrates LCMV NP antigens in GH-producing cells of the anterior lobe of the pituitary in C3H/St mice during persistent infection with LCMV ARM. Mice were infected at birth and studied at day 12. *Panel 2* displays the phenotype of virus-infected and control mice (OLDSTONE et al. 1982, 1984a; VALSAMAKIS et al. 1987; KLAVINSKIS and OLDSTONE 1989). *Panel 3* shows the reassortants made from LCMV ARM (*A*), a strain that causes the GH disease, and LCMV WE (*W*), a strain that does not, to map the relevant viral genes. Reassortants between the long (*L*) and short (*S*) RNAs of these LCMV strains were used to infect C3H/St mice. *Upper panels* display LCMV infections in GH cells in the anterior lobe of the pituitary, and *lower panels* show the phenotypes. The results reveal that viral genes encoded on the S RNA, the GP or NP are responsible for the effect on growth and development (OLDSTONE et al. 1985). *Panel 4* provides evidence that the viral GP is involved by selecting variants of WE that have a single aa mutation in the GP but complete homology in the NP and cause or do not cause GH disease. Here, a single aa mutation in GP residue 153 was associated with GH disease; serine 153 with GH disease⁺ and phenylalanine 153 with GH disease[nil] (BUESA-GOMEZ et al. 1996; TENG et al. 1996a,b). Presumably, the viral GP was required for binding to or entry into the GH-producing cell. In transgenic mice made so that LCMV NP under control of the GH promoter was placed in GH cells, GH disease developed, implicating the viral NP once it is inside the GH cells. In contrast, expressing the viral GP inside the GH cell failed to produce disease. *Panel 5* shows that LCMV infection of cells (PC) that produce both GH and prolactin (PRL) affects GH but not PRL transcription. In these PC cells, each cell transcribes both GH and PRL. *Panel 6* depicts the results from a CAT assay in which the CAT enzyme is under the control of the GH or PRL promoter. The viral effect maps to the GH promoter (DE LA TORRE and OLDSTONE 1992). *Panel 7* shows that, with the use of this GH promoter, deletion mutants map the effect of virus to a 61bp region in the GH promoter that encodes the transcription factor GHF-1 (PIT-1). *Panel 8* shows, at the protein level, the reduced GHF-1 from PC cells in the whole cellular extracted faction (WCF). Mapping of host genes involved in the virus-induced GH disease (TISHON and OLDSTONE 1990; BUREAU et al. 2001) appears in *panels 9 and 10*. On the basis of weight, GH levels and blood glucose measurements, C3H/St mice are susceptible but BALB/W mice are not (*panel 9*). F2 crosses between these two strains and the use of a microsatellite mapping technique to demonstrate host susceptibility gene(s) place the locus on chromosome 17 near the H-2D site between D17 MIT24 and D17 MIT51 (*panel 10*). *Panel 11* demonstrates the reversibility of the viral effect on differentiated function. LCMV infection of PC cells aborts GH transcription; however, use of the anti-arenavirus agent ribavirin (*RBV*) cures the infection and restores GH transcription. NP in the northern blot refers to expression of LCMV NP transcripts (DE LA TORRE and OLDSTONE 1992)

produce both growth retardation and hypoglycemia in C3H/St mice, and LCMV Pasteur does so to a lesser extent, LCMV WE and Traub have only a minimal effect on growth and do not cause hypoglycemia or the related death (OLDSTONE et al. 1985).

When growth is retarded in any murine virus combination, it always correlates with the level of infection in GH-producing cells in the anterior pituitary gland. Most (>90%) GH-producing cells in the anterior pituitary of C3H/St mice infected with LCMV ARM or LCMV E350 become infected and contain viral antigen. With electron microscopy, mature virus particles can be detected budding from the surfaces of these cells (OLDSTONE et al. 1982). In contrast, Traub and WE strains of LCMV infect far fewer GH-producing cells (<15%) in the same mouse strain. Studies with reassortant viruses generated between LCMV ARM and LCMV WE initially mapped the ability to infect GH-producing cells and cause growth retardation in C3H/St mice to the genes encoded on the viral S RNA segment (Fig. 8, panel 3) (OLDSTONE et al. 1985). More recently, LCMV clones have been isolated by selecting plaques from the LCMV WE population that, unlike sister clones or the parental population, retard growth in C3H/St mice (BUESA-GOMEZ et al. 1996; TENG et al. 1996a,b). Sequence comparison of the S RNA segment of WE clones that do and do not cause disease in C3H/St mice has revealed that a single aa change in the viral GP correlates with disease-causing potential (Fig. 8, panel 4). Although this aa difference affects the binding of some LCMV WE clones to their cellular receptor α-DG in vitro (SMELT et al. 2001), the α-DG gene is encoded on mouse chromosome 9, whereas susceptibility to GH disease maps to chromosome 17 (BUREAU et al. 2001). Further, Traub, WE and WE-54 strains, which bind at high affinity to α-DG, do not cause GH disease nor do they infect GH cells in the pituitary, although they replicate to high titers elsewhere. In contrast, LCMV strains like ARM, E-350 and WE2.2 and WE2.5 that bind at low affinity to α-DG infect GH cells in the pituitary. Hence, α-DG is not the relevant receptor on such cells. An alternative, yet to be discovered, receptor may be preferentially expressed on C3H/St and similar susceptible mice.

How does LCMV replication in GH-producing cells activate growth retardation? Oldstone and co-workers (OLDSTONE et al. 1982, 1984) initially reported that LCMV ARM infection of C3H/St mice resulted in a significant reduction (approximately 50% on day 16) in the level of GH in the pituitary, and Valsamakis and colleagues (1987) showed that this decrease correlated with a fivefold reduction in the steady-state level of GH mRNA. In turn, this event was related to a reduction in initiation of transcription of this gene, which appeared to be selective. Transcriptional initiation of another pituitary gene, the precursor of thyroid-stimulating hormone (TSH-β) or of the housekeeping genes, actin and proα2(1) collagen, were only minimally affected (VALSAMAKIS et al. 1987; KLAVINSKIS and OLDSTONE 1989).

A more in-depth characterization of the molecular mechanisms underlying LCMV-induced downregulation of GH mRNA synthesis involved a tissue culture model using a rat pituitary cell line (PC cells) that expresses both GH and prolactin (Fig. 8, panel 5). Persistent infection of these cells (each cell expresses both GH and prolactin) by LCMV markedly downregulated GH mRNA transcription but caused comparatively minimal interference with prolactin transcription (DE LA TORRE and OLDSTONE 1992). Transfection experiments indicated that expression of the reporter gene, chloramphenicol acetyltransferase, in PC cells was significantly decreased by LCMV infection when the reporter gene was expressed under the control of the GH promoter (Fig. 8, panel 6) (DE LA TORRE and OLDSTONE 1992). By contrast, similar levels of chloramphenicol acetyltransferase activity were obtained in uninfected and LCMV-infected cells when chloramphenicol acetyltransferase expression was under control of either a cytomegalovirus immediate-early promoter or a simian virus (SV40) promoter. Next, the use of GH promoter deletion mutants (Fig. 8, panel 7) together with in vitro transcription assays using nuclear fractions from uninfected or LCMV-infected PC cells suggested that the viral effect on GH promoter activity is caused by interference with the GH transactivation factor, GHF1 (PitI). Finally, PitI protein levels were reduced in LCMV-infected PC cells (Fig. 8, panel 8).

Other studies of the molecular basis of the ability of LCMV to downregulate GH mRNA synthesis in PC cells have addressed which viral component(s) mediates the phenomenon. Infection of PC cells with a recombinant vaccinia virus expressing LCMV NP, but not with a control vaccinia virus recombinant or one expressing LCMV GP, significantly decreased the level of GH mRNA, indicating that the viral NP or its mRNA is a sufficient mediator. Confirmation came from results indicating that the interaction of LCMV NP or its mRNA with the PitI protein or its mRNA likely forms the molecular basis of this selective downregulation of GH mRNA transcription and, in turn, GH production, thereby retarding growth. Similarly, in transgenic mice, the GH promoter expressed LCMV NP and mimicked the GH deficiency syndrome caused by persistent LCMV infection.

Recently, the host genes involved in resistance/susceptibility to the virus-induced GH disease were better defined by microsatellite mapping (BUREAU et al. 2001) (Fig. 8, panels 9 and 10). Although C3H/St, BALB/WEHI or CDJ and SWR/J mice infected at birth with LCMV ARM harbored equivalent amounts of virus in their blood, brain, heart, liver, spleen and thymi through life, only C3H/St mice replicated high titers of virus in their anterior lobe of the pituitary gland infecting the majority (>90%) of GH-producing cells (Fig. 8, panels 1 and 3) (OLDSTONE et al. 1985; TISHON and OLDSTONE 1990). In contrast, less than 15% of GH-producing cells became infected in BALB and SWR/J mice. Half the F1 hybrid offspring produced by crossing the susceptible C3H/St GH-deficient strain with the BALB/WEHI GH-resistant mice then developed the disease, but the trait was not sex-linked (TISHON and OLDSTONE 1990). F1 hybrid backcrosses to the susceptible C3H/St parental strain or to the resistant BALB/WEHI parental strain indicated that more than two genes were involved. C3H/St mice have the H-2^k haplotype; even though some other H-2^k strains also developed GH disease after LCMV infection, other H-2^k mice did not (TISHON and OLDSTONE 1990). Further, C3H/Sw mice that have the H-2^b haplotype on the C3H background did develop this disease, further indicating that disease is not related to the H-2^k haplotype but to the C3H background genes. Collectively, these data suggest that the GH deficiency disease induced by LCMV ARM in C3H/St mice is not linked to the mouse MHC haplotype or to sex and is not dependent on a dominant gene. Rather, multiple genes are involved, and these are related to the C3H background (TISHON and OLDSTONE 1990).

Microsatellite mapping across the mouse genome was utilized to identify areas of significant linkage between the clinical findings of growth deficiency induced by the viral infection and host genes, from over 100 individual susceptible C3H/St × resistant BALB/WEHI F1 mice crossed to similar F1 mice. Such studies revealed that the GH deficiency syndrome during persistent LCMV ARM infection maps to a region on chromosome 17 just outside the MHC H-2D site between D17 Mit24 and D17 Mit51. These data linked a region on chromosome 17 encompassing 2.5cM region to the pathogenesis of the GH disease induced by LCMV infection. Further, since murine α-DG, the known receptor for LCMV, residues on chromosome 9 (YOTSUMOTO et al. 1996) not chromosome 17, these findings indicate that an alternative receptor molecule or co-receptor restricted to C3H/St and other GH disease-susceptible mice plays a role in binding and/or entry of LCMV ARM into GH-producing cells.

4.3 Alterations in Behavior and Learning Associated with Persistent LCMV Infection

The link between persistent LCMV infection and clinical signs of severe disease can easily be overlooked. For example, when tested as adults, apparently "normal" mice persistently infected with LCMV can exhibit neurobehavioral abnormalities (HOTCHIN and SEEGAL 1977; DE LA TORRE et al. 1996; GOLD et al. 1994). These include an impaired spatial learning ability, as indicated by a deficit in the acquisition of discriminated avoidance performance and a reduced tendency to explore a novel environment (although their locomotor activity is not affected). During persistent infection of mice with LCMV, viral antigens and nucleic acids in the CNS are localized almost exclusively within neurons (Fig. 9, panels 2 and 3a) (OLDSTONE 1987, 1989). The highest levels of persisting virus are found in the hippocampus, neocortex, limbic system and certain regions of the hypothalamus, with lower levels in the brain stem, thalamus, and basal ganglia (RODRIGUEZ et al. 1983). Virus persistence occurs in the absence of necrosis and inflammation in the CNS. Thus, it is likely that the neurobehavioral alterations seen in mice persistently infected with LCMV are a consequence of direct viral effects on the neuronal populations within which virus persists.

Details of the effects LCMV persistence has on neuronal functioning, and the contributions each of these virus-induced deficits in neuronal functions may make to the neurobehavioral phenotype exhibited by mice persistently infected with LCMV, are beginning to be understood. One series of studies has linked neurochemical abnormalities affecting neurotransmitters to LCMV-mediated neurologic deficits (OLDSTONE et al. 1977; LIPKIN et al. 1988; DE LA TORRE et al. 1996; GOLD et al. 1994). Thus, pharmacologic analysis has shown that mice persistently infected with

LCMV are hypersensitive to the muscarinic cholinergic antagonist, scopolamine, as revealed during their performance in tasks involving learning and motor activity (GOLD et al. 1994). Moreover, in vitro studies (OLDSTONE et al. 1977) demonstrated that persistent infection of murine neuroblastoma cells with LCMV significantly lowered the intracellular levels of choline acetyltransferase and acetylesterase. A cholinergic dysfunction consequent to viral interference with neuronal production of key enzymes involved in neurotransmitter metabolism, therefore, may be one contributor to the learning deficits exhibited by mice persistently infected with LCMV.

The most compelling of these studies (DE LA TORRE et al. 1996) examined whether structural correlates existed for the CNS alterations described in mice persistently infected with LCMV. Specifically sought were alterations in synaptic density and neuronal plasticity, both of which can have profound effects on behavior. The investigators found that, although the overall synaptic density in the neocortex and limbic structures of LCMV-infected mice was preserved, the expression of growth-associated protein-43 (GAP-43), a protein proposed to play an important role in the neuronal plasticity processes accompanying learning and memory, was significantly decreased in the molecular layer of the hippocampus (Fig. 9, panels 3 and 4). In vitro analysis revealed that persistent infection with LCMV of PC12 cells, a cell line that undergoes differentiation from a chromaffin- to a neuron-like phenotype when grown in the presence of neurotrophic growth factor (NGF), prevented NGF induction of GAP-43 upregulation in these cells. NGF-mediated upregulation of amyloid precursor protein in these cells was not affected, indicating the selective viral affect on GAP-43 transcription (Fig. 9, panel 4). Just how LCMV infection affects NGF-mediated upregulation of GAP-43 expression in PC12 cells is not completely clear, but it may interfere with specific pathways of the NGF signal transduction mechanisms, including the protein kinase C-dependent pathway involved in the stabilization of GAP-43 mRNA (CAO et al. 1998). Similarly, LCMV persistence in neurons of mice may interfere with the regulation of GAP-43 expression in response to extracellular signals in the presynaptic terminals of the hippocampal circuitry. The resulting deficit in neuronal plasticity may contribute to the learning defects observed in these mice. Recent studies utilizing a RNA priming technique, hippocampal neurons from LCMV-infected and matched uninfected controls and novel gene array technology (SUTCLIFFE et al. 2000) have uncovered a series of four previously unknown and nine known genes whose transcription is either down- or unregulated with a twofold reproducibly comparing infected and non-infected neurons. Analysis of some of these genes bears the prospect of uncoding the molecular basis of neuronal dysfunction. The cartoon in Fig. 9, panel 1, illustrates the concept of viruses altering differentiated cell functions.

5 Conclusions

The foregoing discussion documents how valuable the study of LCMV infection in its natural rodent host has been to the general understanding of virus-immune

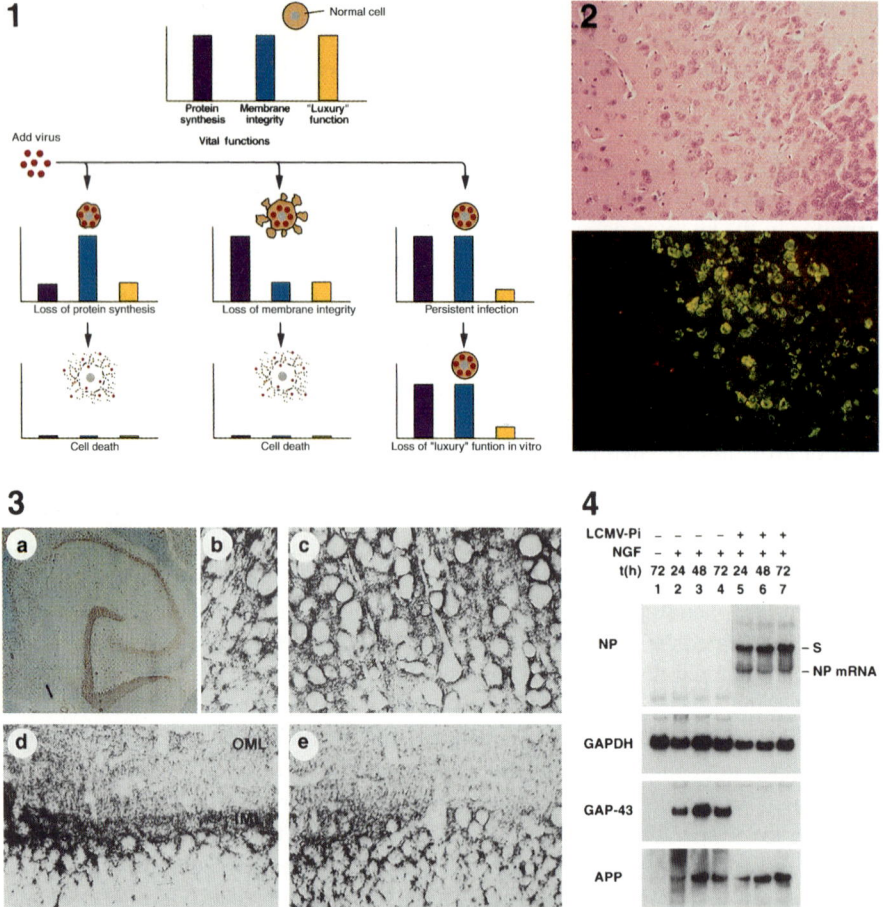

Fig. 9. Virus alters differentiated functions of neuronal cells in vivo and in vitro. *Panel 1* cartoon of virus altering the differentiated (luxury) function of a cell without lysing it. In contrast, viruses can lyse cells by cutting off their protein synthesis or destroying their membranes. *Panel 2* shows that mice persistently infected with LCMV contain virus in neurons of the cerebral cortex without structural destruction of those neurons or an inflammatory response in the area of infection. *Panel 3a* depicts the same phenomenon for hippocampal neurons. In vivo such infected hippocampal neurons undergo a decrease in GAP-43 protein (*panel 3e*) compared to age-matched controls (*panel 3d*). However, other neural markers like synaptophysin are not altered in infected (*panel 3c*) or uninfected neurons (*panel 3b*) (OLDSTONE 1987, 1993; RODRIGUEZ et al. 1983; DE LA TORRE et al. 1993, 1996). These persistently infected mice are defective in cognitive and memory performances (DE LA TORRE et al. 1996; GOLD et al. 1994). *Panel 4* recapitulates in vitro the effect of lowering the GAP-43 level during LCMV infection (LCMV: NP mRNA) of PC-12 neuron-like cells. However, transcription is selective, since amyloid precursor protein (APP), cFOS, etc., transcripts are not altered during the infection

system interactions and viral diseases. As described here and elsewhere in this volume, a remarkable number of key concepts in immunology and virology were first defined in studies of LCMV. Moreover, these conclusions have been extended to many DNA and RNA virus infections of humans. The complex spectrum of variation that marks the outcomes of LCMV infection serves to illustrate the delicate balance governing virus-host interactions, and how seemingly small differences in either host or viral genes can profoundly influence the resolution of infection or the production of end-stage disease. One can safely anticipate that future studies of LCMV infection in mice will yield many more advances in the understanding of viral pathogenesis.

Acknowledgements. This is publication number 13750-NP from the Department of Neuropharmacology, The Scripps Research Institute, La Jolla, CA 92037. This work was supported in part by USPHS grants AI09484 and AI45927. Much of the work presented here was performed in the Viral-Immunobiology Laboratory, TSRI, La Jolla, in collaboration with several outstanding postdoctoral fellows and colleagues over the last 2+ decades including Rafi Ahmed, Christine Biron, Persephone Borrow, Dietmar Berger, Michael Buchmeier, Javier Buesa-Gomez, Wei Cao, Juan Carlos de la Torre, Michael Doyle, Frank Dutko, Michael Eddleston, Claire Evans, Jean Edouard Gairin, Klaus Hahn, Dirk Homann, Etienne Joly, Linda Klavinskis, Stefan Kunz, Hanna Lewicki, W. Ian Lipkin, Dorian McGavern, Lennart Mucke, Denise Naniche, Glenn Rall, Yves Riviere, Moses Rodriguez, Maria Salvato, Noemi Sevilla, Sara Smelt, Peter Southern, Michael Teng, Antoinette Tishon, Alexandra Valsamakis, Raymond Welsh, J. Lindsay Whitton, Matthias von Herrath, and Yusuke Yanagi, and with external collaborators Kevin Campbell, Howard Hughes Medical Institute, University of Iowa; Michel Brahic, Pasteur Institute, Paris; and Anura Rambukkana, Rockefeller Research Institute, New York.

References

Aebischer T, Moskophidis D, Rohrer UH, Zinkernagel RH, Hengartner R (1991) In vitro selection of lymphocytic choriomeningitis virus escape mutants by cytotoxic T lymphocytes. Proc Natl Acad Sci 88:11047–11051
Ahmed R, Byrne JA, Oldstone MBA (1984a) Virus specificity of cytotoxic T lymphocytes generated during acute lymphocytic choriomeningitis virus infection: role of the H-2 region in determining cross-reactivity for different lymphocytic choriomeningitis virus strains. J Virol 51:34–41
Ahmed R, Jamieson BD, Porter D (1987) Immune therapy of a persistent and disseminated viral infection. J Virol 61:3920–3929
Ahmed R, Oldstone MBA (1988) Organ-specific selection of viral variants during chronic infection. J Exp Med 167:1719–1724
Ahmed R, Hahn CS, Somasundaram T, Villarete L, Matloubian M, Strauss JH (1991) Molecular basis of organ-specific selection of viral variants during chronic infection. J Virol 65:4242–4247
Ahmed R, Salmi A, Butler LD, Chiller JM, Oldstone MBA (1984b) Selection of genetic variants of lymphocytic choriomeningitis virus in spleens of persistently infected mice: role in suppression of cytotoxic T lymphocyte response and viral persistence. J Exp Med 160:521–540
Ahmed R, Simon R, Matloubian M, Kohleker S, Southern P, Freedman D (1988) Genetic analysis of in vivo-selected viral variants causing chronic infection: importance of mutation in the L RNA segment of lymphocytic choriomeningitis virus. J Virol 62:3301–3308
Althage A, Odermatt B, Moskophidis D, et al. (1992) Immunosuppression by lymphocytic choriomeningitis virus infection: competent effector T and B cells but impaired antigen presentation. Eur J Immunol 22:1803–1812
Anderson J, Byrne JA, Schreiber R, Patterson S, Oldstone MB (1985) Biology of cloned cytotoxic T lymphocytes specific for lymphocytic choriomeningitis virus: clearance of virus and in vitro properties. J Virol 53:552–560
Baenziger J, Hengartner H, Zinkernagel RM, Cole GA (1986) Induction or prevention of immunopathological disease by cloned cytotoxic T cell lines specific for lymphocytic choriomeningitis virus. Eur J Immunol 16:1237–1242
Battegay M, Moskophidis D, Rahemtulla A, Hengartner H, Mak TW, Zinkernagel RM (1994) Enhanced establishment of a virus carrier state in adult $CD4^+$ T-cell-deficient mice. J Virol 68:4700

Beijerinck MW (1899) Bemerkung zu dem Aufsatz von Herrn Iwanowsky über die Mosaikkrankheit der Tabakspflanze. Zentralbl. Bakteriol. Parasitenkd. Infektionskr. Hyg Abt I Orig 5:310–311

Berger DP, Homann D, Oldstone MBA (2000) Defining parameters for successful immunocytotherapy of persistent viral infection. Virology 266:257–263

Bertoletti A, Costanzo A, Chisari FV, Levrero M, Artini M, Sette A, Penna A, Giuberti T, Fiaccadori F, Ferrari C (1994a) Cytotoxic T lymphocyte response to a wild-type hepatitis B virus epitope in patients chronically infected by variant viruses carrying substitutions within the epitope. J Exp Med 180: 933–943

Bertoletti A, Sette A, Chisari FV, Penna A, Levrero M, De Carli M, Fiaccadori F, Ferrari C (1994b) Natural variants of cytotoxic epitopes are T cell receptor antagonists for anti-viral cytotoxic T cells. Nature 369:407–410

Bjorkman PJ, Saper MA, Samraoui B, Bennett WS, Strominger JL, Wiley DC (1987) Structure of the human class I histocompatibility antigen, HLA-A2. Nature 329:506–511

Blount P, Elder J, Lipkin WI, Southern PJ, Buchmeier MJ, Oldstone MBA (1986) Dissecting the molecular anatomy of the nervous system: Analysis of RNA and protein expression in whole body sections of laboratory animals. Brain Res 382:257–265

Borrow P, Evans CF, Oldstone MBA (1995) Virus-induced immunosuppression: immune system-mediated destruction of virus-infected dendritic cells results in generalized immunosuppression. J Virol 69:1059–1070

Borrow P, Lewicki H, Wei X, Horwitz MS, Peffer N, Meyers H, Nelson JA, Gairin JE, Hahn BH, Oldstone MBA, Shaw GM (1997) Antiviral pressure exerted by HIV-1-specific cytotoxic T lymphocytes (CTLs) during primary infection demonstrated by rapid selection of CTL escape virus. Nature Medicine 3:205–211

Borrow P, Oldstone MBA (1997) Lymphocytic choriomeningitis virus. In: Nathanson N, et al. (eds) Viral Pathogenesis. Lippincott-Raven Publishers, Philadelphia, pp 593–627

Borrow P, Vedovato V, Buesa-Gomez J, de la Torre J, Oldstone M (1993) The in vivo pathogenicity of different LCMV isolates correlates with binding to putative cellular receptors in vitro and the amino acid sequence of the virion attachment protein GP-1 (abstract). IXth International Congress of Virology, Glasgow, Scotland

Buchmeier MJ, Welsh RM, Dutko FJ, Oldstone MBA (1980) The virology and immunobiology of lymphocytic choriomeningitis virus infection. Adv Immunol 30:275–331

Buesa-Gomez J, Teng MN, Oldstone MBA, de la Torre JC (1996) Variants able to cause growth hormone deficiency syndrome are present within the disease-nil WE strain of lymphocytic choriomeningitis virus. J Virol 70:8988–8992

Bureau JF, Le Goff S, Thomas D, Parlow AF, de la Torre JC, Homann D, Brahic M, Oldstone MBA (2001) Disruption of differentiated functions during viral infection in vivo. V. Mapping of a locus involved in susceptibility of mice to growth hormone deficiency due to persistent lymphocytic choriomeningitis virus infection. Virology (in press)

Butz EA, Bevan MJ (1998) Massive expansion of antigen-specific $CD8^+$ T cells during an acute virus infection. Immunity 8:167–175

Byrne JA, Ahmed R, Oldstone MBA (1984) Biology of cloned cytotoxic T lymphocytes specific for lymphocytic choriomeningitis virus. I. Generation and recognition of virus strains and $H-2^b$ mutants. J Immunol 133:433–439

Cao W, Henry MD, Borrow P, Yamada H, Elder JH, Ravkov EV, Nichol ST, Compans RW, Campbell KP, Oldstone MBA (1998) Identification of alpha-dystroglycan as a receptor for lymphocytic choriomeningitis virus and Lassa fever virus. Science 282:2079–2081

Casali P, Rice GPA, Oldstone MBA (1984) Viruses disrupt functions of human lymphocytes: effects of measles virus and influenza virus in lymphocyte-mediated killing and antibody production. J Exp Med 159:1322–1337

Cole GA, Nathanson N, Prendergast RA (1972) Requirement for θ-bearing cells in lymphocytic choriomeningitis virus-induced central nervous system disease. Nature 238:335–337

de Campos-Lima PO, Gavioli R, Zhang QJ, Wallace LE, Dolcetti R, Rowe M, Rickinson AB, Masucci MG (1993) HLA-A11 epitope loss isolates of Epstein-Barr virus from a highly $A11^+$ population. Science 260:98–100

de la Torre JC, Mallory M, Brot M, Gold L, Koob G, Oldstone MBA, Masliah E (1996) Viral persistence in neurons alters synaptic plasticity and cognitive functions without destruction of brain cells. Virology 220:508–515

de la Torre JC, Oldstone MBA (1992) Selective disruption of growth hormone transcription machinery by viral infection. Proc Natl Acad Sci USA 89:9939–9943 de la Torre JC, Rall G, Oldstone C, Sanna P, Borrow P, Oldstone MBA (1993) Replication of lymphocytic choriomeningitis virus is restricted in terminally differentiated neurons. J Virol 67:7350–7359

Dockter J, Evans CF, Tishon A, Oldstone MBA (1996) Competitive selection in vivo by a cell for one variant over another: implications for RNA virus quasispecies in vivo. J Virol 70:1799–1803

Doherty PC, Allan JE (1986) Role of the major histocompatibility complex in targeting effector T cells into a site of virus infection. Eur J Immunol 16:1237–1242

Evans CF, Borrow P, de la Torre JC, Oldstone MBA (1994) Virus-induced immunosuppression: kinetic analysis of the selection of a mutation associated with viral persistence. J Virol 68:7367–7373

Falk K, Rotzschke O, Stevanovic S, Jung G, Rammensee HG (1991) Allele-specific motifs revealed by sequencing of self-peptides eluted from MHC molecules. Nature 351:290–296

Gairin JE, Joly E, Oldstone MB (1991) Persistent infection with lymphocytic choriomeningitis virus enhances expression of MHC class I GP on cultured mouse brain endothelial cells. J Immunol 146:3953–3957

Gairin JE, Mazarguil H, Hudrisier D, Oldstone MBA (1995) Optimal lymphocytic choriomeningitis virus sequences restricted by H-2D^b major histocompatibility complex class I molecules and presented to cytotoxic T lymphocytes. J Virol 69:2297–2305

Garcia KC, Degano M, Stanfield RL, Brunmark A, Jackson MR, Peterson PA, Teyton L, Wilson IA (1996) An αβ T cell receptor structure at 2.5Å and its orientation in the TCR-MHC complex. Science 274:209–219

Gilden DH, Cole GA, Monjan AA, Nathanson N (1972a) Immunopathogenesis of acute central nervous system disease produced by lymphocytic choriomeningitis virus. I. Cyclophosphamide-mediated induction of virus-carrier state in adult mice. J Exp Med 135:860–873

Gilden DH, Cole GA, Nathanson N (1972b) Immunopathogenesis of acute central nervous system disease produced by lymphocytic choriomeningitis virus. II. Adoptive immunization of virus carriers. J Exp Med 135:874–889

Gold LH, Brot MD, Polis I, Schroeder R, Tishon A, de la Torre JC, Oldstone MBA, Koob GF (1994) Behavioral effects of persistent lymphocytic choriomeningitis virus infection in mice. Behav Neural Biol 62:100–109

Goulder PJR, Phillips RE, Colbert RA, McAdam S, Ogg G, Nowak MA, Giangrande P, Luzzi G, Morgan B, Edwards A, McMichael AJ, Rowland-Jones S (1997) Late escape from an immunodominant cytotoxic T-lymphocyte response associated with progression to AIDS. Nature Medicine 3:312–317

Guidotti LG, Borrow P, Brown A, McClary H, Koch R, Chisari FV (1999) Noncytopathic clearance of lymphocytic choriomeningitis virus from the hepatocyte. J Exp Med 189:1555–1564

Henry MD, Campbell KP (1999) Dystroglycan inside and out. Curr Opin Cell Biol 11:602–607

Holtzer H, Biehl J, Yeoh G, Meganathan R, Kaji A (1975) Effect of oncogenic virus on muscle differentiation. Proc Natl Acad Sci USA 72:4051–4055

Holtzer H, Pacifici M, Tapscott S, Bennett G, Payette R, Dlugosz A (1982) Lineages in cell differentiation and in cell transformation. In: Revoltella RF (ed) Expression of differentiated functions in cancer cells. Raven Press, New York, pp 169

Homann D, Teyton L, Oldstone MBA (2001) Loss of antiviral CD4$^+$ T cell memory is associated with reduced bcl-2 and telomerase expression. Nature Medicine (in press)

Hotchin JE, Cintis M (1958) Lymphocytic choriomeningitis infection of mice as a model for the study of latent virus infection. Can J Microbiol 4:149–163

Hotchin JE, Seegal R (1977) Virus-induced behavioral alteration of mice. Science 196:671–674

Hudrisier D, Mazarguil H, Laval F, Oldstone MBA, Gairin JE (1996) Binding of viral antigens to major histocompatibility complex class I H-2D^b molecules is controlled by dominant negative elements at peptide non-anchor residues: implications for peptide selection and presentation. J Biol Chem 271:17829–17836

Ivanovski DI (1899) Ueber die Mosaikkrankheit der Tabakspflanze. Zentralbl. Bakteriol. Parasitenkd. Infektionskr. Hyg Abt II Orig 5:250–254

Jamieson BD, Somasundaram T, Ahmed R (1991) Abrogation of tolerance to a chronic viral infection. J Immunol 147:3521–3529

Joly E, Mucke L, Oldstone MBA (1991) Viral persistence in neurons explained by lack of major histocompatibility complex class I expression. Science 253:1283–1285

Kagi D, Seiler P, Pavlovic J, Ledermann B, Burki K, Zinkernagel RM, Hengartner H (1995) The roles of perforin- and Fas-dependent cytotoxicity in protection against cytopathic and noncytopathic viruses. Eur J Immunol 25:3256–3262

Klavinskis LS, Geckeler R, Oldstone MB (1989a) Cytotoxic T lymphocyte control of acute lymphocytic choriomeningitis virus infection: interferon gamma, but not tumour necrosis factor alpha, displays antiviral activity in vivo. J Gen Virol 70:3317–3325
Klavinskis LS, Oldstone MBA (1989) Lymphocytic choriomeningitis virus selectively alters differentiated but not housekeeping functions: block in expression of growth hormone gene is at the level of transcriptional initiation. Virology 168:232–235
Klavinskis LS, Whitton JL, Oldstone MBA (1989b) Molecularly engineered vaccine which expresses an immunodominant T cell epitope induces cytotoxic T lymphocytes that confer protection from lethal virus infection. J Virol 63:4311–4316
Klenerman P, Rowland-Jones S, McAdam S, Edwards J, Daenke S, Lalloo D, Koppe B, Rosenbrg W, Boyd D, Edwards A, Giangrande P, Phillips RE, McMichael AJ (1994) Cytotoxic T-cell activity antagonized by naturally occurring HIV-1 gag variants. Nature 369:403–407
Koup RA (1994) Virus escape from CTL recognition. J Exp Med 180:779–782
Lampert PW, Oldstone MBA (1974) Pathology of the choroid plexus in spontaneous immune complex disease and chronic viral infections. Virchows Archiv 363:21–32
Leist TP, Zinkernagel RM (1990) Treatment with anti-tumor necrosis factor alpha does not influence the immune pathological response against lymphocytic choriomeningitis virus. Cytokine 2:29–34
Lewicki HA, Tishon A, Borrow P, Evans C, Gairin JE, Hahn KM, Jewell DA, Wilson IA, Oldstone MBA (1995a) CTL escape viral variants. I. Generation and molecular characterization. Virology 210:29–40
Lewicki HA, von Herrath MG, Evans CF, Whitton JL, Oldstone MBA (1995b) CTL escape viral variants. II. Biologic activity in vivo. Virology 211:443–450
Lipkin WI, Battenberg ELF, Bloom FE, Oldstone MBA (1988) Viral infection of neurons can depress neurotransmitter mRNA levels without histologic injury. Brain Res 451:333–339
Lipkin WI, Villarreal LP, Oldstone MBA (1989) Neurotransmitter abnormalities in Borna disease. Brain Res 475:366–370
Loeffler F, Frosch P (1898) Berichte der Kommission zur Erforschung der Maul und Klauenseuche bei dem Institut für Infektionskrankheiten in Berlin. Zentralbl. Bakteriol. Parasitenkd. Infektionskr. Hyg Abt I Orig 23:371–391
Lundstedt C (1969) Interaction between antigenically different cells: virus induced cytotoxicity by immune lymphocytes in vitro. Acta Pathol Microbiol Scand 75:134–147
Marker O, Volkert M (1973) Studies on cell-mediated immunity to lymphocytic choriomeningitis virus in mice. J Exp Med 137:1511–1525
Matloubian M, Concepcion RJ, Ahmed R (1994) $CD4^+$ T cells are required to sustain $CD8^+$ cytotoxic T-cell responses during chronic viral infection. J Virol 68:8056–8063
Matloubian M, Somasundaram T, Kolhekar SR, Selvakumar R, Ahmed R (1990) Genetic basis of viral persistence: single amino acid change in the viral glycoprotein affects ability of lymphocytic choriomeningitis virus to persist in adult mice. J Exp Med 172:1043–1048
Matsumura M, Fremont DH, Peterson PA, Wilson IA (1992) Emerging principles for the recognition of peptide antigens by MHC class I molecules. Science 257:927–934
Medawar PB, Medawar JS (1983) Aristotle to Zeus, a philosophical dictionary of biology. Harvard University Press, Cambridge, Massachusetts, pp 275
Mims CA, Blanden RV (1972) Antiviral action of immune lymphocytes in mice infected with lymphocytic choriomeningitis virus. Infect Immun 6:695–698
Moskophidis D, Zinkernagel RM (1995) Immunobiology of cytotoxic T-cell escape mutants of lymphocytic choriomeningitis virus. J Virol 69:2187–2193
Muller U, Steinhoff U, Reis LFL, et al. (1994) Functional role of type I and type II interferons in antiviral defense. Science 264:1918–1921
Murali-Krishna K, Altman JD, Suresh M, Sourdive DJ, Zajac AJ, Miller JD, Slansky J, Ahmed R (1998) Counting antigen-specific CD8 T cells: a reevaluation of bystander activation during viral infection. Immunity 8:177–187
Odermatt B, Eppler M, Leist TP, Hengartner H, Zinkernagel RM (1991) Virus-triggered acquired immunodeficiency by cytotoxic T-cell-dependent destruction of antigen-presenting cells and lymph follicle structure. Proc Natl Acad Sci USA 88:8252–8256
Oldstone MBA (1975) Virus neutralization and virus-induced immune complex disease: virus-antibody union resulting in immunoprotection or immunologic injury – two sides of the same coin. In: Melnick JL (ed) Progress in medical virology, Vol. 19. S. Karger, Basel, pp 84–119
Oldstone MBA (1984) Virus can alter cell function without causing cell pathology: disordered function leads to imbalance of homeostasis and disease. In: Notkins AL, Oldstone MBA (eds) Concepts in viral pathogenesis. Springer, Heidelberg Berlin New York, pp 269–276

Oldstone MBA (1985) An old nemesis in new clothing: viruses playing new tricks by causing cytopathology in the absence of cytolysis. J Infect Dis 152:665–667
Oldstone MBA (1987) Molecular anatomy of viral disease. Neurology 37:453–460
Oldstone MBA (1989) Viral alteration of cell function. Sci Amer 260:42–48
Oldstone MBA (1991) Molecular anatomy of viral persistence. J Virol 65:6381–6386
Oldstone MBA (1993) Viruses and diseases of the twenty-first century. Amer J Pathol 143:1241–1249
Oldstone MBA (1997) How viruses escape from cytotoxic T lymphocytes: molecular parameters and players. Virology 234:179–185
Oldstone MBA, Ahmed R, Buchmeier MJ, Blount P, Tishon A (1985) Perturbation of differentiated functions during viral infection in vivo. I. Relationship of lymphocytic choriomeningitis virus and host strains to growth hormone deficiency. Virology 142:158–174
Oldstone MBA, Blount P, Southern PJ, Lampert PW (1986) Cytoimmunotherapy for persistent virus infection: unique clearance pattern from the central nervous system. Nature 321:239–243
Oldstone MBA, Dixon FJ (1969) Pathogenesis of chronic disease associated with persistent lymphocytic choriomeningitis viral infection. I. Relationship of antibody production to disease in neonatally infected mice. J Exp Med 129:483–505
Oldstone MBA, Dixon FJ (1972) Disease accompanying in utero viral infection: the role of maternal antibody in tissue injury after transplacental infection with lymphocytic choriomeningitis virus. J Exp Med 135:827–838
Oldstone MBA, Habel K, Dixon FJ (1969) The pathogenesis of cellular injury associated with persistent LCM viral infection. Fed Proc 28:429
Oldstone MBA, Holmstoen J, Welsh RM (1977) Alterations of acetylcholine enzymes in neuroblastoma cells persistently infected with lymphocytic choriomeningitis virus. J Cell Physiol 91:459–472
Oldstone MBA, Rodriguez M, Daughaday WH, Lampert PW (1984a) Viral perturbation of endocrine function: disorder of cell function leading to disturbed homeostasis and disease. Nature 307:278–280
Oldstone MBA, Sinha YN, Blount P, Tishon A, Rodriguez M, von Wedel R, Lampert PW (1982) Virus-induced alterations in homeostasis: alterations in differentiated functions of infected cells in vivo. Science 218:1125–1127
Oldstone MBA, Southern P, Rodriguez M, Lampert P (1984b) Virus persists in beta cells of islets of Langerhans and is associated with chemical manifestations of diabetes. Science 224:1440–1443
Oldstone MBA, Tishon A, Buchmeier MJ (1983) Virus induced immune complex disease: genetic control of C1q binding complexes in the circulation of mice persistently infected with lymphocytic choriomeningitis virus. J Immunol 130:912–918
Oldstone MBA, Tishon A, Chiller J, Weigle W, Dixon FJ (1973) Effect of chronic viral infection on the immune system. I. Comparison of the immune responsiveness of mice chronically infected with LCM virus with that of noninfected mice. J Immunol 110:1268–1278
Oldstone MBA, Whitton JL, Lewicki H, Tishon A (1988) Fine dissection of a nine amino acid glycoprotein epitope, a major determinant recognized by lymphocytic choriomeningitis virus specific class I restricted H-2D^b cytotoxic T lymphocytes. J Exp Med 168:559–570
Oxenius A, Zinkernagel RM, Hengartner H (1998) Comparison of activation versus induction of unresponsiveness of virus-specific CD4$^+$ and CD8$^+$ T cells upon acute versus persistent viral infection. Immunity 9:449–457
Pewe L, Xue S, Perlman S (1998) Infection with cytotoxic T-lymphocyte escape mutants results in increased mortality and growth retardation in mice infected with a neurotropic coronavirus. J Virol 72:5912–5918
Phillips RE, et al. (1991) Human immunodeficiency virus genetic variation that can escape cytotoxic T cell recognition. Nature 354:453–459
Pircher H, Moskophidis D, Rohrer U, Burki K, Hengartner H, Zinkernagel RM (1990) Viral escape by selection of cytotoxic T cell-resistant virus variants in vivo. Nature 346:629–633
Rambukkana A, Oldstone MBA (2001) Manuscript in preparation.
Rammensee HG, Friede T, Stevanovic S (1995) MHC ligands and peptide motifs: first listing. Immunogenetics 41:178–228
Riviere Y, Ahmed R, Southern PJ, Buchmeier MJ, Dutko FJ, Oldstone MBA (1985a) The S RNA segment of lymphocytic choriomeningitis virus codes for the nucleoprotein and glycoproteins 1 and 2. J Virol 53:966–968
Riviere Y, Ahmed R, Southern PJ, Buchmeier MJ, Oldstone MBA (1985b) Genetic mapping of lymphocytic choriomeningitis virus pathogenicity: virulence in guinea pigs is associated with the L RNA segment. J Virol 55:704–709

Rodriguez M, Buchmeier MJ, Oldstone MBA, Lampert PW (1983) Ultrastructural localization of viral antigens in the CNS of mice persistently infected with lymphocytic choriomeningitis virus (LCMV). Amer J Pathol 110:95–100

Rotzschke O, Falk K, Deres K, Schild H, Norda M, Melzger J, Jung G, Rammensee HG (1990) Isolation and analysis of naturally processed viral peptides as recognized by cytotoxic T cells. Nature 348: 252–254

Rowe WP (1954) Studies on pathogenesis and immunity in lymphocytic choriomeningitis infection of the mouse. Research Report NM 005048.14.01. Naval Medical Research Institute, Bethesda, Maryland

Salvato M, Shimomaye E, Southern P, Oldstone MBA (1988) Virus-lymphocyte interactions. IV. Molecular characterization of LCMV Armstrong (CTL$^+$) and that of its variant, Clone 13 (CTL$^-$). Virology 164:517–522

Schrier RD, Oldstone MBA (1986) Recent clinical isolates of cytomegalovirus suppress human cytomegalovirus-specific human leukocyte antigen-restricted cytotoxic T-lymphocyte activity. J Virol 59:127–131

Sevilla N, Kunz S, Holz A, Lewicki H, Homann D, Yamada H, Campbell KP, de la Torre JC, Oldstone MBA (2000) Immunosuppression and resultant viral persistence by specific viral targeting of dendritic cells. J Exp Med 192:1249–1260

Smelt SC, Borrow P, Kunz S, Cao W, Tishon A, Lewicki H, Campbell KP, Oldstone MBA (2001) Differences in affinity of binding of lymphocytic choriomeningitis virus strains to the cellular receptor α-dystroglycan correlate with viral tropism and disease kinetics. J Virol 75:448–457.

Sutcliffe JG, Foye PE, Erlander MG, Hilbush BS, Bodzin LJ, Durham JT, Hasel KW (2000) TOGA: an automated parsing technology for analyzing expression of nearly all genes. Proc Natl Acad Sci USA 97:1976–1981

Teng NM, Oldstone MBA, de la Torre JC (1996a) Suppression of lymphocytic choriomeningitis virus-induced growth hormone deficiency syndrome by disease-negative virus variants. Virology 223: 113–119

Teng NM, Borrow P, Oldstone MBA, de la Torre JC (1996b) A single amino acid change in the glycoprotein of lymphocytic choriomeningitis virus is associated with the ability to cause growth hormone deficiency syndrome. J Virol 70:8438–8443

Tishon A, Borrow P, Evans C, Oldstone MBA (1993a) Virus induced immunosuppression. 1. Age at infection relates to a selective or generalized defect. Virology 195:397–405

Tishon A, Eddleston M, de la Torre JC, Oldstone MBA (1993b) Cytotoxic T lymphocytes cleanse viral gene products from individually infected neurons and lymphocytes in mice persistently infected with lymphocytic choriomeningitis virus. Virology 197:463–467

Tishon A, Lewicki H, Rall G, von Herrath M, Oldstone MBA (1995) An essential role for type 1 interferon-γ in terminating persistent viral infection. Virology 212:244–250

Tishon A, Oldstone MBA (1990) Perturbation of differentiated functions during viral infection in vivo. In vivo relationship of host genes and lymphocytic choriomeningitis virus to growth hormone deficiency. Am J Pathol 137:965–969

Tishon A, Salmi A, Ahmed R, Oldstone MBA (1991) Role of viral strains and host genes in determining levels of immune complexes in a model system: implications for HIV infection. AIDS Res & Human Retrovir 7:963–969

Traub E (1936) Persistence of lymphocytic choriomeningitis virus in immune animals and its relation to immunity. J Exp Med 63:847–861

Valsamakis A, Riviere Y, Oldstone MBA (1987) Perturbation of differentiated functions in vivo during persistent viral infection. III. Decreased growth hormone mRNA. Virology 156:214–220

Varga SM, Welsh RM (1998) Detection of a high frequency of virus-specific CD4$^+$ T cells during acute infection with lymphocytic choriomeningitis virus. J Immunol 161:3215–3218

Villarete L, Somasundaram T, Ahmed R (1994) Tissue-mediated selection of viral variants: correlation between glycoprotein mutation and growth in neuronal cells. J Virol 68:7490–7496

von Herrath MG, Berger DP, Homann D, Tishon T, Sette A, Oldstone MBA (2000) Vaccination to treat persistent viral infection. Virology 268:411–419

von Herrath MG, Dockter J, Nerenberg M, Gairin JE, Oldstone MBA (1994) Thymic selection and adaptability of cytotoxic T lymphocyte responses in transgenic mice expressing a viral protein in the thymus. J Exp Med 180:1901–1910

Walsh CM, Matloubian M, Liu CC, Ueda R, Kurahara CG, Christensen JL, Huang MT, Young JD, Ahmed R, Clark WR (1994) Immune function in mice lacking the perforin gene. Proc Natl Acad Sci USA 91:10854–10858

Whitton JL, Oldstone MBA (2001) The immune response to viruses. In: Knipe D, et al. (eds) Fields Virology, 4th edn. Lippincott Williams & Wilkins, Philadelphia (in press)

Yanagi Y, Maekawa R, Cook T, Kanagawa O, Oldstone MBA (1990) Restricted V-segment usage in T-cell receptors from cytotoxic T lymphocytes specific for a major epitope of lymphocytic choriomeningitis virus. J Virol 64:5919–5926

Yotsumoto S, Fujiwara H, Horton JH, Mosby TA, Wang X, Cui Y, Ko MS (1996) Cloning and expression analyses of mouse dystroglycan gene: specific expression in maternal decidua at the peri-implantation stage. Human Mol Genetics 5:1259–1267

Young AC, Zhang W, Sacchettini JC, Nathenson SG (1994) The three-dimensional structure of H-2D^b at 2.4 Å resolution: implications for antigen-determinant selection. Cell 76:39–50

Young LH, Klavinskis LS, Oldstone MB, Young JD (1989) In vivo expression of perforin by CD8$^+$ lymphocytes during an acute viral infection. J Exp Med 169:2159–2171

Zinkernagel RM, Doherty PC (1974) Restriction of in vitro T cell-mediated cytotoxicity in lymphocytic choriomeningitis within a syngeneic or semiallogeneic system. Nature 248:701–702

Zinkernagel RM, Welsh RM (1976) H-2 compatibility requirement for virus-specific T cell-mediated effector functions in vivo. I. Specificity of T cells conferring antiviral protection against lymphocytic choriomeningitis virus is associated with H-2K and H-2D. J Immunol 117:1495–1502

Contribution of LCMV Transgenic Models to Understanding T Lymphocyte Development, Activation, Tolerance, and Autoimmunity

L.T. Nguyen[1], M.F. Bachmann[2], and P.S. Ohashi[1,3]

1	Thymocyte Development	120
1.1	Timing and Sensitivity of Negative Selection During Thymocyte Development	121
1.2	The Outcome of Positive and Negative Selection Are Determined by the Affinity/Avidity of Thymocyte Interactions	122
1.3	Flexibility of the Interactions that Dictate Positive Selection	123
1.4	Thymocyte Tuning: a Mechanism of Self-Tolerance	124
1.5	Role of Coreceptors During Thymocyte Development	125
1.6	Summary: Thymocyte Selection	126
2	T Cell Activation	127
2.1	The Role of the T Cell Receptor in T Cell Activation	127
2.1.1	The Degree of TCR Internalization Reflects the Strength of the Signal and Induction of Effector Function	127
2.1.2	Why Are TCRs Internalized upon Engagement?	128
2.1.3	TCR Dimerization	128
2.2	The Role of Accessory and Costimulatory Molecules in T Cell Activation	129
2.2.1	Molecules that Enhance T Cell Activation	129
2.2.2	T Cell Molecules that Promote APC Maturation	131
2.3	T Cell Memory	131
2.4	Summary: T Cell Activation	132
3	Peripheral Tolerance	132
3.1	Deletion and Anergy vs. Ignorance	132
3.1.1	Prevention of Diabetes	133
3.1.2	Ignorant/Naive Autoreactive Cells May be Used for Tumor Immunotherapy	134
3.2	Peripheral Tolerance vs. Immunity: the Role of APCs	135
3.3	Peripheral Tolerance: the Role of TNFR Family Members in Apoptosis	136
3.4	Summary: Peripheral Tolerance	137
References		137

T cells play a central role in the elimination of both intracellular and extracellular pathogens. Expression of the $\alpha\beta$ T cell receptor (TCR) heterodimer enables T cells to specifically recognize peptides presented by major histocompatibility complex (MHC) molecules and generate an immune response to those peptides. $CD4^+$

---

[1] Department of Immunology, Ontario Cancer Institute, 610 University Avenue, Toronto, Ontario, Canada, M5G 2M9
[2] Cytos Ag, Wagistr. 21, 8952 Zurich-Schlieren, Switzerland
[3] Department of Medical Biophysics and Immunology, Ontario Cancer Institute, 610 University Avenue, Toronto, Ontario, Canada, M5G 2M9

helper T cells are restricted to MHC class II molecules, and produce cytokines to augment T and B cell responses. $CD8^+$ T cells are restricted to MHC class I molecules, and are generally cytotoxic T lymphocytes (CTLs) whose effector function involves the lysis of infected cells. Broad issues in T cell biology include how mature T cells develop from bone marrow-derived precursors and how T cell activation is regulated. These questions and others have been addressed in many models, including murine TCR transgenic systems, in which the transgenic expression of TCR chains provides a clonal population of antigen-specific T cells that can be readily monitored.

In this chapter, we will outline the contribution of two LCMV-specific transgenic models to various aspects of T cell biology. In P14 TCR transgenic mice, the transgenic Vα2 and Vβ8.1 TCR chains from the CTL clone P14 were cloned and expressed using the H-2K^b promoter and immunoglobulin enhancer (PIRCHER et al. 1987, 1989). Together, this TCR α/β heterodimer recognizes the immunodominant peptide derived from the lymphocytic choriomeningitis virus (LCMV) glycoprotein (gp), called p33. This peptide binds to the MHC class I molecule H-2D^b, and thus the P14 TCR transgenic T cell population is predominantly comprised of $CD8^+$ T cells, with the majority of $CD8^+$ peripheral T cells expressing the transgene (PIRCHER et al. 1989). In another model, we have expressed the LCMV-gp under the control of the rat insulin promoter (RIP), which leads to the expression of the LCMV-gp in the β-islet cells of the pancreas (OHASHI et al. 1991). This model has been independently reported and studied by Oldstone's group (OLDSTONE et al. 1991), and complementary studies using these models are highlighted in other chapters in this book. These models have proven to be powerful tools because they combine a biologically relevant viral system with the ability to follow the differentiation, development and the activation of an immune response of virus-specific T cells. In some studies, we have also used the viral antigen as a self antigen to provide insights into tolerance, autoimmunity and tumor immunity.

1 Thymocyte Development

Maturation of thymocytes through stages of selection and differentiation yields $CD4^+$ and $CD8^+$ T cells that populate the secondary lymphoid organs (SEBZDA et al. 1999). Development of T cells in the thymus begins from progenitor cells derived from the bone marrow. These cells do not yet express TCR chains, nor do they express CD4 or CD8 T cell coreceptors. During this $CD4^-$ $CD8^-$ (double negative, DN) stage, thymocytes randomly rearrange gene segments encoding the variable (V), diversity (D) and joining (J) components of the TCRβ chain. Surface expression of TCRβ with a non-polymorphic pre-TCRα chain allows for selection of those thymocytes with productively rearranged TCRβ loci. Thymocytes then progress to the double positive (DP) stage, characterized by the surface expression of CD4 and CD8. In addition, thymocytes express productively rearranged TCRα and β chains and consequently have the potential to interact with class I or class II major histocompatibility complex (MHC) molecules expressed by thymic epithelial and bone marrow-derived cells. MHC molecules bind and present peptides primarily derived from the thymic environment. By interacting with MHC, the TCRs also encounter a particular peptide bound by the MHC molecule. Therefore, thymocytes at this DP stage are selected by interactions with self peptide/MHC molecules.

Rearrangement of TCR gene loci results in a vast pool of thymocytes, each with a unique receptor. Due to the randomness of this process, each TCR generated may or may not be able to recognize a self peptide/MHC molecule. Experiments using chimeric animals showed that thymocytes are selected for the ability to interact with self MHC molecules (BEVAN 1977; FINK and BEVAN 1978; ZINKERNAGEL et al. 1978a,b). Thus, this process of *positive selection* permits the maturation of thymocytes expressing TCRs that recognize self MHC, and eliminates useless thymocytes that cannot recognize self MHC. However, random rearrangements also produce TCRs that may strongly interact with self-peptides presented by MHC molecules. Another process called *negative selection*, or central tolerance is a mechanism that eliminates potentially autoreactive T cells. TCR transgenic models have demonstrated that central tolerance can be achieved by deletion of self-reactive thymocytes (KISIELOW et al. 1988; SHA et al. 1988), and studies in bone marrow chimeras have shown that self-reactive thymocytes can also undergo clonal inactivation or anergy (BLACKMAN et al. 1990; RAMSDELL et al. 1989).

1.1 Timing and Sensitivity of Negative Selection During Thymocyte Development

The earliest studies using the P14 TCR transgenic system compared clonal deletion induced by two distinct antigens recognized by the P14 TCR: LCMV and Mls-1^a (minor lymphocyte stimulatory) (Mtv-7) (OHASHI et al. 1990; PIRCHER et al. 1989). Because the P14 TCR uses Vβ8.1, this allows this receptor to interact with a superantigen (Mls-1^a) encoded by an endogenous retrovirus (Mtv-7). Tolerance to Mls-1^a was examined by breeding P14 transgenic mice with mice known to express Mtv-7. Tolerance to LCMV was examined by neonatal infection with LCMV, which results in a virus carrier state in which the mice are essentially full of noncytopathic LCMV. P14 TCR transgenic thymocytes were deleted by thymic presentation of either LCMV-gp or Mls-1^a. Negative selection by either LCMV-gp p33 or Mls-1^a resulted in decreased numbers of peripheral $CD8^+$ T cells. However, comparison of thymocyte populations between P14 TCR transgenic LCMV carrier mice and P14 Mls-1^a mice revealed that tolerance induction differed depending on the negatively selecting antigen. P14 TCR transgenic LCMV carrier mice showed deletion of DP thymocytes and no evidence of a positively selected $CD3^{int}$ population, whereas DP thymocytes and a positively selected $CD3^{int}$ population were present P14 Mls-1^a mice.

Two main findings were obtained from these studies. The timing of negative selection depended upon the location of the antigen or the affinity of the self ligand. Negative selection may occur early during the DP stage of thymocyte selection when large amounts of high-affinity antigen are present, as in the case of the LCMV-gp. However, negative selection may also occur late during the DP stage as in the case with Mls-1[a]. Here, Mls is known not to be present in high amounts in the thymic cortex, where DP T cells are found. Accordingly, deletion of the P14 TCR occurs relatively late during development. Further studies examining deletion to variant LCMV viruses demonstrated that interactions that evoke negative selection in the thymus are more sensitive than interactions required for T cell activation in the periphery (PIRCHER et al. 1991).

The second major finding relates to the actual mechanism of how a thymocyte distinguishes between positive and negative selection. At this point, the currently favored "altered ligand" hypothesis of thymocyte development speculated that positive selection occurred first through interactions with a specific population of thymic stromal cells. In this way the TCR was able to send a survival signal, using unique peptide/MHC ligands of the defined "selecting" stromal cell. Subsequently, the "developmental state" of the thymocyte was presumed to change, such that a TCR signal elicited apoptosis. In addition, the developing thymocyte was envisioned to interact with a different stromal cell population, presumably bone marrow-derived cells. In this way, a different subset of peptide/MHC complexes were now eliciting negative selection and death of thymocytes. However, these studies with the P14 model now demonstrated that negative selection could occur before or after positive selection (OHASHI et al. 1990). These findings clearly disputed the "altered ligand" model of thymocyte selection (MARRACK and KAPPLER 1988), and supported an affinity model of selection in which low-affinity interactions promoted positive selection, while high-affinity interactions led to negative selection (SPRENT et al. 1988).

1.2 The Outcome of Positive and Negative Selection Are Determined by the Affinity/Avidity of Thymocyte Interactions

Further studies were done using the P14 TCR transgenic model that solidified and modified the affinity model of thymocyte selection. In an attempt to elucidate the mechanisms underlying positive and negative selection, studies have examined the contribution and importance of defined peptides in thymocyte development. Fetal thymic organ cultures (FTOCs) were done using thymii from various class I deficient mice as in vitro systems to study thymocyte development. β_2 microglobulin (β_2m) is an essential non-polymorphic component of the class I molecule. In the absence of β_2m, class I molecules remain improperly folded, and consequently impaired $CD8^+$ T cell development was observed (KOLLER et al. 1990). Another molecule, TAP (transporter associated with antigen processing) is also important for peptide transport to the ER. In the absence of TAP there is a lack of class I expression on the surface, due to a lack of peptides associated with the class I molecule (VAN KAER et al. 1992). Various transgenic TCR have been bred onto genetic backgrounds that do not express β_2m or TAP. FTOCs from such TCR β_2m$^{-/-}$ or TCR TAP$^{-/-}$ mice can be used to address the role of peptides in thymocyte selection. A variety of defined peptides may be added in FTOCs (plus exogenous β_2m for β_2m$^{-/-}$ cultures) to restore the expression of MHC class I, permitting the evaluation of the effect of a specific peptide on selection.

Using P14 TCR transgenic TAP1$^{-/-}$ and P14 TCR transgenic β_2m$^{-/-}$ FTOCs, it was shown that the same peptide (p33) could mediate both positive and negative selection of the P14 TCR (ASHTON-RICKARDT et al. 1994; SEBZDA et al. 1994). Whether the thymocytes underwent positive or negative selection was solely dependent on the concentration of the p33 peptide, with low concentrations inducing positive selection and high concentrations, negative selection. This demonstrated that the avidity of the interactions clearly played a role in determining the fate of thymocytes, and it was not necessary to have two distinct peptides for positive and negative selection. Further studies have also shown that a spectrum of ligands from antagonists to agonist ligands for a given TCR may promote positive selection. Together many studies using the P14 model (CHIDGEY and BOYD 1997, 1998; MARIATHASAN et al. 1998; SEBZDA et al. 1996) and other models (COOK et al. 1997; DELANEY et al. 1998; FUKUI et al. 1997; HOGQUIST et al. 1994; JAMESON et al. 1994) support the affinity/avidity model of thymocyte selection, where both the affinity and avidity of the interaction between a TCR and its peptide/MHC ligand determines the selection outcome.

Later studies comparing TCR downregulation in thymocytes after positive and negative selection signals further support the affinity/avidity model (MARIATHASAN et al. 1998). Engagement of TCRs by their cognate peptide/MHC ligands is known to result in the downregulation of the TCRs and initiation of intracellular signaling pathways including those dependent on increased intracellular Ca^{2+} levels. FTOCs of P14 TCR transgenic β_2m$^{-/-}$ lobes with peptides that were shown to induce suboptimal TCR downregulation and Ca^{2+} flux resulted in positive selection of P14 TCR transgenic thymocytes. Conversely, when peptides that were shown to induce strong TCR downregulation and Ca^{2+} flux were added to β_2m$^{-/-}$ FTOCs, negative selection of P14 TCR transgenic thymocytes was observed. Thus, the intensity of TCR signaling correlated with the outcome of thymocyte selection.

1.3 Flexibility of the Interactions that Dictate Positive Selection

Studies over the years have suggested that a given TCR has relatively degenerate interactions that can induce positive selection. Models of TCR transgenics with the "bm" mutant MHC molecules revealed that slightly different MHC could promote positive selection of defined TCRs (OHASHI et al. 1993b; SHA et al. 1990). This flexibility in selecting signals was supported by early experiments examining the P14 TCR transgenic on bm mutant MHC backgrounds (OHASHI et al. 1993b). P14 TCR transgenic thymocytes presented with the mutant H-2D^b molecule H-2^{bm13} underwent enhanced positive selection, despite inefficient presentation of the

P14-specific peptide, p33, by H-2D^{bm13}. Thus, the flexibility in positively selecting signals may lie in the conformation of MHC or in the identity of the MHC-bound peptide. Addition of peptides to FTOC showed that a range of different peptides were able to mediate selection of one particular TCR (Table 1) (ASHTON-RICKARDT et al. 1994; HOGQUIST et al. 1994; HOGQUIST et al. 1997; HU et al. 1997; MARIA-THASAN et al. 1998; SEBZDA et al. 1996; SEBZDA et al. 1994). Hu et al. have used the P14 TCR transgenic to identify a self peptide that can direct positive selection (HU et al. 1997). They identified an abundant self-peptide eluted from H-2D^b molecules on thymic epithelial cells that was able to mediate positive selection of functional P14 TCR transgenic thymocytes when added to TAP1$^{-/-}$ FTOCs.

Together these findings demonstrate that each TCR can interact with more than one ligand, and that positive selection can occur within a range of affinities between the TCR and its peptide/MHC ligands (ALAM et al. 1996). Thus, flexibility in peptide recognition for positive selection provides a means by which the limited number of thymic self-peptides can select a diverse population of T cells that can respond to foreign antigens. Notably, the ability of self-peptides to select transgenic thymocytes did not correlate with their ability to stabilize self-MHC molecules, further supporting the role of peptide in direct interaction with TCR rather than stabilizing MHC surface expression.

1.4 Thymocyte Tuning: a Mechanism of Self-Tolerance

In addition to inducing positive and negative selection, TCR-peptide/MHC interactions can also result in clonal inactivation of thymocytes. Clonal inactivation has been observed in several models, including functional tolerance to the Mls-1^a antigen observed in P14 TCR transgenic T cells selected in the presence of Mls-1^a (KAWAI and OHASHI 1995). The Vβ8.1 TCR chain expressed by the P14 TCR can

Table 1. Flexibility in TCR-peptide/MHC interactions mediating positive selection

TCR specificity (model)	Peptide	Type of peptide	Concentration [M]	Reference
Ova/K^b (β$_2$m−/−)	SIINFEKL	Wild-type		
	RGYNYKEL	Antagonist	−5	(HOGQUIST et al. 1994)
	KIINFEKL	Partial ant	−5	(HOGQUIST et al. 1994)
	SIIRFEKL	Antagonist	−5	(HOGQUIST et al. 1994)
Ova/K^b (TAP−/−)	ISFKFDHL	Natural	−3–4	(HOGQUIST et al. 1997)
LCMV/D^b (TAP−/−)	KAVYNFATC	Wild-type		
	KAVYNFATM	Agonist	−5	(ASHTON-RICKARDT et al. 1994)
	KAMYNFATM	Weak agonist	−5	(ASHTON-RICKARDT et al. 1994)
LCMV/D^b (β$_2$m−/−)	KAVYNFATM	Agonist	−12	(SEBZDA et al. 1994)
	KAVANFATM	Weak agonist	−5–9	(SEBZDA et al. 1996)
	KAVYNLATM	Weak agonist	−7–8	(MARIATHASAN et al. 1998)
	KAVWNFATM	Weak agonist	−7–8	(MARIATHASAN et al. 1998)
LCMV/D^b (TAP−/−)	FQIVNPHLL	Natural	−4	(HU et al. 1997)

recognize Mls-1^a in the H-2^k MHC haplotype. In H-2$^{b/k}$ mice, partial deletion of P14 thymocytes was observed. The remaining Vβ8.1$^+$ mature peripheral T cells did not proliferate in response to Mls-1^a, and thus selection in the presence of Mls-1^a resulted in tolerance to this self-antigen without deleting all self-reactive T cells. However, these T cells retained some functionality, as they proliferated normally when stimulated with other cognate antigens such as p33 or the superantigen staphylococcal enterotoxin B (SEB). Because p33 and SEB are more strongly stimulating antigens than Mls-1^a for the P14 TCR, a "tuning" model for thymocyte maturation was proposed, where lower affinity/avidity interactions during selection result in a higher threshold for activation such that mature T cells do not respond to the selecting self-antigen, but are still able to respond to foreign antigens with higher affinities. This model is supported by other studies that demonstrate unresponsiveness to selecting peptide (GIRAO et al. 1997; HOGQUIST et al. 1995; JAMESON et al. 1994). Thus, tuning of thymocytes results in a broader repertoire of peripheral TCRs that are unable to respond to their selecting self-antigen.

Tuning of thymocytes during selection has also been demonstrated in the P14 TCR transgenic model using the weak agonist peptide variant of p33, A4Y (SEBZDA et al. 1996), other related peptides (MARIATHASAN et al. 1998) and in other models (LUCAS et al. 1999). Addition of A4Y to P14 TCR transgenic β$_2$m$^{-/-}$ FTOCs induced positive selection of P14 TCR transgenic T cells. These selected T cells were unresponsive to A4Y, but when stimulated with p33, proliferated at levels similar to T cells selected by endogenous peptides. Thus, tuning can occur in response to negatively selecting (Mls-1^a) and positively selecting (A4Y) stimuli.

By what mechanism can selecting peptides alter the activation threshold required by mature T cells? Studies by Chidgey et al. in adult stromal cell suspension cultures indicate that increasing strengths of positively selecting signals result in thymocytes with increased downregulation of the CD8β chain (CHIDGEY and BOYD 1998). These studies also showed that thymocytes expressing CD8αα homodimers were less responsive than those expressing CD8αβ heterodimers, suggesting that CD8β downregulation may be a potential tuning mechanism. Other groups have also reported a correlation between unresponsiveness and CD8 downregulation (JAMESON et al. 1994; ROCHA and VON BOEHMER 1991). Alternatively or concurrently, tuning may involve the altered expression of other molecules that regulate TCR signal strength, or altered subcellular localization or activation state of intracellular signaling molecules. Future studies are required to address these possibilities.

1.5 Role of Coreceptors During Thymocyte Development

Upon TCR binding to peptide/MHC ligands, the CD8 and CD4 coreceptors interact with MHC class I and II, respectively. This interaction may stabilize TCR-peptide/MHC binding, and may also augment signaling by providing the TCR complex with intracellular kinases. The role of CD8 in thymocyte selection has been investigated in several models. Although the association of the CD8 cytoplasmic domain with the intracellular protein tyrosine kinase lck is not required for selection (CHAN et al. 1993), other studies have shown that selection is impaired in the absence of the CD8α cytoplasmic domain, as well as when CD8-MHC class I interactions are disrupted (ALDRICH et al. 1991; FUNG-LEUNG et al. 1993; INGOLD et al. 1991; KILLEEN et al. 1992).

Studies in FTOCs using P14 TCR transgenic lobes lacking CD8 expression indicated the CD8 coreceptor is not necessary for positive selection, as positive selection was restored in these cultures when a p33 peptide variant A4Y was added (SEBZDA et al. 1997). Addition of this peptide variant enhanced the signal transmitted to the TCR from peptide/MHC, and was able to overcome the requirement for CD8. Thus, the physiological role of CD8 in positive selection may be to increase the net signal transmitted by suboptimal peptides, thus expanding the repertoire of positively selected thymocytes to include those with lower affinity TCRs.

The findings from the above study also address the mechanism by which $CD4^+ CD8^+$ double positive thymocytes differentiate into either the $CD4^+$ or $CD8^+$ single positive T cell lineage. The two main models for T cell lineage commitment are the instructive model and the stochastic model (VON BOEHMER 1996), while other models have also been proposed (BASSON et al. 1998; MATECHAK et al. 1996; SUZUKI et al. 1995). In the instructive model, DP thymocytes that bear MHC class I-restricted TCRs downregulate their CD4 coreceptors, whereas DP thymocytes that bear MHC class II-restricted TCRs downregulate their CD8 coreceptors. The stochastic model proposes that each thymocyte randomly downregulates either CD4 or CD8, and subsequent maturation only occurs in thymocytes where the MHC-restriction of the TCR corresponds to expression of the appropriate coreceptor. The above study and another by Goldrath et al. (GOLDRATH et al. 1997), reported a system in which mature MHC class I-restricted T cells committed to the cytotoxic T cell lineage can be generated despite the absence of CD8. These findings suggest that the coreceptors do not send a critical signal for commitment to the $CD8^+$ T cell lineage. Current models suggest that the timing and intensity of the signals received by the thymocyte contribute to commitment to the CD4 or CD8 lineage (BASSON et al. 1998; HERNÁNDEZ-HOYOS et al. 2000; MATECHAK et al. 1996). However, the precise mechanism of how a T cell decides to become a helper $CD4^+$ T cell versus a cytotoxic $CD8^+$ T cell remains unclear.

1.6 Summary: Thymocyte Selection

The P14 TCR transgenic model has been used to address questions concerning several aspects of thymocyte development. This model has been pivotal in supporting the current affinity/avidity model of thymocyte selection. Studies using the P14 TCR transgenic have also contributed to our understanding of the flexibility in recognition of positively selecting ligands, the properties of peptides in positive versus negative selection signals, and has provided some insights into $CD8^+$ T cell lineage commitment. Data from this model have also indicated that selection "tunes" the activation threshold of thymocytes as a mechanism of concurrently maintaining self unresponsiveness and broadening the peripheral T cell repertoire.

Many questions concerning thymocyte development remain unanswered. The P14 TCR transgenic model may be well suited to explore questions such as how and when TCR signaling pathways diverge to result in positive versus negative selection, given the affinity/avidity model for selection. Various altered peptide ligands of p33 that induce various strengths of signal through the P14 TCR have recently been identified (MARIATHASAN et al. 1998). Use of these ligands, together with the ability to follow a synchronized, clonal population of thymocytes, will allow for comparisons of biochemical events and gene transcription induced by different selecting conditions.

2 T Cell Activation

Activation of naive T cells is a carefully orchestrated process involving antigen-specific recognition by the T cell receptor (TCR) and a multitude of adhesion and costimulatory molecules. In addition, it has emerged more recently that the signal flow during T cell activation is not unidirectional from APCs to T cells, but rather a dialogue between the T cells and APCs. In the first step, T cells recognize peptides bound to MHC molecules on APCs. In the second step, this TCR-triggering activates the expression of various membrane bound and secreted molecules by the T cells, such as CD40L. These factors finally enhance the immunogenicity of APCs by promoting their survival and upregulating the expression of costimulatory molecules such as the B7 family members in the third step (GREWAL et al. 1996; YANG and WILSON 1996). Thus, the interaction between T cells and APCs may be viewed as a series of communication events, where APCs activate T cells leading to the production of T cell factors that trigger the activation of APCs further enhancing T cell activation. Using P14 TCR transgenic mice, these various interactions could be dissected at the molecular level.

2.1 The Role of the T Cell Receptor in T Cell Activation

2.1.1 The Degree of TCR Internalization Reflects the Strength of the Signal and Induction of Effector Function

Engagement of TCRs by peptides bound to MHC molecules rapidly leads to the phosphorylation of ITAM motifs in the CD3 complex, recruitment and activation of ZAP-70 and the induction of more downstream signaling cascades such as the ERK1/2 pathway and a rise in intracellular free Ca^{2+} (CANTRELL 1996; CRABTREE and CLIPSTONE 1994; QIAN and WEISS 1997). In addition, by a not yet fully understood mechanism, TCR-engagement leads to rapid (within seconds to minutes) internalization and lysosomal degradation of TCRs (VALITUTTI et al. 1995, 1997). Using T cell clones, a study by Valitutti et al., showed that only specifically engaged TCRs are internalized (VALITUTTI et al. 1995). This observation offered the possibility to quantitatively measure TCR-triggering by assessing TCR-surface expression upon antigen-specific stimulation. Using cytotoxic T cells derived from TCR-transgenic mice, these findings were subsequently confirmed for naive T cells (BACHMANN et al. 1997c; CAI et al. 1997). Moreover, it was shown that the extent and in particular the rate of TCR-triggering on naive T cells was critical for the efficiency of T cell activation (BACHMANN et al. 1997b; VIOLA and LANZAVECCHIA 1996). Peptides that induced greater than 90% TCR downregulation within 5h were able to induce strong and extended Ca^{2+}-fluxes and T cells stimulated with these peptides efficiently proliferated, produced high amounts of cytokines and differentiated to lytic effector cells. In contrast, peptides that induced only intermediate TCR downregulation triggered only transient Ca^{2+}-fluxes and limited T cell proliferation. Moreover, these peptides failed to induce generation of lytic T cell effector cells. Interestingly, however, target cells pulsed with these peptides were efficiently recognized in ^{51}Cr-release assays (BACHMANN et al. 1996, 1997a,c). Peptides that triggered only minimal TCR downregulation failed to induce a measurable response in naive T cells. In fact, they were only able to induce lysis of target cells by pre-activated effector cells. Interestingly, the ability of the peptides to induce TCR downregulation inversely correlated with their ability to act as T cell antagonists (BACHMANN et al. 1998b). Thus, peptides that triggered minimal TCR downregulation were able to maximally inhibit T cell responses triggered by stronger peptides. Taken together, the efficiency of various peptides to trigger TCR downregulation directly predicted their agonist and antagonist properties (reviewed by BACHMANN and OHASHI 1999).

2.1.2 Why Are TCRs Internalized upon Engagement?

Several explanations come to mind. It is possible that the rapid downregulation of triggered TCRs increases the signal to noise ratio. Interactions between TCRs and specific MHC/peptide complexes occur fast and MHC/peptide complexes that are released by a triggered and internalized TCR will rapidly engage another TCR, allowing continuous signaling. By contrast, non-specific reactions usually occur slowly and MHC/peptide complexes that bind a TCR "non-specifically" will fail to rapidly engage a number of TCRs, leading to signal-extinction. Thus, TCR downregulation may allow for screening of rapid and hence specific TCR-MHC/peptide interactions (BACHMANN and OHASHI 1999). In addition, T cells that have downregulated a sizable fraction of their TCRs may become desensitized to the antigen. If TCR transgenic T cells were stimulated with a high-affinity ligand inducing massive TCR downregulation, they failed to respond subsequently to a low-affinity ligand. In contrast, normal T cells that had not been pretreated responded efficiently to stimulation with the low-affinity ligand (MARTIN and BEVAN 1998).

2.1.3 TCR Dimerization

The question of whether TCRs oligomerize during T cell activation has not been definitively resolved. There is some indirect evidence that TCR-oligomerization does occur. It has been shown that soluble TCRs bound to peptide pulsed MHC molecules spontaneously oligomerize in vitro (ALAM et al. 1999; REICH et al. 1997). Moreover, dimeric and multimeric MHC/peptide complexes can activate T cells in vitro (ABASTADO et al. 1995; BONIFACE et al. 1998; MEUER et al. 1984; RÖTZSCHKE et al. 1997). However, MHC molecules are not oligomeric on APCs and the aforementioned experiments therefore suggest, but do not demonstrate, that TCRs oligomerize under physiological conditions. In fact, the question of whether TCRs oligomerize during T cell-APC interactions has rarely been experimentally investigated (BACHMANN et al. 1998a). To directly address this question, naive cytotoxic T cells were isolated from TCR transgenic animals and stimulated with peptide pulsed macrophages to assess TCR downregulation. Using a similar mathematical formalism as used to investigate enzyme kinetics, a model was generated that described the interaction between T cells and APCs. This model allowed assessment of TCR-oligomerization by measuring the rate of TCR downregulation as a function of TCR-density. Assuming that each TCR that productively engaged a MHC molecule was internalized, the model predicted a first order reaction and consequently that the rate of TCR downregulation was directly proportional to the TCR density. In contrast, under the assumption that TCRs dimerized before internalization, the model predicted a second order reaction. Under these conditions, the rate of TCR downregulation was predicted to be proportional to the square of the TCR-density. The results experimentally obtained clearly demonstrated a parabolic relation, indicating that TCRs underwent dimerization before internalization during antigen-specific activation of naive cytotoxic T cells by peptide pulsed macrophages (BACHMANN et al. 1998a).

Using a similar mathematical formula, it could be shown that this TCR-dimerization step is essential for T cell specificity since it serves as a proof-reading system (SALZMANN and BACHMANN 1998). Moreover, TCR-dimerization as an essential step for T cell activation was able to explain the phenomenon of T cell antagonism, since engagement of TCR molecules by MHC/antagonist complexes may lead to a "poisoning" of TCR-dimers. Specifically, if the antagonist/MHC complex falls off the TCR within a dimer before a signal could effectively be generated, presence of MHC/antagonist complexes within an TCR-MHC/peptide dimer may be able to severely inhibit T cell activation (SALZMANN and BACHMANN 1998). Thus, MHC/antagonist complexes may be able to inhibit TCR-triggering at the level of dimer formation (reviewed by BACHMANN and OHASHI 1999).

2.2 The Role of Accessory and Costimulatory Molecules in T Cell Activation

2.2.1 Molecules that Enhance T Cell Activation

Although it is possible to stimulate T cells to proliferate by triggering the TCR alone, T cell activation may be inefficient in the absence of additional signals.

Although many of these molecules are usually called costimulatory ligands, it may be useful to distinguish mechanistically how they promote T cell activation. The 2 signal theory postulates two different signals required for T cell activation. An antigen-specific signal 1 that can conveniently be assessed by measuring TCR downregulation, and a costimulatory signal 2 required for full T cell activation (BRETSCHER and COHN 1970; COHN and LANGMAN 1990; SCHWARTZ 1990). Accordingly, at least two types of accessory molecules can be defined: those that affect T cell activation by increasing signal 1 (i.e., TCR-triggering) and those that promote T cell activation without affecting TCR-triggering (costimulatory molecules delivering signal 2). However, in the absence of defined monoclonal T cell populations, it remained difficult to experimentally distinguish between these possibilities. Thus, only the use of TCR transgenic mice allowed a mechanistic assessment of the role of various accessory molecules in promoting activation of naive T cells. CD28 clearly emerged as a true costimulatory molecule that enhanced T cell activation without changing TCR-internalization (BACHMANN et al. 1997b; KÜNDIG et al. 1996). These properties of CD28 were reflected in vivo, since CD28-deficient mice failed to mount T cell responses against abortively replicating viruses (KÜNDIG et al. 1996) and low-affinity ligands (BACHMANN et al. 1996). In contrast, LFA-1 and CD2, often also called costimulatory molecules, enhanced T cell activation by increasing signal 1. The dose response of T cell activation was shifted in the absence of CD2 and/or LFA-1 and the minimal amount of antigen required for TCR-internalization (i.e., signal 1) was similarly increased in the absence of these molecules (BACHMANN et al. 1999a,b). Interestingly, CD2 and LFA-1 acted additively: while the absence of either CD2 or LFA-1 alone shifted the dose response by about a factor of ten, absence of both molecules resulted in a 100-fold shift. The enhanced TCR-triggering observed in the presence of CD2 and LFA-1 was in part due to enhanced T cell-APC conjugation (BACHMANN et al. 1999a,b). In addition, both CD2 and LFA-1 help in orchestrating the molecular arrangement of the various molecules in the T cell-APC contact site, thereby facilitating TCR-triggering (DAVIS and VAN DER MERWE 1996; DUSTIN et al. 1998). Interestingly, despite the reduced T cell responses observed in vitro upon stimulation with low antigen concentrations, both CD2 and LFA-1 deficient mice mounted normal anti-viral CTL responses. Surprisingly, not even the absence of both molecules significantly altered the T cell response upon infection with LCMV. However, CTL responses were clearly reduced in vivo if antigens were introduced to the immune system at low concentration, for example, by injecting limiting amounts of peptide. In addition, CTL responses induced by cross-priming were also dependent on functional CD2 expression, a finding which is compatible with the notion that cross-presentation does not yield high peptide densities on the APC surface. Thus, under conditions where the amounts of the peptides presented are low, full in vivo T cell responses require the presence of CD2 or LFA-1 (BACHMANN et al. 1999a).

In addition to CD28, LFA-1 and CD2, there is a large population of additional accessory molecules, such as HSA, 4-1BB, CD44, OX40, and ICOS (WATTS and DEBENEDETTE 1999). The molecular mechanism by which these molecules facilitate T cell activation remains, however, to be elucidated.

2.2.2 T Cell Molecules that Promote APC Maturation

As already mentioned, T cell activation is not a unidirectional process where information flows exclusively from the APC to the T cell. In contrast, the process seems to be bi-directional, and T cells signal back to APCs to increase their potential to mediate T cell activation. Thus, there is a third class of molecules assisting T cell activation – those that activate APCs. Interestingly, there is little evidence for the aforementioned accessory molecules such as CD28, LFA-1 or CD2 to induce activation of APCs and a different set of molecules seems to be responsible for APC stimulation. Two members of the TNF family, CD40L and TRANCE (OPGL, RANKL) are the best described examples of proteins that can trigger activation, maturation and survival of APCs (ANDERSON et al. 1997; GREWAL and FLAVELL 1998; JOSIEN et al. 1999; WONG et al. 1997a,b). As a consequence, many T cell responses, in particular $CD4^+$ T cell responses, are dependent to a high degree on functional CD40L-CD40 and TRANCE-RANK interactions (BACHMANN et al. 1999d; GREWAL and FLAVELL 1998). In addition, secreted cytokine-like factors may also mediate activation of APCs (CELLA et al. 1997). Some of these factors are secreted with suprisingly high kinetics and induce the full maturation program in APCs within 24h after initial T cell-APC contact (RUEDL et al. 1999). The relative contribution of membrane bound molecules versus secreted cytokine-like factors for APC activation remains to be elucidated.

2.3 T Cell Memory

When compared with primary T cell responses, memory responses are both stronger and faster. To a large degree, these enhanced memory T cell responses can be explained by the presence of increased numbers of specific T cells in primed individuals. However, it is also possible that memory T cells show a better performance at the single cell level. While this question has been debated for a long time, it was difficult to assess the problem experimentally until recently. Naive T cells are present at very low frequencies in normal mice and are therefore difficult to study. Moreover, while memory T cells may be present at frequencies high enough to directly study their functional properties, these memory T cells may express a different set of TCRs than primary T cells, due to the high selective pressure placed on them during the establishment of T cell memory. P14 TCR transgenic mice expressing a TCR specific for LCMV showed themselves ideal for the study this question. Untreated P14 TCR transgenic mice harbor large numbers of naive T cells which can easily be studied in vitro. Moreover, upon adoptive transfer of naive P14 TCR transgenic T cells and subsequent LCMV infection of recipient mice, the transferred T cells become activated and expand massively within a week (MOSKOPHIDIS et al. 1993; ZIMMERMANN et al. 1996a). After elimination of the virus, frequencies decline but high numbers of memory T cells expressing a defined TCR remain present in the recipient mice (ZIMMERMANN et al. 1996a). This system can therefore be used to directly compare the performance of naive and memory T cells (BACHMANN et al. 1999b; ZIMMERMANN et al. 1999). A quite interesting picture emerged. In some respects, memory T cells behaved very similarly to naive T cells. They required the same amounts of antigen for activation as naive T cells, gained lytic activity with similar kinetics, required a similar time span of about 24h to start proliferating, and divided with similar doubling times in vivo. In some respects, however, memory T cells were clearly different from naive T cells. Memory T cells secreted cytokines, such as IL-2 or IFN-γ within 1–2h after stimulation while naive T cells required 24h. In addition, although memory T cells started to proliferate upon antigenic stimulation only after about 24h, they were committed for proliferation within about 2h of contact with antigen, which is in marked contrast to the 12–24h required by naive T cells. Thus, memory T cells were able to respond to antigenic stimulation with greatly enhanced kinetics, in particular if the secretion of cytokines was assessed (BACHMANN et al. 1999b; ZIMMERMANN et al. 1999). In addition to these kinetic differences, memory T cells were also observed to be less dependent upon costimulation for activation than naive T cells (BACHMANN et al. 1999c). This reduced requirement for CD28 correlated with a rearrangement of the TCR signaling cascade. While lck, a key signaling molecule, exhibited a cytoplasmic distribution in naive T cells, in memory T cells lck was targeted to the cell membrane and associated with CD8 molecules. Thus, in memory but not in naive T cells, lck is reshuffled to a strategically optimized position, rendering the TCR-mediated signaling cascade more efficient in memory T cells (BACHMANN et al. 1999c).

2.4 Summary: T Cell Activation

The use of transgenic mice expressing a TCR specific for LCMV allowed the study of the activation of naive T cells in great detail. It was possible to correlate the rate of TCR downregulation with the degree of T cell activation using defined altered peptide ligands. It was also possible to experimentally demonstrate that TCRs dimerize during T cell activation and dissect the mechanisms of enhanced T cell activation in the presence of various accessory molecules such as CD28, LFA-1 and CD2. Moreover, it could be demonstrated that memory T cells respond with enhanced kinetics to antigenic stimulation at the single cell level and that they have improved signaling machinery due to a relocalization of lck.

3 Peripheral Tolerance

3.1 Deletion and Anergy vs. Ignorance

When mature T cells are exposed to high doses of a tolerogenic antigen, studies have clearly demonstrated that the specific T cells undergo an expansion phase, followed by deletion and the induction of anergy (RAMMENSEE et al. 1989; WEBB et al. 1990). This has also been repeated with the LCMV P14 model, demonstrating that these observations also hold true for $CD8^+$ mature T cells (KYBURZ et al. 1993; MOSKOPHIDIS et al. 1993). For $CD8^+$ T cells, the anergic state can be reversed by the addition of IL-2 and IL-4 (KYBURZ et al. 1993). Experiments using transgenic mice have also demonstrated that deletion may occur to self antigens that are cross-presented in vivo (ADLER et al. 1998; CARBONE et al. 1998; KURTS et al. 1997; MORGAN et al. 1999).

Studies have also been done to address mechanisms of tolerance to tissue-specific antigens using the LCMV model. RIP-gp transgenic mice express the LCMV-gp in the β-islet cells of the pancreas, but not in the thymus (OHASHI et al. 1991). Studies using RIP-gp single transgenic mice and RIP-gp/P14 double transgenic mice have clearly shown that systemic LCMV-gp-specific CTL were naive and unaware of the LCMV-gp that is present on the islet cells. Infection with LCMV led to the activation of gp-specific CTL, infiltration into the pancreas, and the induction of CD8 mediated diabetes. This model has become a "classic" model which demonstrates that not all self reactive T cells are tolerized to all self antigens. Instead, some self reactive cells remain "ignorant" or unaware of sequestered self antigens. Many other studies have demonstrated that ignorance does occur towards some self peptides (BOEHME et al. 1989; HEATH et al. 1995; KURTS et al. 1999; MAMULA 1993; MURPHY et al. 1989; SCHILD et al. 1990; SOLDEVILA et al. 1995), and in some models spontaneous autoimmunity is observed (GEIGER et al. 1992; GOVERMAN et al. 1993; KATZ et al. 1993; LAFAILLE et al. 1994; ROMAN et al. 1990; SARUKHAN et al. 1998; SCOTT et al. 1994).

Further studies have characterized parameters that influence the induction of diabetes or CD8 mediated pathology in this model. By breeding the same RIP-gp transgene to a variety of strains, the onset of diabetes was shown to be affected by the genetics of the mice (OHASHI et al. 1993a). Backcrossing the RIP-gp transgene two generations to the B10.BR strain ($H-2^k$), demonstrated that diabetes was induced after LCMV infection in 67% of the animals with a mean onset time of 24 days, compared to 10 days in the C57Bl/6 background. Backcrossing both the RIP-gp/P14 transgenes onto a 129/J ($H-2^b$) background leads to the induction of spontaneous diabetes in approximately 30% of the mice (GARZA et al. 2000b). Spontaneous disease was rare in RIP-gp/P14 mice maintained on the C57Bl/6 background (less than 3%). Studies using gene deficient mice have shown that the induction of diabetes is dependent upon perforin (KÄGI et al. 1996) but not TNF-R1 (tumor necrosis factor receptor-1) (MCKALL-FAIENZA et al. 1998).

3.1.1 Prevention of Diabetes

Using this model, several approaches have been tested to examine ways to tolerize 'ignorant' self reactive cells and prevent diabetes. Tolerance to LCMV-gp has been induced by generating virus carrier mice (OEHEN et al. 1992) or high dose peptide administration (AICHELE et al. 1994). Both approaches successfully prevented diabetes. In addition, a novel approach was also successful (ALLY et al. 1995). The LCMV-gp was expressed in a retroviral vector, which was subsequently used to infect bone marrow-derived cells. These BM cells expressing LCMV-gp, were used to reconstitute an irradiated RIP-gp host. After several weeks of hematopoetic cell reconstitution, the mice were analyzed to determine whether tolerance to LCMV-gp expressed by the BM-derived cells occurred. Thymic tolerance by clonal deletion was the predominant mechanism for eliminating the response to LCMV-gp, and preventing virus induced diabetes.

3.1.2 Ignorant/Naive Autoreactive Cells May be Used for Tumor Immunotherapy

Recent work by many groups has begun to define tumor associated antigens (VAN PEL et al. 1995). Experiments have shown that the immune response may be activated against a variety of tumor antigens to promote tumor rejection. In order to investigate the requirements for and effectiveness of tumor immunotherapy, we have developed a tumor model based on LCMV. We have bred the RIP-gp mice with mice provided by Hanahan that express the SV40 large T antigen under the control of the rat insulin promoter (HANAHAN 1985). In this way, we have a spontaneously arising tumor of the pancreatic β-islet cells, with a defined tumor associated antigen, the LCMV-gp (SPEISER et al. 1997). The novel aspect of this model is that the tumor arises naturally in the host, and is not a transplanted, in vitro cultivated tumor cell line. The main advantages are that these are naturally arising tumors with normal vasculature, such that relevant questions may be asked regarding the development of tolerance to tumor antigens. In addition, because these tumors are not maintained in vitro, they do not possess unnatural mutations acquired during culture, nor are they administered in high cell numbers in an immunogenic fashion.

The tumor burden of RIP-gp/SV40 double transgenic mice is reliably measured by following the blood glucose levels. A low blood glucose reading reflects a large number of β-islet cells producing insulin, and has been correlated with the presence of insulinomas. By following the blood glucose, we can also determine the effectiveness of immunotherapy against the LCMV-gp antigen, after administration of LCMV or other agents to induce an LCMV-gp-specific response. Our studies have shown that the presence of LCMV-gp on the islet tumors is not sufficient to induce an anti-tumor response by the "naive" LCMV-gp-specific T cells. Again, infection with virus is required to induce effective LCMV-gp-specific T cells, and prolong survival in tumor bearing mice. Interestingly, after activation with LCMV, an anti-tumor response was not sustained by the tumor, despite the presence of memory LCMV-gp-specific T cells. Therefore, several boosts will be required to induce and maintain tumor immunity. Importantly, there was also no chronic, long lasting immune response against the islet cells. This demonstrates that an anti-tumor response may be induced by activating naive 'ignorant' T cells, using a tissue-specific self ligand as a target molecule, without the hazard of initiating a chronic long-term autoimmune disease against a particular tissue. However, there is always the danger that normal tissue expressing the self antigen is destroyed (LUDEWIG et al. 2000). Ideally, perhaps breast or prostate cancer may be suitable for these types of approaches for tumor immunotherapy.

3.2 Peripheral Tolerance vs. Immunity: the Role of APCs

When a mature T cell encounters antigen, the cell must decide whether an immune response will be evoked, or whether tolerance will occur. For many years, the predominant hypothesis was known as the two signal model, where signal 1, through the TCR alone, lead to anergy. Signal 1 together with a costimulatory signal 2 lead to the induction of a full immune response (BRETSCHER and COHN 1970; SCHWARTZ 1990). Over the years, costimulatory signals have been defined that include members of the immunoglobulin superfamily such as CD28/B7, ICOS/B7-RP-1, and other members of the TNF/TNFR superfamily such as CD40/CD40L, TRANCE (OPGL)/RANK, 41BB/41BBL, OX40/OX40L and many others (LENSCHOW et al. 1996; WATTS and DEBENEDETTE 1999). Studies using the RIP-gp model have shown the important role for B7-1 in vivo (HARLAN et al. 1994; VON HERRATH et al. 1995). More recently, the two signal model has been revised and reconsidered at the level of the APC.

Seminal studies by Lafferty and coworkers have described the importance of the APC as the "Passenger Leukocyte Concept" (LAFFERTY et al. 1983). In their studies, they found that if various grafts were depleted of APCs, then a graft would be likely survive an allogeneic transplant. Many models have now speculated that the APC is the crucial mediator of T cell fate, and determines whether the induction of tolerance or immunity occurs (BANCHEREAU and STEINMAN 1998). Encounter with antigen (self or non-self) on resting APC would lead to tolerance, while the interaction between T cells and activated APCs leads to the induction of immunity (Fig. 1).

Studies with the RIP-gp mice have supported the importance of APC activation. The induction of diabetes required viral infection (OHASHI et al. 1991; OLDSTONE et al. 1991; VON HERRATH et al. 1994). Because LCMV is a natural mouse pathogen, it is likely that this led to the activation of APCs via the induction of a normal immune response. As a consequence, the induction of autoimmunity also occurred. Other similar models, expressing a defined autoantigen in the pancreas using the rat insulin promoter, found it difficult to induce diabetes (HEATH et al. 1995; SOLDEVILA et al. 1995). It is likely that this was because none of the "activating" antigens were from a natural mouse pathogen, and hence an effective immune response, including APC activation did not occur. Accordingly, reports have shown that spontaneous autoimmunity is associated with chronic infection and the cleanliness of the animal colony (GOVERMAN et al. 1993), pointing towards an important role for the activation of APCs. However, other factors like the ability to induce a sufficient number of CTL to cause pathology are also critical (OHASHI et al. 1993a).

Experiments have also shown the importance of DCs in the initiation of the immune and autoimmune response in this model. In vivo administration of DCs that express the LCMV-gp have been shown to be sufficient for the induction of diabetes (LUDEWIG et al. 1998). In addition, non-specific bystander activation of P14 T cells was not sufficient to induce disease (EHL et al. 1997). Interestingly, treatment of RIP-gp/P14 mice with the LCMVgp peptide p33 was shown to induce transient CTL function, followed by tolerance. These mice did not develop diabetes, despite the induction of LCMV-gp-specific CTL activity. However, activation of the APC with anti-CD40 Ab together with the administration of peptide, lead to insulitis and the onset of diabetes (GARZA et al. 2000a). Together these data support the importance of APCs in determining whether immunity or tolerance is induced (Fig. 1).

3.3 Peripheral Tolerance: the Role of TNFR Family Members in Apoptosis

Another mechanism proposed to govern the induction and maintenance of peripheral tolerance is the deletion of antigen-specific T cells, whereby T cells undergo transient activation and subsequently apoptosis following encounter with self antigens in the periphery (GREEN and SCOTT 1994). Several members of the tumor necrosis factor receptor (TNFR) superfamily, including TNFR1 and CD95 (Fas), are able to signal apoptosis (SCREATON and XU 2000), and have been investigated as potential mediators of T cell deletion in the downregulation of T cell responses to pathogens, as well as peripheral tolerance.

The role of CD95L-CD95 interactions in T cell deletion in vivo has been investigated in several different systems. While some studies have found a requirement for CD95 signaling in the induction of T cell deletion (RUSSELL et al. 1993; SCOTT et al. 1993; SINGER and ABBAS 1994), others have demonstrated that deletion can occur independently of CD95 (EHL et al. 1996; TEH et al. 1996; ZIMMERMANN et al. 1996b). The requirement for CD95 in the regulation of $CD8^+$ T cell homeostasis has been evaluated using the LCMV system by Pircher's group (ZIMMERMANN et al. 1996b). In this study, the deletion that followed the expansion phase in response to p33 peptide administration occurred with normal kinetics in P14 TCR transgenic mice on a CD95-deficient background. In addition, downregulation of virus-specific CTLs following LCMV infection and tolerance of peptide-specific T cells induced by high dose p33 administration were not altered in the absence of CD95. A later study by the same group showed that in the absence of CD95, deletion of P14 TCR transgenic T cells was impaired only in response to low doses of peptide (REICH et al. 2000). Overall, these studies indicate that CD95 is only required for homeostatic regulation of $CD8^+$ T cells following encounter with a low peptide dose.

The role of TNFR1 signaling in deletion of $CD8^+$ T cells has also been examined using the LCMV system (SPEISER et al. 1996). An impairment of deletion was demonstrated in TNFR1-deficient P14 TCR transgenic mice following p33 administration. However, this requirement for TNFR1 only occured with low doses of peptide, indicating that other mechanisms mediate deletion in response to higher peptide doses.

The data has indicated that under many conditions of antigenic stimuli, deletion of T cells can proceed normally in the absence of either TNFR1 or CD95, but this begs the question of whether TNFR1 and CD95 might cooperate to mediate deletion. This question has been addressed by two groups using the LCMV system by creating mice deficient for both molecules (NGUYEN et al. 2000; REICH et al. 2000). Both groups found that deletion following activation of T cells by LCMV infection was not impaired in the absence of both TNFR1 and CD95. In addition, they observed that TNFR1 and CD95 do not cooperate to induce the deletion of P14 TCR transgenic T cells in response to p33 administration. However, there was a profound defect in tolerance induced by the administration of p33 emulsified in incomplete Freund's adjuvant (IFA) in non-TCR transgenic mice deficient in both TNFR1 and CD95 (NGUYEN et al. 2000). These results show that under defined conditions, TNFR1 and CD95 contribute to $CD8^+$ T cell deletion.

Together, the data support the possibility that although TNFR1 and CD95 are not important in maintaining T cell homeostasis following viral infection, they may contribute to peripheral tolerance. In the LCMV system, the only conditions where TNFR1 and CD95 have been found to play a role in deletion are those which more closely mimic self antigens present in a non-inflammatory milieu and non-pathogenic stimuli, i.e., low dose peptide administration of TCR transgenic animals and chronic peptide stimulation (by emulsion in IFA) of non-TCR transgenic animals. Thus, the data is consistent with a hypothesis in which TNFR1 and CD95 play a role in the induction of peripheral tolerance to self antigens.

3.4 Summary: Peripheral Tolerance

The LCMV model has made significant contributions to the study of peripheral T cell tolerance and autoimmunity. This model clearly demonstrated that self reactive T cells can exist in a naive, "ignorant" state. These T cells do not generally become activated in vivo and spontaneous autoimmunity is rare. This "ignorant" population of T cells must remain unreactive to avoid autoimmunity, however, they may be utilized in immunotherapy against cancer. Further studies using the LCMV RIP-gp model have solidified the importance of APCs in determining whether tolerance or immunity/autoimmunity is evoked, and has provided insights for the role of TNFR family members in apoptosis and peripheral tolerance.

References

Abastado J-P, Lone Y-C, Casrouge A, Boulot G, Kourilsky P (1995) Dimerization of soluble major histocompatibility complex-peptide complexes is sufficient for activation of T cell hybridoma and induction of unresponsiveness. J Exp Med 182:439–447

Adler AA, Marsh DW, Yochum GS, Guzzo JL, Nigam A, Nelson WG, Pardoll DM (1998) CD4+ T cell tolerance to parenchymal self-antigens requires presentation by bone marrow-derived antigen-presenting cells. J Exp Med 187:1555–1564

Aichele P, Kyburz D, Ohashi PS, Odermatt B, Zinkernagel RM, Hengartner H, Pircher H (1994) Peptide-induced T-cell tolerance to prevent autoimmune diabetes in a transgenic mouse model. Proc Natl Acad Sci USA 91:444–448

Alam SM, Davies GM, Lin CM, Zal T, Nasholds W, Jameson SC, Hogquist KA, Gascoigne NR, Travers PJ (1999) Qualitative and quantitative differences in T cell receptor binding of agonist and antagonist ligands. Immunity 10:227–237

Alam SM, Travers PJ, Wung JL, Nasholds W, Redpath S, Jameson SC, Gascoigne NRJ (1996) T-cell-receptor affinity and thymocyte positive selection. Nature 381:616–620

Aldrich CJ, Hammer RE, Jones-Youngblood S, Koszinowski U, Hood L, Stroynowski I, Forman J (1991) Negative and positive selection of antigen-specific cytotoxic T lymphocytes affected by the $\alpha 3$ domain of MHC I molecules. Nature 352:718–721

Ally BA, Hawley TS, MacKall-Faienza K, Kündig TM, Oehen SU, Pircher H, Hawley RG, Ohashi PS (1995) Prevention of autoimmune disease by retroviral-mediated gene therapy. J Immunol 155:5404–5408

Anderson DM, Maraskovsky E, Billingsley WL, Dougall WC, Tometsko ME, Roux ER, Teepe MC, DuBose RF, Cosman D, Galibert L (1997) A homologue of the TNF receptor and its ligand enhance T-cell growth and dendritic-cell function. Nature 390:175–179

Ashton-Rickardt PG, Bandeira A, Delaney JR, Van Kaer L, Pircher HP, Zinkernagel RM, Tonegawa S (1994) Evidence for a differential avidity model of T cell selection in the thymus. Cell 76:651–663

Bachmann MF, Barner M, Kopf M (1999a) CD2 sets quantitative thresholds in T cell activation. J Exp Med 190:1383–1392

Bachmann MF, Barner M, Viola A, Kopf MF (1999b) Distinct kinetics of cytokine production and cytolysis in effector and memory T cells after viral infection. Eur J Immunol 29:291–299

Bachmann MF, Gallimore A, Linkert S, Cerundolo V, Lanzavecchia A, Kopf M, Viola A (1999c) Developmental regulation of Lck targeting to the CD8 coreceptor controls signaling in naive and memory T cells. J Exp Med 189:1521–1530

Bachmann MF, Mariathasan S, Bouchard D, Speiser DE, Ohashi PS (1997a) Four types of Ca^{2+} signals in naive $CD8^+$ cytotoxic T cells after stimulation with T cell agonists, partial agonists and antagonists. Eur J Immunol 27:3414–3419

Bachmann MF, McKall-Faienza K, Schmits R, Bouchard D, Beach J, Speiser DE, Mak TW, Ohashi PS (1997b) Distinct roles for LFA-1 and CD28 during activation of naive T cells: adhesion versus costimulation. Immunity 7:1–20

Bachmann MF, Ohashi PS (1999) The role of T cell receptor dimerization in T cell activation. Immunol Today 20:568–576

Bachmann MF, Oxenius A, Speiser DE, Mariathasan S, Hengartner H, Zinkernagel RM, Ohashi PS (1997c) Peptide-induced T cell receptor down-regulation on naive T cells predicts agonist/partial agonist properties and strictly correlates with T cell activation. Eur J Immunol 27:2195–2203

Bachmann MF, Salzmann M, Oxenius A, Ohashi PS (1998a) Formation of TCR dimers/trimers as a crucial step for T cell activation. Eur J Immunol 28:2571–2579

Bachmann MF, Sebzda E, Kündig TM, Shahinian A, Speiser DE, Mak TW, Ohashi PS (1996) T cell responses are governed by avidity and costimulatory thresholds. Eur J Immunol 26:2017–2022

Bachmann MF, Speiser DE, Zakarian A, Ohashi PS (1998b) Inhibition of TCR-triggering by a spectrum of altered peptide ligands suggests the mechanism for TCR-antagonism. Eur J Immunol 28:3110–3119

Bachmann MF, Wong BR, Josien R, Steinman RM, Oxenius A, Choi Y (1999d) TRANCE, a tumor necrosis factor family member critical for CD40 ligand-independent T helper cell activation. J Exp Med 189:1025–1031

Banchereau J, Steinman RM (1998) Dendritic cells and the control of immunity. Nature 392:245–252

Basson MA, Bommhardt U, Cole MS, Tso JY, Zamoyska R (1998) CD3 ligation on immature thymocytes generates antagonist-like signals appropriate for CD8 lineage commitment, independently of T cell receptor specificity. J Exp Med 187:1249–1260

Bevan MJ (1977) In a radiation chimera, host H-2 antigens determine immune responsiveness of donor cytotoxic cells. Nature 269:417–418

Blackman MA, Gerhard-Burgert H, Woodland DL, Palmer E, Kappler JW, Marrack P (1990) A role for clonal inactivation in T cell tolerance to Mls-1a. Nature 345:540–542

Boehme J, Haskins K, Stecha P, van Ewijk W, LeMeur M, Gerlinger P, Benoist C, Mathis D (1989) Transgenic mice with I-A on islet cells are normoglycemic but immunologically intolerant. Science 244:1179–1183

Boniface JJ, Rabinowitz JD, Wulfing C, Hampl J, Reich Z, Altman JD, Kantor RM, Beeson C, McConnell HM, Davis MM (1998) Initiation of signal transduction through the T cell receptor requires the multivalent engagement of peptide/MHC ligands [corrected] [published erratum appears in Immunity 1998 Dec;9(6):891]. Immunity 9:459–466

Bretscher P, Cohn M (1970) A theory of self-nonself discrimination. Science 169:1042–1049

Cai Z, Kishimoto H, Brunmark A, Jackson MR, Peterson PA, Sprent J (1997) Requirements for peptide-induced T cell receptor downregulation on naive $CD8^+$ T cells. J Exp Med 185:641–651
Cantrell D (1996) T cell antigen receptor signal transduction pathways. Annu Rev Immunol 14:259–274
Carbone FR, Kurts C, Bennet SRM, Miller JFAP (1998) Cross-presentation: a general mechanism for CTL immunity and tolerance. Immunol Today 19:368
Cella M, Sallusto F, Lanzavecchia A (1997) Origin, maturation and antigen presenting function of dendritic cells. Curr Opin Immunol 9:10–16
Chan IT, Limmer A, Louie MC, Bullock ED, Fung-Leung WP, Mak TW, Loh DY (1993) Thymic selection of cytotoxic T cells independent of CD8α-lck association. Science 261:1581–1584
Chidgey A, Boyd R (1997) Agonist peptide modulates T cell selection thresholds throgh qualitative and quantitative shifts in CD8 co-receptor expression. Int Immunol 9:1527–1536
Chidgey AP, Boyd RL (1998) Positive selection of low responsive potentially autoreactive T cells induced by high avidity, non-deleting interactions. Int Immunol 10:999–1008
Cohn M, Langman RE (1990) The unit of humoral immunity selected by evolution. Immunol Rev 115: 7–142
Cook JR, Wormstall E-M, Hornell T, Russell J, Connolly JM, Hansen TH (1997) Quantitation of the cell surface level of L^d resulting in positive versus negative selection of the 2 C transgenic T cell receptor in vivo. Immunity 7:233–241
Crabtree GR, Clipstone NA (1994) Signal transmission between the plasma membrane and nucleus of T lymphocytes. Annu Rev Biochem 63:1045–1083
Davis SJ, van der Merwe PA (1996) The structure and ligand interactions of CD2: implications for T cell function. Immunol Today 17:177–187
Delaney JR, Sykulev Y, Eisen HN, Tonegawa S (1998) Differences in the level of expression of class I major histocompatibility complex proteins on thymic epithelial and dendritic cells influence the decision of immature thymocytes between positive and negative selection. Proc Natl Acad Sci USA 95:5235–5240
Dustin ML, Olszowy MW, Holdorf AD, Li J, Bromley S, Desai N, Widder P, Rosenberger F, van der Merwe PA, Allen PM, Shaw AS (1998) A novel adaptor protein orchestrates receptor patterning and cytoskeletal polarity in T-cell contacts. Cell 94:667–677
Ehl S, Hoffmann-Rohrer U, Nagata S, Hengartner H, Zinkernagel R (1996) Different susceptibility of cytotoxic T cells to CD95 (Fas/Apo-1) ligand-mediated cell death after activation in vitro versus in vivo. J Immunol 156:2357–2360
Ehl S, Hombach J, Aichele P, Hengartner H, Zinkernagel RM (1997) Bystander activation of cytotoxic T cells: studies on the mechanism and evaluation of in vivo significance in a transgenic mouse model. J Exp Med 185:1241–1251
Fink PJ, Bevan MJ (1978) H-2 antigens of the thymus determine lymphocyte specificity. J Exp Med 148:766–775
Fukui Y, Ishimoto T, Utsuyama M, Gyotoku T, Koga T, Nakao K, Hirokawa K, Katsuki M, Sasazuki T (1997) Positive and negative CD4+ thymocyte selection by a single MHC class II/peptide ligand affected by its expression level in the thymus. Immunity 6:401–410
Fung-Leung WP, Louie MC, Limmer A, Ohashi PS, Ngo K, Chen L, Kawai K, Lacy E, Loh DY, Mak TW (1993) The lack of CD8 alpha cytoplasmic domain resulted in a dramatic decrease in efficiency in thymic maturation but only a moderate reduction in cytotoxic function of CD8+ T lymphocytes. Eur J Immunol 23:2834–2840
Garza KM, Chan SM, Suri R, Nguyen LT, Odermatt B, Schoenberger SP, Ohashi PS (2000a) Role of antigen-presenting cells in mediating tolerance and autoimmunity. J Exp Med 191:2021–2027
Garza KM, McKall-Faienza KJ, Zakarian A, Odermatt B, Ohashi PS (2000b) Enhanced T cell responses contribute to the genetic predisposition of CD8-mediated spontaneous autoimmunity. Eur J Immunol, in press
Geiger T, Gooding LR, Flavell RA (1992) T-cell responsiveness to an oncogenic peripheral protein and spontaneous autoimmunity in transgenic mice. Proc Natl Acad Sci USA 89:2985–2989
Girao C, Hu Q, Sun J, Ashton-Rickardt PG (1997) Limits to the Differential Avidity Model of T Cell Selection in the Thymus. J Immunol 159:4205–4211
Goldrath AW, Hogquist KA, Bevan MJ (1997) CD8 lineage commitment in the absence of CD8. Immunity 6:633–642
Goverman J, Woods A, Larson L, Weiner LP, Hood L, Zaller DM (1993) Transgenic mice that express a myelin basic protein specific T cell receptor develop spontaneous autoimmunity. Cell 72:551–560
Green DR, Scott DW (1994) Activation-induced apoptosis in lymphocytes. Curr Opin Immunol 6: 476–487

Grewal IS, Flavell RA (1998) CD40 and CD154 in cell-mediated immunity. Annu Rev Immunol 16: 111–135

Grewal IS, Foellmer HG, Grewal KD, Xu J, Hardardottir F, Baron JL, Janeway Jr CA, Flavell RA (1996) Requirement for CD40 ligand in costimulation induction, T cell activation, and experimental allergic encephalomyelitis. Science 273:1864–1867

Hanahan D (1985) Heritable formation of pancreatic β-cell tumours in transgenic mice expressing recombinant insulin/simian virus 40 oncogenes. Nature 315:115–122

Harlan DM, Hengartner H, Huang ML, Kang YH, Abe R, Moreadith RW, Pircher H, Gray GS, Ohashi PS, Freeman GJ, Nadler LM, June CH, Aichele P (1994) Mice expressing both B7 and viral glycoprotien on pancreatic beta cells along with glycoprotein-specific transgenic T cell develop diabetes due to a breakdown of T lymphocyte unresponsiveness. Proc Natl Acad Sci USA 91:3137–3141

Heath WR, Karamalis F, Donoghue J, Miller JFAP (1995) Autoimmunity caused by ignorant $CD8^+$ T cell is transient and depends on avidity. J Immunol 155:2339–2349

Hernández-Hoyos G, Sohn SJ, Rothenberg EV, Alberola-Ila J (2000) Lck activity controls CD4/CD8 T cell lineage commitment. Immunity 12:313–322

Hogquist KA, Jameson SC, Bevan MJ (1995) Strong agonist ligands for the T cell receptor do not mediate positive selection of functional $CD8^+$ T cells. Immunity 3:78–86

Hogquist KA, Jameson SC, Heath WR, Howard JL, Bevan MJ, Carbone FR (1994) T cell receptor antagonist peptides induce positive selection. Cell 76:17–27

Hogquist KA, Tomlinson AJ, Kieper WC, McGargill MA, Hart MC, Naylor S, Jameson SC (1997) Identification of a naturally occurring ligand for thymic positive selection. Immunity 6:389–399

Hu Q, Bazemore Walker CR, Girao C, Opferman JT, Sun J, Shabanowitz J, Hunt DF, Ashton-Rickardt PG (1997) Specific recognition of thymic self-peptides induces the positive selection of cytotoxic T lymphocytes. Immunity 7:221–231

Ingold AL, Landel C, Knall C, Evans GA, Potter TA (1991) Co-engagement of CD8 with the T cell receptor is required for negative selection. Nature 352:721–723

Jameson SC, Hogquist KA, Bevan MJ (1994) Specificity and flexibility in thymic selection. Nature 369:750–752

Josien R, Wong BR, Li H-L, Steinman RM, Choi Y (1999) TRANCE, a TNF family member, is differentially expressed on T cell subsets and induces cytokine production in dendritic cells. J Immunol 162:2562–2568

Katz JD, Wang B, Haskins K, Benoist C, Mathis D (1993) Following a diabetogenic T cell from genesis through pathogenesis. Cell 74:1089–1100

Kawai K, Ohashi PS (1995) Immunological function of a defined T-cell population tolerized to low-affinity self antigens. Nature 374:68–69

Kägi D, Odermatt B, Ohashi PS, Zinkernagel RM, Hengartner H (1996) Development of insulitis without diabetes in transgenic mice lacking perforin-dependent cytotoxicity. J Exp Med 183:2143–2152

Killeen N, Moriarty A, Teh HS, Littman DR (1992) Requirement for CD8-Major Histocompatibility Complex class I interaction in positive and negative selection of developing T cells. J Exp Med 176:89–97

Kisielow P, Blüthmann H, Staerz UD, Steinmetz M, von Boehmer H (1988) Tolerance in T cell receptor transgenic mice involves deletion of nonmature $CD4^+8^+$ thymocytes. Nature 333:742–746

Koller BH, Marrack P, Kappler JW, Smithies O (1990) Normal development of mice deficient in beta 2 M, MHC class I proteins, and $CD8+$ T cells. Science 248:1227–1230

Kurts C, Kosaka H, Carbone FR, Miller JFAP, Heath WR (1997) Class I-restricted cross-presentation of exogenous self-antigens leads to deletion of autoreactive $CD8^+$ T cells. J Exp Med 186:239–245

Kurts C, Sutherland RM, Davey G, Li M, Lew AM, Blanas E, Carbone FR, Miller JFAP, Heath WR (1999) CD8 T cell ignorance or tolerance to islet antigens depends on antigen dose. Proc Natl Acad Sci USA 96:12703–12707

Kündig TM, Shahinian A, Kawai K, Mittruecker HW, Sebzda E, Bachmann MF, Mak TW, Ohashi PS (1996) Duration of TCR stimulation determines costimulatory requirements of T cells. Immunity 5:41–52

Kyburz D, Aichele P, Speiser DE, Hengartner H, Zinkernagel RM, Pircher H (1993) T cell immunity after a viral infection versus T cell tolerance induced by soluble viral peptides. Eur J Immunol 23:1956–1962

Lafaille JJ, Nagashima K, Katsuki M, Tonegawa S (1994) High incidence of spontaneous autoimmune encephalomyelitis in immunodeficient anti-myelin basic protein T cell receptor transgenic mice. Cell 78:399–408

Lafferty KJ, Prowse SJ, Simeonovic CJ (1983) Immunobiology of tissue transplantation: a return to the passenger leukocyte concept. Annu Rev Immunol 1:143–173

Lenschow DJ, Walunas TL, Bluestone JA (1996) CD28/B7 system of T cell costimulation. Annu Rev Immunol 14:233–258

Lucas B, Stefanova I, Yasutomo K, Dautigny N, Germain RN (1999) Divergent changes in the sensitivity of maturing T cells to structurally related ligands underlies formation of a useful T cell repertoire. Immunity 10:367–376

Ludewig B, Ochsenbein AF, Odermatt B, Paulin D, Hengartner H, Zinkernagel RM (2000) Immunotherapy with dendritic cells directed against tumor antigens shared with normal host cells results in severe autoimmune disease. J Exp Med 191:795–803

Ludewig B, Odermatt B, Landmann S, Hengartner H, Zinkernagel RM (1998) Dendritic cells induce autoimmune diabetes and maintain disease via De Novo formation of local lymphoid tissue. J Exp Med 188:1493–1501

Mamula MJ (1993) The inability to process a self-peptide allows autoreactive T cells to escape tolerance. J Exp Med 177:567–571

Mariathasan S, Bachmann MF, Bouchard D, Ohteki T, Ohashi PS (1998) Degree of TCR internalization and Ca2+ flux correlates with thymocyte selection. J Immunol 161:6030–6037

Marrack P, Kappler J (1988) The T-cell repertoire for antigen and MHC. Immunol Today 9:308–315

Martin S, Bevan MJ (1998) Transient alteration of T cell fine specificity by a strong primary stimulus correlates with T cell receptor down-regulation. Eur J Immunol 28:2991–3002

Matechak EO, Killeen N, Hedrick SM, Fowlkes BJ (1996) MHC class II-specific T cells can develop in the CD8 lineage when CD4 is absent. Immunity 4:337–347

McKall-Faienza K, Kawai K, Kündig TM, Odermatt B, Bachmann MF, Zakarian A, Mak TW, Ohashi PS (1998) Absence of TNFRp55 influences virus-induced autoimmunity despite efficient lymphocytic infiltration. Int Immunol 10:405–412

Meuer SC, Hussey RE, Cantrell DA, Hodgdon JC, Schlossman SF, Smith KA, Reinherz EL (1984) Triggering of the T3-Ti antigen-receptor complex results in clonal T-cell proliferation through an interleukin 2-dependent autocrine pathway. Proc Natl Acad Sci USA 81:1509–1513

Morgan DJ, Kreuwel HTC, Sherman LA (1999) Antigen concentration and precursor frequency determine the rate of CD8+ T cell tolerance to peripherally expressed antigens. J Immunol 163:723–727

Moskophidis D, Lechner F, Pircher H, Zinkernagel RM (1993) Virus persistence in acutely infected immunocompetent mice by exhaustion of antiviral cytotoxic effector T cells. Nature 362:758–761

Murphy KM, Weaver CT, Elish M, Allen PM, Loh DY (1989) Peripheral tolerance to allogeneic class II histocompatibility antigens expressed in transgenic mice: evidence against a clonal-deletion mechanism. Proc Natl Acad Sci USA 86:10034–10038

Nguyen LT, McKall-Faienza K, Zakarian A, Speiser DE, Mak TW, Ohashi PS (2000) TNF receptor 1 (TNFR1) and CD95 are not required for T cell deletion after virus infection but contribute to peptide-induced deletion under limited conditions. Eur J Immunol 30:683–688

Oehen S, Ohashi PS, Aichele P, Hengartner H, Zinkernagel RM (1992) Vaccination or tolerance to prevent diabetes. Eur J Immunol 22:3149–3153

Ohashi PS, Oehen S, Aichele P, Pircher H, Odermatt B, Herrera P, Higuchi Y, Buerki K, Hengartner H, Zinkernagel RM (1993a) Induction of diabetes is influenced by the infectious virus and local expression of MHC class I and TNF-α. J Immunol 150:5185–5194

Ohashi PS, Oehen S, Bürki K, Pircher H, Ohashi CT, Odermatt B, Malissen B, Zinkernagel R, Hengartner H (1991) Ablation of "tolerance" and induction of diabetes by virus infection in viral antigen transgenic mice. Cell 65:305–317

Ohashi PS, Pircher H, Bürki K, Zinkernagel RM, Hengartner H (1990) Distinct sequence of negative or positive selection implied by thymocyte T cell receptor densities. Nature 346:861–863

Ohashi PS, Zinkernagel RM, Luescher IF, Hengartner H, Pircher H (1993b) Enhanced positive selection of a transgenic TCR by a restriction element that does not permit negative selection. Int Immunol 5:131–138

Oldstone MBA, Nerenberg M, Southern P, Price J, Lewicki H (1991) Virus infection triggers insulin-dependent diabetes mellitus in a transgenic model: role of anti-self (virus) immune response. Cell 65:319–331

Pircher H, Bürki K, Lang R, Hengartner H, Zinkernagel R (1989) Tolerance induction in double specific T-cell receptor transgenic mice varies with antigen. Nature 342:559–561

Pircher H, Hoffmann Rohrer U, Moskophidis D, Zinkernagel RM, Hengartner H (1991) Lower receptor avidity required for thymic clonal deletion than for effector T cell function. Nature 351:482–485

Pircher H, Mak TW, Lang R, Ballhausen W, Rüedi E, Hengartner H, Zinkernagel RM, Bürki K (1989) T cell tolerance to Mls^a encoded antigens in T cell receptor β-chain transgenic mice. EMBO J 8:719–727

Pircher H, Michalopoulos EE, Iwamoto A, Ohashi PS, Baenziger J, Hengartner H, Zinkernagel RM, Mak TW (1987) Molecular analysis of the antigen receptor of virus-specific cytotoxic T cells and identification of a new V-alpha family. Eur J Immunol 17:1843–1846

Qian D, Weiss A (1997) T cell antigen receptor signal transduction. Curr Opin Cell Biol 9:205–212

Rammensee H-G, Kroschewski R, Frangoulis B (1989) Clonal anergy induced in mature V β 6+ T lymphocytes on immunizing Mls-1^b mice with Mls-1^a expressing cells. Nature 339:541–544

Ramsdell F, Lantz T, Fowlkes BJ (1989) A nondeletional mechanism of thymic self tolerance. Science 246:1038–1041

Reich A, Korner H, Sedgwick JD, Pircher H (2000) Immune down-regulation and peripheral deletion of CD8 T cells does not require TNF receptor-ligand interactions nor CD95. Eur J Immunol 30:678–682

Reich Z, Boniface JJ, Lyons DS, Borochov N, Wachtel EJ, Davis MM (1997) Ligand-specific oligomerization of T-cell receptor molecules. Nature 387:617–620

Rocha B, von Boehmer H (1991) Peripheral selection of the cell repertoire. Science 251:1225–1228

Roman LM, Simons LF, Hammer RE, Sambrook JF, Gething M-JH (1990) The expression of influenza virus hemagglutinin in the pancreatic β cells of transgenic mice results in autoimmune diabetes. Cell 61:383–396

Rötzschke O, Falk K, Stromingher JL (1997) Superactivation of an immune response triggered by oligomerized T cell epitopes. Proc Natl Acad Sci USA 94:14642–14647

Ruedl C, Kopf M, Bachmann MF (1999) CD8(+) T cells mediate CD40-independent maturation of dendritic cells in vivo. J Exp Med 189:1875–1884

Russell JH, Rush B, Weaver C, Wang R (1993) Mature T cells of autoimmune lpr/lpr mice have a defect in antigen-stimulated suicide. Proc Natl Acad Sci USA 90:4409–4413

Salzmann M, Bachmann MF (1998) The role of T cell receptor dimerization in T cell antagonism and T cell specificity. Mol Immunol 53:271–277

Sarukhan A, Lanoue A, Franzke A, Brousse N, Buer J, von Boehmer H (1998) Changes in function of antigen-specific lymphocytes correlating with progression towards diabetes in a transgenic model. EMBO J 17:71–80

Schild HJ, Rötzschke O, Kalbacher H, Rammensee H-G (1990) Limit of T cell tolerance to self proteins by peptide presentation. Science 247:1587–1589

Schwartz RH (1990) A cell culture model for T lymphocyte clonal anergy. Science 248:1349–1356

Scott B, Liblau R, Degermann S, Marconi LA, Ogata L, Caton AJ, McDevitt HO, Lo D (1994) A role for non-MHC genetic polymorphism in susceptibility to spontaneous autoimmunity. Immunity 1:73–82

Scott DE, Kisch WJ, Steinberg AD (1993) Studies of T cell deletion and T cell anergy following in vivo administration of SEB to normal and lupus-prone mice. J Immunol 150:664–672

Screaton G, Xu X-N (2000) T cell life and death signalling via TNF-receptor family members. Curr Opin Immunol 12:316–322

Sebzda E, Choi M, Fung-Leung WP, Mak TW, Ohashi PS (1997) Peptide-induced positive selection of TCR transgenic thymocytes in a coreceptor-independent manner. Immunity 6:643–653

Sebzda E, Kündig TM, Thomson CT, Aoki K, Mak SY, Mayer J, Zamborelli TM, Nathenson S, Ohashi PS (1996) Mature T cell reactivity altered by a peptide agonist that induces positive selection. J Exp Med 183:1093–1104

Sebzda E, Mariathasan S, Ohteki T, Jones R, Bachmann MF, Ohashi PS (1999) Selection of the T cell repertoire. Annu Rev Immunol 17:829–874

Sebzda E, Wallace VA, Mayer J, Yeung RSM, Mak TW, Ohashi PS (1994) Positive and negative thymocyte selection induced by different concentrations of a single peptide. Science 263:1615–1618

Sha WC, Nelson CA, Newberry RD, Kranz DM, Russell JH, Loh DY (1988) Positive and negative selection of an antigen receptor on T cells in transgenic mice. Nature 336:73–76

Sha WC, Nelson CA, Newberry RD, Pullen JK, Pease LR, Russell JH, Loh DY (1990) Positive selection of transgenic receptor-bearing thymocytes by K^b antigen is altered by K^b mutations that involve peptide binding. Proc Natl Acad Sci USA 87:6186–6190

Singer GG, Abbas AK (1994) The Fas antigen is involved in peripheral but not thymic deletion of T lymphocytes in T cell receptor transgenic mice. Immunity 1:365–371

Soldevila G, Geiger T, Flavell RA (1995) Breaking immunologic ignorance to an antigenic peptide of simian virus 40 large T antigen. J Immunol 155:5590–5600

Speiser DE, Miranda R, Zakarian A, Bachmann MF, McKall-Faienza K, Odermatt B, Hanahan D, Zinkernagel RM, Ohashi PS (1997) Self antigens expressed by solid tumors do not efficiently stimulate naive or activated T cells: implications for immunotherapy. J Exp Med 186:645–653

Speiser DE, Sebzda E, Ohteki T, Bachmann MF, Pfeffer K, Mak TW, Ohashi PS (1996) Tumor necrosis factor receptor p55 mediates deletion of peripheral cytotoxic T lymphocytes in vivo. Eur J Immunol 26:3055–3060

Sprent J, Lo D, Gao E, Ron Y (1988) T cell selection in the thymus. Immunol Rev 101:173–190

Suzuki H, Punt JA, Granger LG, Singer A (1995) Asymmetric signalling requirements for thymocyte commitment to the $CD4^+$ versus $CD8^+$ T cell lineages: a new perspective on thymic commitment and selection. Immunity 2:413–425

Teh SJ, Dutz JP, Motyka B, Teh HS (1996) Fas (CD95)-independent regulation of immune responses by antigen-specific $CD4^-CD8^+$ T cells. Int Immunol 8:675–681

Valitutti S, Muller S, Cella M, Padovan E, Lanzavecchia A (1995) Serial triggering of many T-cell receptors by a few peptide-MHC complexes. Nature 375:148–151

Valitutti S, Müller S, Salio M, Lanzavecchia A (1997) Degradation of T cell receptor (TCR)-CD3-zeta complexes after antigenic stimulation. J Exp Med 185:1859–1864

Van Kaer L, Ashton-Rickardt PG, Ploegh HL, Tonegawa S (1992) TAP1 mutant mice are deficient in antigen presentation, surface class I molecules, and $CD4-CD8^+$ T cells. Cell 71:1205–1214

Van Pel A, van der Bruggen P, Coulie PG, Brichard VG, Lethe B, Van den Eynde B, Uyttenhove C, Renauld J-C, Boon T (1995) Genes coding for tumor antigens recognized by cytolytic T lymphocytes. Immunol Rev 145:229–250

Viola A, Lanzavecchia A (1996) T cell activation determined by T cell receptor number and tunable thresholds. Science 273:104–106

von Boehmer H (1996) CD4/CD8 lineage commitment: back to instruction? J Exp Med 183:713–715

von Herrath MG, Dockter J, Oldstone MBA (1994) How virus induces a rapid or slow onset insulin-dependent diabetes mellitus in a transgenic model. Immunity 1:231–242

von Herrath MG, Guerder S, Lewicki H, Flavell RA, Oldstone MBA (1995) Coexpression of B7-1 and viral ("self") transgenes in pancreatic β cells can break peripheral ignorance and lead to spontaneous autoimmune diabetes. Immunity 3:727–738

Watts TH, DeBenedette MA (1999) T cell co-stimulatory molecules other than CD28. Curr Opin Immunol 11:286–293

Webb S, Morris C, Sprent J (1990) Extrathymic tolerance of mature T cells: clonal elimination as a consequence of immunity. Cell 63:1249–1256

Wong BR, Josien R, Lee SY, Sauter B, Li H-L, Steinman RM, Choi Y (1997a) TRANCE (tumor necrosis factor [TNF]-related activation-induced cytokine), a new TNF family member predominantly expressed in T cells, is a dendritic cell-specific survival factor. J Exp Med 186:2075–2080

Wong BR, Rho J, Arron J, Robinson E, Orlinick J, Chao WN, Kalachikov S, Cayani E, Bartlett FS, Frankel WN, Lee SY, Choi Y (1997b) TRANCE is a novel ligand of the tumor necrosis factor receptor family that activates c-Jun N-terminal kinase in T cells. J Biol Chem 272:25190–25194

Yang Y, Wilson JM (1996) CD40 ligand-dependent T cell activation: requirement of B7-CD28 signaling through CD40. Science 273:1862–1864

Zimmermann C, Brduscha-Riem K, Blaser C, Zinkernagel RM, Pircher H (1996a) Visualisation, characterization and turnover of CD8+ memory T cells in virus-infected hosts. J Exp Med 183:1367–1375

Zimmermann C, Prevost-Blondel A, Blaser C, Pircher H (1999) Kinetics of the response of naive and memory CD8 T cells to antigen: similarities and differences. Eur J Immunol 29:284–290

Zimmermann C, Rawiel M, Blaser C, Kaufmann M, Pircher H (1996b) Homeostatic regulation of $CD8^+$ T cells after antigen challenge in the absence of Fas (CD95). Eur J Immunol 26:2903–2910

Zinkernagel RM, Callahan GN, Althage A, Cooper S, Klein PA, Klein J (1978a) On the thymus in the differentiation of "H-2 self-recognition" by T cells: evidence for dual recognition? J Exp Med 147:882–896

Zinkernagel RM, Callahan GN, Klein J, Dennert G (1978b) Cytotoxic T cells learn specificity for self H-2 during differentiation in the thymus. Nature 271:251–253

Regulation of Virally Induced Autoimmunity and Immunopathology: Contribution of LCMV Transgenic Models to Understanding Autoimmune Insulin-Dependent Diabetes Mellitus

M.G. VON HERRATH

1   Introduction... 145
2   Section 1: Paradigms for Virally Induced Autoimmune Diabetes 146
2.1 The RIP-LCMV Mouse Model for Virally Induced Autoimmune Diabetes:
    Role of the Thymus in Autoimmunity... 148
2.2 Paradigm I: Breaking of "Peripheral Tolerance" Through Viral Infections –
    Role of the Thymus.. 150
2.3 Paradigm II: The Local "Danger" Factor in Autoimmune Diabetes –
    Need for Co-Stimulation or Viral Presence in the Islets........................ 150
2.4 Paradigm III: Numbers of Autoreactive Lymphocytes Correlate with Severity
    of Autoimmune Disease; After a Critical "Cut-Off" Level, Disease Progression Ceases 152
2.5 Paradigm IV: How β-Cells Die – Essential Role of Inflammatory Cytokines 154
2.6 Paradigm V: Immunotherapy – How Autoreactive Regulatory Lymphocytes
    can Prevent or Abrogate Autoimmunity... 155
2.7 How do Regulatory, Autoreactive Lymphocytes Work?
    The Concept of Bystander Suppression 157
2.8 Induction of Regulatory Cells by Oral Antigens and DNA Vaccination 158
2.9 Summary and Open Questions... 161
3   Section 2: Regulation of Virally Induced Immunopathologic Effects 162
3.1 Numbers of Anti-Viral Lymphocytes... 162
3.2 Costimulation and Infection of Antigen-Presenting Cells....................... 164
3.3 Cytokines and Chemokines as Inducers and/or Regulators of Immunopathologic Effects ... 165
3.4 Conclusions... 166
4   Future Considerations – Is it Possible to Model and Predict Immunopathology
    in Autoimmunity or Viral Infections?... 166
References... 169

1 Introduction

Viral infections have the capacity to damage host organisms in multiple ways. First, direct lytic effects of the virus can destroy cells or whole organ systems. Second, as discussed in this chapter, the anti-viral immune response can be harmful, in con-

---

Division of Virology, Department of Neuropharmacology, IMM6, The Scripps Research Institute, 10550 N. Torrey Pines Road, La Jolla, CA 92037 USA

trast and addition to its beneficial effect of eliminating or decreasing the viral load. For example, cytokines and chemokines secreted by cells and other components of the immune system can have detrimental side effects on particularly sensitive cell types and organs. Via this mechanism, viral infections are excellent candidates for causing chronic self-destructive processes that we think of as autoimmune. A third possibility is that viral infection directly activates antigen-presenting cells (APCs) or inflammatory mediators, ultimately enhancing uptake and presentation of foreign as well as self-antigens and, in this way, producing "true" autoimmunity. It is evident from these considerations that the fine-tuned regulation of anti-viral immunity is crucial and that viral elimination, although it is, of course, the final and most desirable goal, should not be achieved at any cost or through too strong and too rapid an immune response which can damage the host. However, too weak and protracted an anti-viral response may lead to persistent infection with chronic activation of the immune system, which would be harmful in the long-term. In this chapter, I will discuss experimental scenarios illustrating the foregoing considerations. For these studies, we have used the arenavirus, lymphocytic choriomeningitis virus (LCMV) and a transgenic mouse model of virally (LCMV)-induced autoimmune diabetes.

2 Section 1: Paradigms for Virally Induced Autoimmune Diabetes

Insulin-dependent diabetes mellitus (IDDM – type 1 diabetes) (BACH 1994a) and multiple sclerosis (MS) (ANDERSEN et al. 1993; MARTIN et al. 1998) are thought to be autoimmune diseases, but their precise etiologies are still unknown. In both of these disorders T cells reactive to islet or myelin components have been detected systemically (humans) and in lesions (animal models) (TUOHY et al. 1999) and, therefore, implicated in the related pathogenesis (LERNMARK et al. 1985). However, to date no definite proof exists for the concept that autoimmune disorders in humans are really caused by autoaggressive T cells (TUOHY et al. 1999). Genetic factors are definitely involved in their pathogenic activities (BACH 1994b; HAFLER 1999; KURTZKE et al. 1997, 1998; WICKER et al. 1995), but the concordance for disease is only 30%–40% in monozygotic twins at risk (REDONDO et al. 1999). Thus, the participation of an additional environmental trigger (HARRISON and HONEYMAN 1999) or infectious agent must be postulated. Viruses are "prime" candidates (LEITER and HAMAGUCHI 1990), because they can selectively infect target organs or cells and induce strong T and B cell-mediated immune responses. Several mechanisms could underlie a viral etiology of autoimmune diseases: First, lytic viruses could directly infect target cells or organs (the pancreas/islets in type 1 diabetes, the brain or oligodendrocytes in MS) leading to their damage (JENSON et al. 1980; NOTKINS et al. 1981; PRABHAKER et al. 1982; PRINCE et al. 1978; YOON et al. 1984). In this scenario, the resulting disease would NOT be primarily autoimmune in nature, although autoreactive cells could be generated secondarily after organ or cellular damage has been initiated (BENNETT et al. 1997; KATZ-LEVY et al. 1999). Second, non-lytic viruses could infect the targeted organ, and the immune response that clears the viral infection could lead to "bystander" damage of the organ (HORWITZ et al. 1998). Such immunopathologic effects could be severe enough to destroy sufficient cells to produce clinically manifested disease. Again, during the progression of organ or cellular destruction secondary autoimmune phenomena are possible. In both these situations, however, the primary cause for organ destruction is not autoimmunity but anti-viral immunity, and the secondary autoreactive responses could either be aggressive or even regulatory. For "true" autoimmunity induced by viral infections or other environmental triggers, a primarily pathogenic autoaggressive lymphocyte or antibody response would be required. Two mechanisms are possible. First, viral and self-antigens could share cross-reactive determinants that lead to recognition of self-antigens during the anti-viral immune response, a mechanism termed molecular mimicry (OLDSTONE 1987, 1989b). Second, in the absence of shared determinants, autoaggressive lymphocytes could be activated by inflammatory cytokines or APC-related mechanisms as bystanders during strong systemic anti-viral responses (HORWITZ et al. 1998; OLDSTONE 1989a; VON HERRATH and OLDSTONE 1996). Which of these two scenarios is implicated in human type 1 diabetes, MS or other autoimmune disorders is completely unclear. Since these possibilities are extremely difficult to analyze in humans at this point, the use of animal models is an indispensable tool if we are to have any hope of understanding the underlying mechanisms in vivo. Multiple models, each unique in some respect, are required to "piece together" the puzzle. Moreover, each model has a lesson to teach that might apply to some facet of human "autoimmune" disease. The goal of this section, then, is to introduce and provide background for one such model – transgenic animals with virally induced autoimmune diabetes (OHASHI et al. 1991; OLDSTONE et al. 1991) – and develop the important paradigms that this model provides for understanding the immunopathogenesis of related diseases in humans.

Murine models, in general, take a reductionist in vivo approach in that they limit the number of variables by using genetically defined mouse strains and, frequently, clearly defined self-antigens (BACH and MATHIS 1997; VON HERRATH et al. 1994b). This experimental design has the advantage that autoreactive cells can be tracked easily and that disease incidence is reproducibly high and not as diverse as in the "out-bred" population that humans represent. Consequently, the principles learned should point us in the right direction as we seek related answers in humans. An additional advantage is that potential interventions (CHATENOUD et al. 1994; SEMPE et al. 1994; VON HERRATH et al. 1996a) can be tested in such models, and their validity can be established in multiple different but well-defined situations. In other words, although it would be a mistake to assume that genes or viral infections capable of propagating disease in mice will always do the same in humans, the causative mechanisms might be very similar, even if the "players" are exchanged.

These presumptions led to the establishment of the RIP-LCMV transgenic mouse model of virally induced autoimmune diabetes (OHASHI et al. 1991; OLDSTONE et al. 1991). Since these studies began in the late 1980s, results with

RIP-LCMV mice have validated the concept that peripheral tolerance/unresponsiveness to a defined self-antigen (transgene) can be broken by a systemic viral infection, leading to attack of β-cells and their eventual destruction. For the "success" of the autoreactive aggressive response, viral infection of the exocrine pancreas, or a continuous 'driving' of autoreactive lymphocytes by activated dendritic cells (LUDEWIG et al. 1998) is required. That step leads to local priming of APCs, which subsequently (after systemic viral clearance) helps to propagate an autoimmune reaction in the pancreas and islets (VON HERRATH and HOLZ 1997). Further, destruction of most β-cells occurs via the action of inflammatory cytokines, in this case mostly interferon (IFN)-γ made by autoreactive lymphocytes (VON HERRATH and OLDSTONE 1997), but somewhat surprisingly not by perforin-mediated killing of β-cells by cytotoxic T-lymphocytes (CTL), although β-cells are sensitive to lysis by autoreactive CTL in vitro (HOMANN et al. 1999). Given this (and other, see below) knowledge about the pathogenic process leading to IDDM in RIP-LCMV mice, we believe that this powerful model exhibits the most salient features of human type 1 diabetes.

In the following paragraphs I will describe in more detail the pathogenically relevant observations made by our group using the RIP-LCMV model and the therapeutic strategies explored so far.

2.1 The RIP-LCMV Mouse Model for Virally Induced Autoimmune Diabetes: Role of the Thymus in Autoimmunity

RIP-LCMV transgenic mice were developed simultaneously in the laboratories of Oldstone and Zinkernagel in 1990 (OHASHI et al. 1991; OLDSTONE et al. 1991) and have since been used by multiple laboratories as an animal model of virally induced type 1 diabetes. The underlying idea was to create mice that express a well-defined target autoantigen exclusively in pancreatic β-cells but not in other organs (Fig. 1). Using the rat insulin promoter (RIP) was a very successful approach for generating transgenic lines of mice whose pancreatic islets expressed as a self-antigen either the glycoprotein (GP) or nucleoprotein (NP) of LCMV (OLDSTONE et al. 1996; VON HERRATH et al. 1994b). Expression of these viral transgenes did not lead to β-cell destruction, and the majority of such transgenic mice (>95%) remained healthy throughout their full life span without signs of islet dysfunction or inflammatory cell infiltration. These mice thus constituted an ideal tool for manipulating the immune response to a single, defined self-antigen and for testing which type of autoreactive responses engendered clinically overt disease. Indeed, most dominant and subdominant T cell (CD4 and CD8) epitopes for LCMV have been mapped for the mouse H-2^b and H-2^d haplotypes (GAIRIN et al. 1995; OXENIUS et al. 1995; VAN DER MOST et al. 1998; WHITTON et al. 1993). The immune response has been quantified precisely in many different laboratories, and most "tools of the trade" such as tetramers (MURALI-KRISHNA et al. 1998), FACS used for intracellular cytokine production (WHITMIRE et al. 1998) and ELISPOT assays have been established. Consequently, this experimental model is uniquely useful in that the

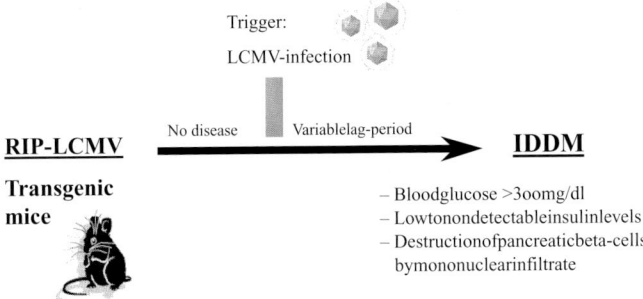

Fig. 1. The RIP-LCMV mouse as model for autoimmune diabetes

autoreactive immune response can be tracked, manipulated and defined, because the disease of interest does not occur spontaneously and the initiating self-antigen is known.

Analysis of several different RIP-LCMV transgenic murine lines resulted in the discovery of an interesting difference: Some lines developed rapid-onset IDDM that occurred 10–18 days after infection with LCMV, whereas other lines developed slow-onset IDDM occurring 4–6 months post-infection (VON HERRATH et al. 1994b). Two lines were selected for further studies, the RIP-LCMV-GP rapid-onset line and the RIP-LCMV-NP slow-onset line, both with MHC H-2^b (C57Bl6/ J) or MHC H-2^d (Balb/c) genetic backgrounds. Direct comparison showed that, whereas both lines expressed about equivalent levels of the transgene in the pancreas while the slow-onset RIP-NP mice additionally expressed the viral (NP) transgene in the thymus. Further analysis showed that thymic expression of the transgene led to the deletion of most LCMV-NP-specific (anti-self) high affinity CTL, but fewer and lower affinity CTL were still detectable in the periphery of these mice. These CTL were able to trigger IDDM, but the development was much slower. In addition, IDDM did not develop at all in the absence of CD4 helper lymphocytes. In contrast, rapid-onset IDDM in RIP-GP mice did not depend on the participation of $CD4^+$-help (VON HERRATH et al. 1994b). Thus, thymic expression of a self-antigen does not necessarily lead to the elimination of all autoreactive lymphocytes, and the remaining CTL are quite capable of causing slowly progressive self-destruction and autoimmune disease, a process that depends on CD4 help. The CD4-dependent RIP-LCMV model therefore seems to reproduce a scenario likely for human autoimmune diabetes and, because of the long lag phase preceding the onset of disease, offers a suitable "window" for testing therapeutic interventions. Further, recent evidence (HOMANN et al. 1999; T. Dyrberg, M. Oldstone, unpublished data) suggests that the autoimmune attack initially directed only to the LCMV transgene, spreads to other islet cell antigens before the onset of IDDM. This is reflected in the generation of antibodies to GAD in RIP-LCMV slow-onset mice and parallels the findings of islet cell antibodies preceding human IDDM.

2.2 Paradigm I: Breaking of "Peripheral Tolerance" Through Viral Infections – Role of the Thymus

The first and foremost insight gained with this model was that peripheral "tolerance" can be overcome and broken. When infected with LCMV, more than 90% of RIP-LCMV mice develop autoimmune diabetes that is characterized by extensive islet infiltration containing CD4 and CD8 lymphocytes, dendritic cells, activated macrophages and some B cells. Thus, RIP-LCMV mice are not really tolerant to the pancreatic self-transgene, because good numbers of naive, non-activated autoreactive (anti-viral) lymphocytes do not undergo thymic negative selection but reach the periphery where they circulate and can be activated by LCMV infection. Therefore, it is better to use the term "unresponsiveness", which encompasses the fact that autoreactive cells are present but cannot react with the β-cell viral (self-) transgene. Why are these autoreactive (LCMV specific) lymphocytes unable to attack β-cells expressing the viral transgene and why does spontaneous autoimmunity fail to occur? Two possibilities come to mind. First, non-activated lymphocytes never reach the pancreatic islets and therefore do not have an opportunity to interact with the transgene, a situation termed "ignorance". Second, since the self-antigen is expressed only on non-"professional" APCs (the β-cells) that are incapable of providing the necessary co-stimulatory signals for T-cell activation, which the autoreactive lymphocytes circulating in the periphery would require to become fully activated. It turned out that both are true. As will become clear from studies described in the following paragraphs, very few peripheral lymphocytes have the opportunity to interact directly with β-cell antigens (VON HERRATH et al. 1995b). Further, co-stimulatory signals provided by activated professional APCs, such as dendritic cells or macrophages, are essential for the development of type 1 diabetes (VON HERRATH et al. 1995b).

2.3 Paradigm II: The Local "Danger" Factor in Autoimmune Diabetes – Need for Co-Stimulation or Viral Presence in the Islets

To test whether autoantigens must be expressed or presented by professional APCs that are co-stimulation competent, we did a series of experiments. Co-stimulatory signals were provided by the interaction of B7.1 or B7.2 molecules expressed by activated APCs and CD28 molecules expressed on T cells (HEROLD et al. 1997). During engagement of the T cell receptor (TcR) with the fitting MHC peptide complex on an activated APC, these signals are necessary to generate a fully activated T cell capable of effector functions [cytotoxicity – (SIGAL et al. 1998), cytokine/chemokine secretion, T cell help – (HARDING and ALLISON 1993; ZHENG and LIU 1997)]. In the absence of such signals, engagement of the TcR alone might even lead to the anergy of T cells (MILLER et al. 1997; SALOJIN et al. 1998) and their ultimate elimination.

In the first study, B7 molecules were co-expressed with the LCMV transgenes on β-cells. Interestingly, our studies showed that B7.1 when expressed on β-cells led to local activation and amplification of CTL. As a result, we observed that IDDM occurred spontaneously without LCMV infection in a fraction (30%) of RIP-B7.1 × RIP-GP mice (without thymic expression of the GP) and that the virus-induced slow-onset IDDM normally observed in RIP-LCMV-NP mice with thymic expression of the transgene was markedly accelerated. These different incidences of IDDM correlated well with the numbers of CTL precursors (pCTL) isolated locally from the pancreas or the spleen. Local expression of B7.1 led to a tenfold increase of pCTL in the pancreas of RIP-LCMV thymic expressor mice, which correlated with the accelerated incidence of IDDM (VON HERRATH et al. 1995b). Similar findings were shown for the spontaneous IDDM in RIP-LCMV non-thymic expressor mice: Expression of B7.1 induced local activation and amplification of CTL in the pancreas, resulting in numbers comparable to those found in RIP-LCMV single transgenic mice with IDDM (VON HERRATH et al. 1995b).

In the second study, we analyzed, whether LCMV infection of the exocrine pancreas is a prerequisite for the development of type 1 diabetes. Indeed, very shortly (1–4 days) after intra-peritoneal inoculation with LCMV, replicating virus was present in the pancreas, but not the islets (VON HERRATH and HOLZ 1997). This was associated with clear signs of APC activation (such as upregulation of MHC class II). We attempted to induce disease by transferring in vitro activated LCMV (self-) antigen-specific lymphocytes into uninfected RIP-LCMV transgenic recipients and observed that transfer, even of large numbers of highly activated LCMV(self-transgene)-specific CTL, never yielded IDDM. Insulitis developed routinely, but β-cells were not destroyed (see Table 1 and VON HERRATH and HOLZ 1997). This outcome prompted us to look for unique pathogenic changes in islets of LCMV infected transgenic mice, i.e., changes that preceded infiltration and accumulation of autoreactive LCMV-specific CTL. Indeed, these studies showed that upregulation of MHC class II and activation of macrophages occurred as soon as 4days post-infection around the islets, clearly before the first CTL and CD4 lymphocytes appeared in the pancreas (day 7). Unquestionably, these changes resulted from the presence of infectious virus in the pancreas (VON HERRATH and HOLZ 1997). Once activated lymphocytes reached the target organ, professional antigen presentation of viral antigens followed, resulting in their expansion. During this

Table 1. Transfer of syngeneic peptide-activated CTL does not induce IDDM in RIP-LCMV transgenic recipients unless B7.1 is present locally on islet cells

CTL activity (% specific ^{51}CR release) of donor splenocytes in vitro prior to transfer (H-2^b)				Recipient	Incidence	
	LCMV	NP peptide	GP peptide			
Stimulated with GP-1 and GP2 peptides	45+/−8	4+/−1	44+/−7	RIP-GP	1/5	5/5
Stimulated with GP-1 and GP2 peptides	45+/−8	4+/−1	44+/−7	RIP-NP	0/5	0/5
Stimulated with NP peptide	55+/−11	67+/−10	9+/−2	RIP-NP	0/5	4/5
Stimulated with NP peptide	55+/−11	63+/−10	4+/−2	RIP-NP/B7.1	4/6	5/6
Stimulated with TNFR peptide	12+/−1	13+/−2	1+/−1	RIP-NP	5/5	5/5

process, the viral transgene expressed on β-cells was also recognized by these lymphocytes, likely causing the direct or indirect killing of β-cells by LCMV CTL. It is not clear at this point, how much of a role cross-presentation of β-cell antigens, including the LCMV transgenes, plays in the continuation of the autoimmune process (CARBONE et al. 1998). At later stages, around the time when clinical IDDM develops and most β-cells are destroyed, islet infiltration is organized similarly to a "mini-lymph node" around a network of dendritic cells, and activated macrophages are only found in the periphery. Similar findings have been reported with the RIP-HA transgenic mouse model (SCOTT et al. 1994).

In conclusion, it appears that viral presence in the pancreas is required to locally activate APCs (VON HERRATH and HOLZ 1997). This occurs in normal non-transgenic mice as well as RIP-LCMV transgenics and must precede the "arrival" of activated autoreactive lymphocytes (that could have been primed somewhere else). Once autoreactive (LCMV-specific) lymphocytes arrive, virus is cleared (in non-transgenic as well as RIP-LCMV transgenic mice). In RIP-LCMV transgenics, these "autoreactive" LCMV specific cells continue to assault the islet cells, because the β-cells persistently express viral antigen. This process is driven by local APCs and continues until most β-cells have been destroyed and diabetes is clinically manifest. Indeed, repeated transfer of dendritic cells can "mimic" the viral presence and continuously activate enough autoreactive lymphocytes to cause diabetes in RIP-LCMV mice (LUDEWIG et al. 1998). Two important lessons can be learned from these findings. On the one hand, our study implies that local infection of an organ targeted for autoimmune attack might be required to establish an inflammatory environment (i.e., including activated APCs) that can propagate autoreactive lymphocytes once they "arrive". Second, autoimmune diseases, if triggered by a viral infection, can occur as a hit-and-run event, even after all infectious virus has been cleared by the time clinical disease is diagnosed.

2.4 Paradigm III: Numbers of Autoreactive Lymphocytes Correlate with Severity of Autoimmune Disease; After a Critical "Cut-Off" Level, Disease Progression Ceases

One important and yet unresolved issue is how many autoreactive lymphocytes it takes to generate clinical disease (STEINMAN 1996, 1999). In theory, very few antigen-specific cells might be able to selectively attack a specific organ or cell type. When this initial event leads to the release of sufficient self-antigens that are then taken up and presented by APCs in the area, the result would be the propagation of a chronic autoimmune process with antigenic and epitope spreading. This type of scenario has been observed in several animal models and is a very attractive concept to explain the development of disease (STEINMAN 1999; YANG et al. 1996; YOON et al. 1999). If only one or very few self-antigens are involved in the earliest disease process, interventions could be designed to tolerize against these antigens in a targeted way. However, the difficult feat has been to define the early or initiating autoantigens in pre-diabetic patients or in the non-obese diabetic (NOD) mouse model of spontaneous IDDM. Since the primary self-antigen is the viral transgene in RIP-LCMV mice, we attempted to determine how many antigen-specific, activated lymphocytes would be sufficient to cause type 1 diabetes. Several studies were undertaken and evidence was obtained from four different experiments:

In the first, RIP-LCMV mice were infected with vaccinia virus recombinants that expressed the LCMV proteins NP or GP. Infection with these recombinants induces about 10–50-fold fewer LCMV NP- or GP-specific CTL than regular LCMV infection. None of the mice infected with vaccinia/LCMV recombinants developed IDDM, but insulitis was regularly observed (OLDSTONE et al. 1991). A second piece of evidence emerged from studies of RIP-LCMV × RIP-B7.1 double transgenic mice. The virus-induced slow-onset IDDM typical of RIP-LCMV-NP mice with thymic expression of the transgene accelerated markedly. These different incidences of IDDM correlated well with the numbers of pCTL isolated locally from the pancreas or the spleen. Local expression of B7.1 led to a tenfold increase of pCTL in the pancreata of RIP-LCMV thymic expressor mice, which correlated with the accelerated incidence of IDDM (VON HERRATH et al. 1995b). The third set of experiments used several viral strains of LCMV, each of which was characterized by inducing high or low numbers of anti-self (LCMV-NP) pCTL in RIP-NP mice. Thereafter, the incidence of IDDM correlated very well with the amount of precursors generated. For example, IDDM occurrence was 90%–100% in transgenic mice infected with our regularly used strain, LCMV-Armstrong, which induces 1/50–1/300 LCMV-NP pCTL 7days post-infection. In contrast, LCMV Pasteur induced pCTL levels around 1/2,000, and only 50% of the RIP-NP mice developed IDDM (SEVILLA et al. 1999). When fewer than 1/6,000 pCTL were generated, no diabetes developed, as observed after infection with the LCMV-WE strain. Last, this type of "cut-off" for the number of autoreactive pCTL required to produce disease was also reflected by comparing the incidence of IDDM in RIP-LCMV thymic expressors and non-expressors on the $H-2^b$ background. Non-expressors developed IDDM within 2 weeks (1/100 pCTL), whereas thymic expressor lines did not show signs of IDDM until 2–6 months post-infection (VON HERRATH et al. 1994a,b, 1998).

Thus, a critical cut-off level (VON HERRATH et al. 1998), which depends on the numbers of autoreactive lymphocytes, appears to determine whether disease will develop or not. Several studies in animal models as well as humans have confirmed that low numbers of naive, non-activated autoreactive cells are present in the peripheral blood (BACH 1994a; BIEGANOWSKA et al. 1999; GENAIN et al. 1994; SCHOLZ et al. 1998). Their quantity or degree of activation is presumably too low to cause disease. However, if external factors [such as a virus-induced cellular event (crossactivation)] or predisposing genetic factors collide, this fragile equilibrium can be tilted toward a pathogenic process resulting in autoimmunity (KATZ et al. 1995).

The fact that the number of autoreactive lymphocytes correlates well with the incidence and severity of IDDM, which has been shown in this but also other animal models (FORSTER et al. 1995; MORGAN et al. 1996, 1998), is a finding that can be exploited therapeutically. As we found recently, MHC class I blocking peptides that bond with high affinity to the MHC D^b allele but did not engage the

TcR of LCMV H-2D^b restricted lymphocytes could, when given at high concentrations in vivo, prevent the expansion of autoreactive lymphocytes and avert IDDM in RIP-LCMV transgenic mice. LCMV-CTL levels dropped below 1/6,000 7days after infection in blocking peptide-treated mice (VON HERRATH et al. 1998). However, the design and development of peptide analogs that do not undergo rapid degradation in vivo will be necessary to make such an approach clinically feasible (GAIRIN and OLDSTONE 1992).

2.5 Paradigm IV: How β-cells Die – Essential Role of Inflammatory Cytokines

The participation of viruses in causing IDDM could take two routes. First, cross-reactivity that may involve molecular mimicry between viral and "self-" (islet) antigens could lead to direct, perforin-mediated killing of β-cells (KAGI et al. 1996, 1997). Alternatively or in addition, inflammatory cytokines secreted by such CTL and/or anti-viral CD4 lymphocytes could lead to INF-γ, tumor necrosis factor (TNF)-α and interleukin (IL)-1β-mediated death of islet cells (CAMPBELL et al. 1988a). Historically, autoreactive CTL (DILORENZO et al. 1998; VON HERRATH et al. 1994b), FAS/FAS ligand-mediated killing (AMRANI et al. 1999; CHERVONSKY et al. 1997; ITOH et al. 1997; KANG et al. 1997; KIM et al. 1999; KING et al. 1998b; RABINOVITCH et al. 1999) as well as direct effects of cytokines (CAMPBELL et al. 1988a; HORWITZ et al. 1998; PAKALA et al. 1999; RABINOVITCH et al. 1999; THOMAS et al. 1999) have been shown to destroy β cells . However, not much is known about the precise relative contributions of these factors in vivo or their importance at different stages of IDDM pathogenesis. Previous studies demonstrated the requirement for IFN-γ in type I diabetes and its role in upregulating MHC class I on β-cells in vivo (CAMPBELL et al. 1988b; THOMAS et al. 1998). The direct β-cell toxicity of IFN-γ, as well as other cytokines such as TNF-α, has also been shown in vitro (CAMPBELL et al. 1988a). CTL can rapidly produce these cytokines upon activation (SLIFKA et al. 1999), but their other task is to kill target cells expressing the appropriate MHC class I/peptide complex via the perforin granule exocytosis pathway (KAGI et al. 1994). Various lines of evidence have implicated CTL in initiation as well as effector phases of the pathogenesis typical of type I diabetes (DILORENZO et al. 1998; WONG et al. 1999). We reason that knowing their precise effector contributions in IDDM will be extremely important for judging the potential efficacy of antigen-specific interventions and for precise targeting of CTL during the optimal susceptible phase(s) in that pathogenic process.

Based on these considerations, and the fact that MHC class I-restricted CTLs capable of killing target cells through secretion of perforin granules are one of the main components of most anti-viral immune responses (KAGI et al. 1995), our goal was to evaluate the relative contribution of perforin-competent, lytic CTL and inflammatory cytokines in an antigen-specific model of autoimmune diabetes. In doing so, we demonstrated that even large numbers of autoreactive CTL are unable to lyse β-cells unless MHC class I is upregulated on islets. This necessity requires the presence of inflammatory cytokines (mainly IFN-γ) induced by viral infection of the exocrine pancreas, but not of the β-cells. Unexpectedly, in preliminary experiments, we found that the resulting perforin-mediated killing of β-cells by autoreactive CTL was not sufficient to produce clinically overt diabetes in vivo, although it was a prerequisite for the development of insulitis. In turn, the destruction of most β-cells required a direct effect of IFN-γ produced by islet-infiltrating $CD4^+$ and $CD8^+$ lymphocytes. Thus, although our studies in perforin-deficient mice clearly indicate that autoreactive CTL are needed to initiate disease in a perforin-dependent manner, destruction of most β-cells in this model occurs as cytokine-mediated death. If so, this finding has great conceptual importance in the design of suitable interventions in pre-diabetic individuals at risk of developing IDDM.

2.6 Paradigm V: Immunotherapy – How Autoreactive Regulatory Lymphocytes can Prevent or Abrogate Autoimmunity

Regulation of Immunity by Cytokines. Cytokines are important messenger and regulatory molecules that constitute a major part of the "cross-talk" network that connects lymphocytes, APCs and other cells of the immune system. In the recent past, the so-called "Th_1/Th_2 paradigm" has gained increasing importance as well as causing controversy in establishing regulators of immune processes (CHARLTON and LAFFERTY 1995; LIBLAU et al. 1995). This scheme divides cytokines into two groups: Th_1 or T_{C1} cytokines, which include IFN-γ, TNF-α, IL-12 and others, classified as inflammatory cytokines. Secretion of these cytokines results in upregulation of MHC molecules on APCs, enhancement of CTL killing, increased activation-induced cell death and proliferation and recruitment of other cells that secrete Th_1 cytokines. The main immunoglobulin subclass induced by a Th_1 response is IgG2a. In contrast, the Th_2/T_{C2} group of cytokines, which includes IL-4, IL-5, IL-6 and IL-13, is thought to have mainly a regulatory function. The latter group's foremost property is to induce dampening of Th_1 processes, but they have other tasks such as activation of B cells to secrete IgE and IgG1, enhancement of cellular proliferation and lengthening the life spans of naive lymphocytes. Already, the foregoing findings have established that the "Th_1/Th_2 paradigm" is not an entirely "black and white" situation and that a "gray zone" exists in which Th_2 cytokines might enhance or maintain an inflammatory process and Th_1 cytokines might, conversely, lead to the termination of an inflammation through induction of an activation-dependent cell death (KATZ et al. 1995; LIBLAU et al. 1995). For example, Marrack and her colleagues noted that IL-4 can significantly increase the life span of naive lymphocytes in vitro (VELLA et al. 1997). Elsewhere, increasing IL-2 levels lowered the incidence of IDDM in NOD mice due to increased apoptosis of autoreactive lymphocytes. Therefore, extensive evaluation of cytokine-dependent effects is essential because of their variable behavior from one autoimmune disease to another and among the organs affected. In the NOD mouse, Th_1 and Th_2 clones have been isolated from the pancreas and, in some instances, Th_2 clones conferred protection, whereas in other studies they were diabetogenic (HASKINS and WEGMANN 1996; D. Wegmann, unpublished data).

From this starting point, we explored the role of several cytokines in the RIP-LCMV model of virus-induced IDDM. Two routes were chosen for these experiments: First, various cytokines were expressed well above basal levels locally in the β-cells by generating RIP-cytokine transgenic mice that were then crossed with the RIP-LCMV transgenic mice. Second, cytokine "knock-out" mice were back-crossed to the RIP-LCMV transgenic mice. The incidence of spontaneous and LCMV-induced IDDM was then monitored in both groups of mice and compared. When **IFN-γ** was over-expressed in β-cells, IDDM occurred spontaneously without LCMV infection in all RIP-LCMV × RIP-IFN-γ double transgenic mice. Massive inflammation and upregulation of MHC class I molecules permeated the pancreas, and anti-LCMV CTL activated spontaneously without LCMV infection were found in the islets (LEE et al. 1995). In contrast, when RIP-LCMV IFN-γ deficient mice were infected with LCMV, IDDM did not develop (VON HERRATH and OLDSTONE 1997). Our studies showed that this was the case despite the generation of good levels of anti-LCMV (autoreactive) CTL that infiltrated the pancreas. However, the MHC class I upregulation on β-cells usually seen in RIP-LCMV transgenic mice after LCMV infection was not detectable. From these studies, we conclude that (a) IFN-β has a strong inflammatory effect by upregulation of MHC molecules and that (b) this cytokine is a prerequisite for β-cell destruction, since CTL against a self-antigen on β-cells cannot successfully destroy them in the absence of IFN-γ-induced MHC class I upregulation. These studies also illustrate the "built-in" safety mechanisms organisms have to prevent autoimmune disease: Basal MHC class I levels on β-cells appear to be insufficient to support β-cell destruction by autoreactive CTL.

IL-2 was evaluated using a double transgenic RIP-LCMV × RIP-IL-2 model. Interestingly enough, and in contrast to IFN-γ, spontaneous IDDM did not occur despite the presence of local insulitis in the absence of LCMV infection (VON HERRATH et al. 1995a). However, virus-induced IDDM was enhanced in these double transgenic mice.

Several Th$_2$ cytokines were also evaluated as described in the previous paragraphs for IFN-γ and IL-2. Interestingly, *IL-10* promoted IDDM, showing clearly that the Th$_{1/2}$ paradigm does not always apply (LEE et al. 1996). In contrast, *IL-4* prevented IDDM in our preliminary studies using the RIP-LCMV model. These findings resemble those described using NOD mice (MUELLER and SARVETNICK 1996). IL-4 prevented islet destruction under certain but not all conditions, whereas IL-10 did not prevent IDDM. Thus, the regulation of local autoimmunity is remarkably complex. In addition to the amount of the cytokine produced, the timing in relation to the stage of the autoimmune process also seems to be very important in order to an escape from or onset of disease (COPE et al. 1997; GREEN et al. 1998).

A "third class" of cytokines termed "Th$_3$", of which the main candidate is *TGF*-β, is produced predominantly by lymphocytes in the gut and may have a regulatory function, particularly in response to orally administered antigens

(BRIDOUX et al. 1997; KING et al. 1998b). Several studies by Weiner's group and members of our laboratory have found a predominance of TGF-β producing lymphocytes in the pancreata of mice treated orally with insulin, which prevented IDDM (MILLER et al. 1993) (see also the paragraph on treatment of autoimmune diseases). However, upstream factors such as IL-4 probably precede the involvement of TGF-γ in regulating autoimmunity (HOMANN et al. 1999; KING et al. 1998a).

When direct immune-histochemical or ELISPOT analyses of cytokines were performed in the islets during these studies, an interesting scheme emerged. Mice that were pre-diabetic or had received successful immune-intervention therapy had more IL-4 than γ-IFN in their islets (VON HERRATH et al. 1995b, 1996b). In contrast, mice with IDDM had a predominance of γ-IFN. This outcome indicates a potential regulatory or islet-protective role for certain cytokines such as IL-4 and provides a mechanism for implementing immune intervention to influence the local inflammatory process in pancreatic islets, as discussed more fully in the following paragraphs.

2.7 How do Regulatory, Autoreactive Lymphocytes Work? The Concept of Bystander Suppression

This apparent regulation of type 1 diabetes by cytokines is amenable to therapeutic exploitation. Although autoreactive lymphocytes are usually thought of as destructive mediators of autoimmune disease, some effectively ameliorate or prevent it – the so-called regulatory cells. These autoreactive lymphocytes are frequently characterized by the secretion of IL-4, IL-10 and TGF-β. In type 1 diabetes, splenic β cell-reactive CD4+ T cells as well as thymic regulatory cells prevented diabetes after adoptive transfer into prediabetic NOD mice (AKHTAR et al. 1995; HERBELIN et al. 1998; SEMPE et al. 1994). A promising strategy to induce and expand populations of such regulatory cells in vivo for therapeutic purposes has been called "oral tolerance", which refers to the prevention and treatment of autoimmune diseases by the oral administration of antigens related to autoimmune target tissue-specific antigens. Although high-dose regimens resulted in the deletion or anergy of autoreactive lymphocytes specific for the administered antigen, low to medium dosages represented a form of mucosal immunization, since their success relied on the induction of antigen-specific regulatory cells (WEINER 1997). In several animal models of autoimmune disease, including type 1 diabetes, the generation of such regulatory cells by oral immunization ("oral tolerance") was described: Initially, these cells were reported to be CD8 T cells (LIDER et al. 1989; MILLER et al. 1991; NUSSENBLATT et al. 1990); however, subsequent publications documented disease suppression by CD8 and CD4 T cells in experimental autoimmune encephalitis (EAE) (CHEN et al. 1994, 1995) as well as regulatory CD4 T cells in collagen-induced arthritis (TADA et al. 1996), diabetes (BERGEROT et al. 1994) and tolerance to ovalbumin (GARSIDE and MOWAT 1997). Intestinal γδTcR T cells (reviewed by KAPP and KE 1997) have also mediated oral tolerance in the course of experimental autoimmune uveitis (WILDNER et al. 1996). CD8/γδTCR T cells capable of preventing diabetes in non-obese diabetic (NOD) mice were also noted after aerosol insulin treatment (HARRISON et al. 1996). However, not much is known about their antigenic specificities or effector mechanism(s).

The special immune environment for lymphocyte priming at large mucosal surfaces may be critical in the generation of regulatory cells characterized by their ability to secrete cytokines such as IL-4, IL-10 and TGF-β. Such cells can act in a suppressive regulatory manner if they home to a site targeted for autoimmune attack and, thereby, prevent disease. This mechanism has been termed "*bystander suppression*" if the specificity of regulatory cells differs from that of the autoreactive destructive lymphocytes (Fig. 2). "Bystander suppressor" cells, which by definition must recognize an autoantigen different from the initiating self-antigen (Fig. 2), have been suggested to act in EAE (MILLER et al. 1991) and arthritis (YOSHINO 1996). They were also found in our virus-induced diabetes model (HOMANN et al. 1999; VON HERRATH et al. 1996a). This and further observations in a transgenic TcR system (CHEN et al. 1994) argue that antigen-specific autoreactive T cells can differentiate in vivo into either autoaggressive or regulatory cells depending on the site and context of the initial priming event.

2.8 Induction of Regulatory Cells by Oral Antigens and DNA Vaccination

Oral Self-Antigens Prevent Autoimmune Diabetes. To address the potential mechanism of "bystander suppression" in type 1 diabetes, we took advantage of the well-defined primary self (LCMV) antigen in the RIP-LCMV transgenic system of virus-induced autoimmunity (VON HERRATH et al. 1994b). In mice of the transgenic

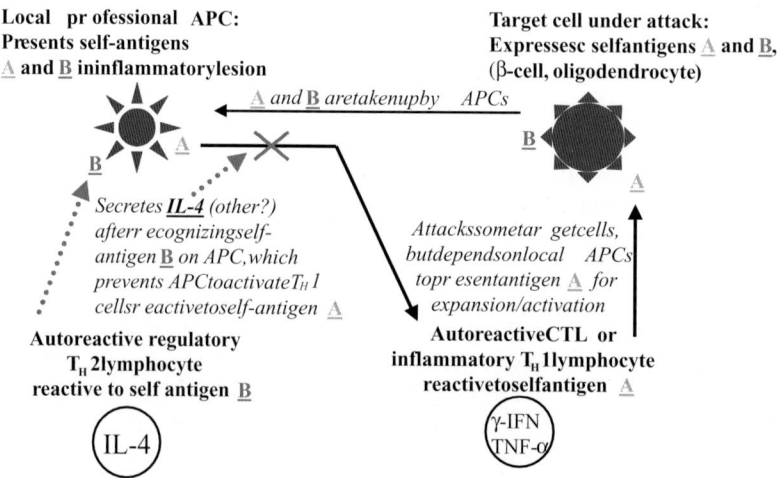

Fig. 2. The concept of bystander suppression

RIP-NP line used for these studies, the development of disease is dependent on NP specific CD8+ T cells, CD4+ T helper cells as well as IFN-γ (VON HERRATH et al. 1996a). We showed that orally administered insulin can prevent diabetes in these mice. This protection was associated with a "mixed" "T_H1/T_H2" type cytokine profile in the pancreas and a local (pancreas) but not systemic abrogation of NP-specific ("anti-self") CTL activity. We next derived cell lines from pancreatic draining lymph nodes of RIP-NP mice protected from type 1 diabetes by porcine insulin feedings. These lymphocytes were then cultured in the presence of the immune-dominant insulin B-chain (HOMANN et al. 1999) or B-chain peptides. Analysis of these established cell lines revealed a cytokine profile similar to that reported for other regulatory cells, namely production of IL-4, IL-10 and some IFN-γ (HOMANN et al. 1999). When transferred intravenously into pre-diabetic RIP-NP mice, the regulatory cells proliferated only in the pancreatic draining nodes of recipients (Fig. 3) and completely protected them from autoimmune diabetes (Table 2) by locally suppressing the virus (self)-specific, "diabetogenic" T cell responses (Fig. 4). IL-4 and stat6, a critical component of the IL-4 signaling pathway, were essential for the induction of regulatory cells as demonstrated by the failure of oral insulin treatment to efficiently protect RIP-NP X IL-4$^{-/-}$ or RIP-NP X stat6$^{-/-}$ mice from diabetes. In this instance, the prevention of diabetes likely occurred through IL-4/stat6-mediated modulation of APCs resulting in decreased expansion of and cytokine secretion by autoreactive CD8+ and CD4+ T cells. These findings provide direct evidence for "bystander" suppression in a dual self-antigen specific model.

Immunization with Plasmid DNA Expressing Self-Antigens. The next approach was vaccination with DNA plasmids expressing self-antigens. Mice expressing LCMV NP as a transgene in their β-cells developed IDDM only after LCMV infection. Inoculation of plasmid DNA encoding the insulin B-chain reduced the incidence of virally induced autoimmune diabetes (IDDM) by 50% in this model (Fig. 5).

CD4+ regulatory T cells from protected donor

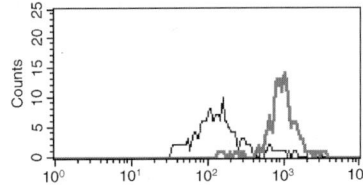

CD4+ regulatory T cells from diabetic donor

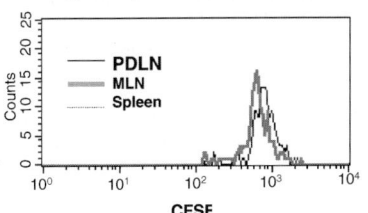

Fig. 3. Regulatory T cells will only proliferate in the pancreatic draining lymph node but not any other lymphoid sites after transfer into prediabetic RIP-LCMV recipients. For the experiment shown, regulatory T cells were labeled in vitro with CSFE dye and their divisions were tracked in vivo by FACS analysis. A shift to the *left* indicates proliferation. *PDLN*, pancreatic draining lymph node; *MLN*, mesenteric lymph node

Table 2. Adaptive transfer of insulin-B specific CD4 regulatory lymphocytes that produce IL-4 but low amounts of interferon-gamma can protect RIP-LCMV recipients from autoimmune diabetes[a]

RIP-NP mice receiving transfer 5 days post LCMV	Cells stimulated in the presence of	Cytokines produced by transferred cells		Outcome (% IDDM)
		IFN-γ	IL-4	
No transfer	N/A	N/A	N/A	100%
5×10^5 CD4 + *lymphocytes* from pancreatic LN of protected mouse	Porcine-B + IL-4	0.12	0.1	0%
	Porcine-B no IL-4	0.12	0.0	100%
5×10^6 *splenocytes* from protected mouse (oral insulin)	Porcine B-chain	0.33	0.9	50%
	LCMV-NP (118–126)	3.2	0.16	100%
5×10^6 splenocytes from non-protected mouse (oral insulin)	Porcine B-chain	1.4	0.11	83%
	LCMV-NP (18–126)	2.8	0.13	100%

[a] Note that propagation of regulatory lymphocytes on the "regulatory" insulin-B autoantigen resulted in protective cell lines, whereas their propagation on the "aggressive" LCMV NP (118–126) antigen resulted in non-protective cultures.

Much like the success of oral insulin treatments, vaccination with insulin B-chain DNA was effective through induction of anti-self, regulatory CD4 lymphocytes that reacted with the insulin B-chain, secreted IL-4 and locally reduced the activity of LCMV-NP autoreactive CTL in the pancreatic draining lymph nodes (COON et al. 1999). In contrast, similar vaccination with plasmids expressing the LCMV viral ("self-") protein did not prevent IDDM, because no such regulatory cells were induced (Fig. 5). Thus, DNA immunization with plasmids expressing self-antigens might constitute novel and attractive therapies for preventing autoimmune diseases, if the antigens are carefully pre-selected for the ability to induce regulatory lymphocytes in vivo. On the basis of our findings, those self-antigens that are primarily involved in the destructive inflammatory autoimmune process are probably not suited for induction of regulatory lymphocytes (such as the LCMV self-antigens). In contrast, antigens not involved in this primary autore-

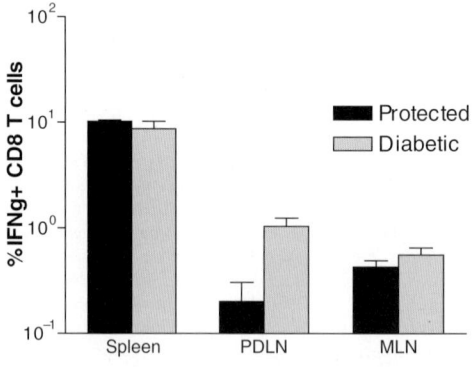

Fig. 4. Regulatory cells only suppress the autoaggressive, LCMV specific response in RIP-LCMV prediabetic mice in the pancreatic draining lymph node, but not in spleen or other lymphoid organs. Thus, the figure shows a selective reduction in auto-aggressive CD8 lymphocytes as evidenced by their antigen specific interferon-γ production, in the PDLN of a RIP-LCMV recipient that was protected from T1D after transfer of regulatory lymphocytes

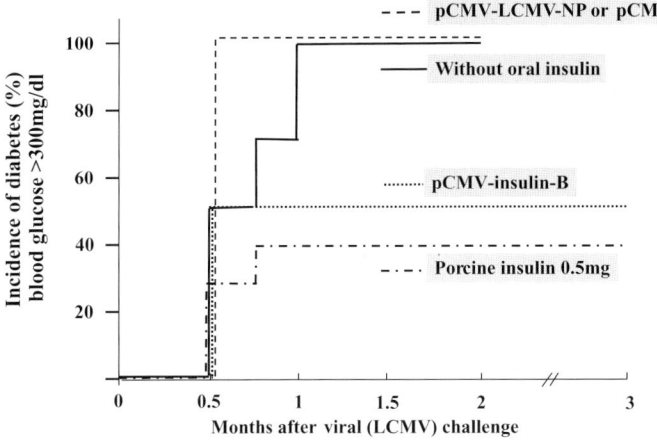

Fig. 5. DNA vaccination with the insulin B-chain but not LCMV-NP expressing plasmids prevents IDDM in RIP-LCMV mice

active process, such as insulin, are good candidates for the induction of bystander suppressor or regulatory cells (COON et al. 1999).

2.9 Summary and Open Questions

The RIP-LCMV transgenic mouse used as a model of virally induced autoimmune diabetes enabled us to define important pathogenic activities that could resemble those in human type 1 diabetes. The pathogenic process leading to IDDM can be divided into at least three distinct phases:

Phase 1: Local APC activation caused by viral infection and/or enhanced by genetic predisposition. No or very minor β-cell destruction (2 days post-infection).

Phase 2: Entry of autoreactive CD4 and/or CD8 cells into the pancreas and islets. Initiation of islet destruction potentially involving perforin. No clinically overt diabetes yet (7 days post-infection).

Phase 3: Propagation of islet infiltration by activated APCs in the islets and draining lymph node. Progressive β-cell destruction mainly through inflammatory cytokines leading to overt IDDM (10 days up to months post-infection).

First, APC activation including upregulation of MHC class II molecules must occur before autoreactive lymphocytes reach the pancreas. This might be achieved through an acute or chronic viral infection or enhanced by genetic predisposition. In the second phase, activated, autoreactive lymphocytes enter the pancreas and the islets. Their activation can occur systemically or locally in the pancreas and draining node via molecular mimicry or bystander activation. Third, local APC can then expand the population of autoreactive lymphocytes, promoting chronic infiltration and β-cell destruction. The autoimmune process next spreads to islet autoantigens other than the initiating one. As a consequence, most β-cells die, likely through the presence of inflammatory cytokines, not perforin- or FAS-mediated lysis by CTL. Treatment of pre-diabetic mice that have already entered phase 3 is possible by activating autoreactive regulatory lymphocytes. These cells expand selectively in the pancreatic-draining node, where they encounter activated APCs expressing islet antigens and act as "bystander suppressors": Secretion of IL-4 and probably other mediators by such cells dampens the expansion and activation of LCMV-specific autoaggressive CD4 and CD8 lymphocytes locally, which leads to abrogation of the autoimmune process. As outlined in this chapter, such immune interventions are an attractive possibility, but many questions must be answered before such a strategy reaches the clinic.

3 Section 2: Regulation of Virally Induced Immunopathologic Effects

Each viral infection, even if the virus is not lytic and causes only limited direct damage to infected cells, induces a certain degree of immunopathologic effects. Factors influencing the amount of such immune-mediated damage, as discussed in the following section, include the magnitude of an anti-viral response, the cytokines and chemokines produced and the numbers and types of cells infected by the virus. Since many immune functions are controlled and restricted by host genetic factors, every individual responds uniquely to an infection. The outcome, therefore, may vary greatly. Understanding the factors that determine the degree of immunopathologic effects resulting from a given viral encounter would be of great value for devising appropriate immune interventions to modify/improve the outcome. In principle, such interventions would be guided by paradigms much like those described in the previous section. However, the clinical situation is potentially even more complex, because even though dampening the anti-viral response should lower the immune-mediated damage, such suppression might also impede viral elimination. No reliable data yet exist to predict the potential of such interventions in humans. Studies of experimental animals are just beginning to teach us how to influence the course of viral infections toward a beneficial outcome. As this section portends, in the long term such immunomodulatory treatments might have a good chance to achieve clinical reality.

3.1 Numbers of Anti-Viral Lymphocytes

In resolving an acute viral infection, therapies that generate a rapid response of high magnitude are generally the most beneficial. For example, most anti-viral vaccines derive their protective effect from priming anti-viral memory lymphocytes that, upon re-encountering the same antigen, respond more rapidly than naive unprimed lymphocytes (ALEXANDER et al. 1998; BONA et al. 1998; BRANDER and WALKER 1999; FU et al. 1999; LUDEWIG et al. 1999; RODRIGUEZ et al. 1997). This connection has been proven in numerous situations. However, once a virus has produced a prolonged and persistent infection, the paradigm shifts (GALLIMORE et al. 1998; MOSKOPHIDIS et al. 1995). At this point, one cannot easily predict whether a strong response will be beneficial or detrimental. For example, infection by the human immunodeficiency virus (HIV) induces a very strong CTL response that cannot eliminate the virus but induces considerable immunologic damage (BORROW et al. 1997; CAO et al. 1997; GANDHI et al. 1998; ZINKERNAGEL and HENGARTNER 1994). Further, the location of the viral infection is important. In this respect, the brain and perhaps other specialized organs or cells, such as the pancreatic islets, apparently do not tolerate well any type of immune response (GAMBLE et al. 1973; GLADISCH et al. 1976; KATZ-LEVY et al. 1999; PROBERT et al. 1997; YOKOYAMA et al. 1995; YOON et al. 1979). Hosts infected at these sites might benefit most from lowering the persisting anti-viral response to limit excessive immunopathologic effects (OLDSTONE et al. 1999). Indeed, such intervention has drastically improved the outcome of a cerebral viral infection or peripheral persisting viral infection in some experimental animals. In those studies, done in the LCMV model, persistent infection of tolerant mice (infected at birth) did not cause excessive immunologic damage (ZAJAC et al. 1998), whereas persistent infection in the presence of an ongoing immune response (e.g., as in mouse strains lacking perforin or IFN-γ) (BINDER et al. 1998; VON HERRATH et al. 1997) eventually led to the animals' death (SOUTHERN and OLDSTONE 1986). One can, therefore, assume that for LCMV, and probably other infections, a balance is struck between viral load, ongoing immunity, the class of immune response and the outcome of infection, as depicted in the following table, which shows issues in antigen-specific immunotherapy used to treat or prevent autoimmune diseases:

1. Exogenous antigen dose	High dose	Deletion/anergy
	Intermediate	Immunization/regulation
	Low	No effect
2. Endogenous self-antigen level	High	Earlier response, destructive
	Low	Later response, regulatory
3. Route and mode of immunization	Oral	Regulation, Th2
	DNA	Regulation/destruction
	Adjuvants	Destructive (?), Th1
4. Autoreactive T-cell repertoire	High avidity	Destructive (?)
	Low avidity	Regulatory (?)
	Thymic deletion	Magnitude

Based on these considerations, one can hypothesize that dampening the antiviral response might be of benefit in situations involving proven excessive immunopathologic effects caused by anti-viral lymphocytes. Indeed, some experimental examples from animal models demonstrate that eliminating or blocking the activity of cytotoxic anti-viral lymphocytes can drastically improve the outcome of an otherwise destructive infection. In these studies, a blocking peptide for one

MHC class I allele (D^b) was given daily to mice that were suffering from the normally lethal intracerebral infection with LCMV (OLDSTONE et al. 1999). These animals' survival rate increased significantly due to a reduction in the CTL response to LCMV (from blocking of the MHC class I peptide/TcR interaction) and consequent prevention of cytokine-mediated CNS inflammation. Of course, these findings not only illuminate the potentially beneficial aspect of reducing an antiviral response, but at the same time make us understand the finely tuned kinetics that govern most if not all such immune responses. For this reason, such interventions are still relatively distant from any clinical application. First we must develop markers that signal when overly strong suppression of an anti-viral response is detrimental. Additionally, the persistence of a viral infection could, by itself, lead to chronic inflammation (DE LA TORRE et al. 1991; OLDSTONE 1990). However, in the above-mentioned experimental scenario, clearance of LCMV was delayed but not abrogated, which was the desired outcome.

Another study from our laboratory employed a different strategy to achieve a similar outcome (reduction of immunopathologic effects). Perforin-deficient mice, when persistently infected with LCMV, usually succumb to lethal immunopathologic effects within 2–3 months after infection. In these experiments one viral protein, LCMV NP, was expressed in the thymus of perforin-deficient mice leading to the elimination of most but not all high-affinity anti-LCMV-NP CTL (VON HERRATH et al. 1999). As an intriguing consequence, immune responses to other viral proteins, i.e., LCMV GP, were not augmented as compensation, and the ensuing reduction in immunopathologic effects resulted in a 50% increased survival rate. Of course, one cannot yet express viral proteins reliably in the thymus as a therapeutic asset, but future technologies using special gene delivery vehicles might make this feasible. Additionally, the engineering of special APCs, i.e., previously described "killer" dendritic cells (MATSUE et al. 1999), might even enable us to eliminate a lymphocyte subset via an antigen-specific agent during ongoing infection to reduce immunopathologic effects.

3.2 Costimulation and Infection of Antigen-Presenting Cells

The extent of "professional" APCs infection is definitely of great importance in determining the outcome of a viral infection. For example, infection with certain strains of LCMV (i.e., the LCMV variant Clone 13) causes more CD11c positive dendritic cells to become infected than do other strains (i.e., LCMV Armstrong). When more dendritic cells are infected, the anti-viral response is lost over time, because these cells are compromised by the infection or killed by anti-viral CTL (BORROW et al. 1995). Persistent viral infection results. In contrast, infection of fewer dendritic cells results in a more 'gentle' immune response with comparatively slower kinetics that can still eliminate the infection but at a lower 'cost': Losing fewer dendritic cells, in this situation, enables the host organism to sustain the anti-viral response until every viral particle has been eliminated (BORROW and OLDSTONE 1997). Directly or even indirectly influencing how many APCs are infected by a given virus would be very difficult, because so many factors engage in this event. Minute differences in a virus' genomic sequence (LCMV Armstrong and Clone 13 strains differ in only a few amino acids, of which aa260 in the glycoprotein is the essential residue to which the differing infection kinetics are mapped) can translate into huge biological differences (AHMED et al. 1984; WONG et al. 1996). Further, predicting the quantity and efficacy of receptor binding for most viruses is still difficult, because alternative receptors and multiple binding sites most often play a role (CAO et al. 1998). Further, once APCs (or other cells) become infected, it is very difficult, if not impossible, to predict how well viral antigens or peptides will be presented. Most viruses possess more than one gene or protein that can interfere with antigen presentation at various sites. For example, herpesviruses, which are 'experts' in persisting, have multiple genes that affect peptide transport and can, in this way, lower recognition of infected cells by CTL (FRUEH et al. 1995). The concept of the "Trojan horse", a metaphor for viruses that use certain cells (FINZI et al. 1999) as a vehicle for traveling to sequestered sites (e.g., in HIV infection of the brain), further complicates any therapeutic approach that targets antigen presentation to reduce immunopathologic effects. Generalized systemic immunosuppression is not an attractive strategy, because responses to other pathogens are compromised, and an already weakened organism would become even more vulnerable to infectious attacks. One important issue with selective (see previous paragraph) as well as systemic immunosuppression during viral infections is that, generally speaking, increasing viral titers predict a poor outcome of infection. Thus, lowering viral titers concurrently with selective immunesuppressive interventions will probably be important for reducing the anti-viral response enough to benefit patients in whom an over-zealous immune response is harmful. Other types of therapy might be directed toward establishing a peaceful coexistence between host and virus in those situations when the infection has very few or no adverse effects on cells, e.g., for some non-lytic viruses. Then, a decrement of viral load might not be as necessary as for more aggressive infections that can interfere with basic or "luxury" cellular functions.

3.3 Cytokines and Chemokines as Inducers and/or Regulators of Immunopathologic Effects

Certainly, pro-inflammatory cytokines such as IFN-γ and TNF-α can induce substantial degrees of tissue and organ damage, via mechanisms much like those operative in autoimmune diabetes (GUIDOTTI et al. 1996; VON HERRATH et al. 1999). However, counter-regulation with regulatory cytokines such as IL-4 appears to be much less efficient, possibly because the magnitude and commitment of the anti-viral response are frequently high when IFN and TNF are involved. For example, we attempted to exert a positive influence on pulmonary infection with influenza virus (a mouse-adapted PR8 isolate) by over-expressing IL-4 in the lung (BOT et al. 2000). Although the generation of anti-viral memory CTL and Th_1 lymphocytes decreased, the survival rate of these infected mice did not improve. Viral clearance was marginally delayed. Comparing these findings to the extremely beneficial effects of IL-4 on the course and severity of autoimmune diabetes when expressed locally in islets, one must note the lack of efficacy in the latter situation (HOMANN et al. 1999). Perhaps the Th_1 or Tc_1 lymphocytes, once committed to their phenotype, were not easily (if at all) 'turned around' to assume a Th_2 or Tc_2 cytokine profile (HUANG and PAUL 1998; MOSMANN and COFFMAN 1989; ROCKEN et al. 1996; SALLUSTO et al. 1998), a notion based on the responses of Th_1 and Th_2 cell lines exposed to various cytokines in vitro. The other and more likely explanation is that anti-viral responses are not usually governed by simple regulatory circuits such as the Th_1/Th_2 paradigm. Indeed, many viral infections induce a mixed cytokine profile that consists of a diverse *Gemisch* of Th_1, Th_2 and other cytokines/chemokines. No firm evidence has yet confirmed that counter-regulatory mechanisms play a large role. In contrast, cellular (Th_1-driven) as well as humoral (Th_2-driven) antibody responses must frequently combine to efficiently combat an infection making both indispensable. A last consideration would be to selectively reduce immunopathologic effects by identifying the most predominantly harmful cytokines and systemically antagonize their generation or receptor binding. This practice, although already in clinical use for rheumatoid arthritis (TNF receptor antagonists) (MAINI et al. 1997), has not been attempted in models of viral infection and little is known about its potential. However, systemic reduction of one or several cytokines is likely to have side effects, and because of the inherent redundancy and diversification of anti-viral cytokine responses, would bear few benefits.

3.4 Conclusions

The components of anti-viral responses resemble those of other autoimmune processes. However, the relative influence of these "players" on the outcome of immunopathologic effects differs. Apparently, the reduction of antigen-specific lymphocytes, which is difficult to achieve in autoimmune diseases because the pathogenically important self-antigens are frequently unknown, can have very positive effects. Precise understanding of the underlying kinetics is, however, required before directing such therapy to the small margin between improvement of immunopathologic effects and viral persistence. In contrast, cytokine modulatory approaches, which appear to be beneficial for autoimmune diseases, would not improve the outcome of most viral infections, because such a variety of regulatory and inflammatory cytokines is required to achieve effective cellular as well as humoral immune responses.

4 Future Considerations – Is it Possible to Model and Predict Immunopathology in Autoimmunity or Viral Infections?

As described in this chapter, our work has shown that the amount of immunopathology or tissue injury is determined not only by the magnitude of a localized

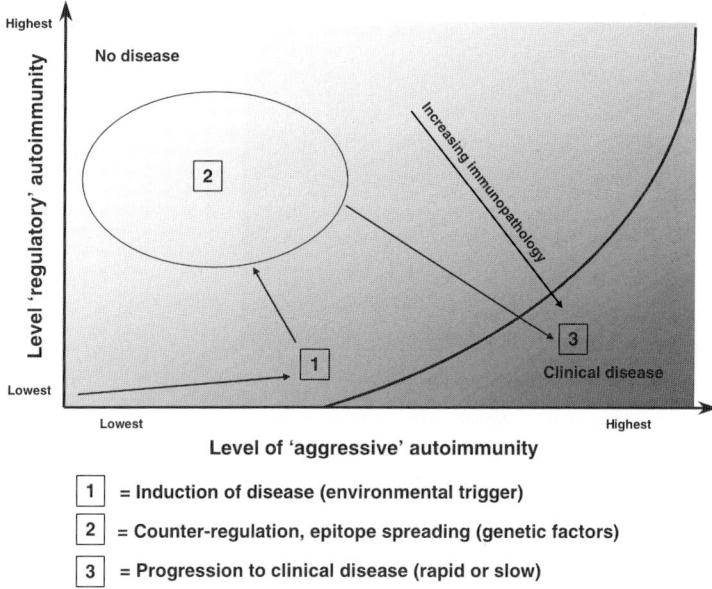

Fig. 6. Regulation of autoimmunity as a function of autoaggressive and autoreactive regulatory responses. The goal of immune-based interventions is to preserve stage 2 and prevent its progressing towards the clinical stage 3. Molecules known to be instrumental in this decision are inflammatory and regulatory cytokines, chemokines, adhesion molecules, and the activation profile of autoreactive lymphocytes as well as antigen-presenting cells. Many of these molecules can have beneficial or detrimental effects based on the time and level of expression in relation to the ongoing disease process. Due to the complexity of this situation, it has therefore been very difficult to make good predictions about the safety and efficacy of a given approach. Importantly, many of these molecules can be assessed as markers in the peripheral blood and could potentially be used as a basis for a predictive model that would allow a rating of the success of immune interventions

or systemic immune process, but also, to a large extent, by its components or the class(es) of responses it encompasses. Thus, each immune or autoimmune reaction has at least a more aggressive and a more regulatory component that balance each other and in this way have a strong effect on the duration or magnitude of the response and the resulting tissue injury (Fig. 6). In autoimmune diseases, it is possible to take therapeutic advantage of this paradigm and generate autoreactive regulatory cells by targeted immunization with self-antigens. We have shown that such cells can be induced by oral immunization, DNA vaccines and peripheral immunization and are able to selectively suppress an ongoing autoimmune reaction, because these cells are preferentially retained in the draining lymph node closest to the target organ where they exert their regulatory function. Similarly, dampening the anti-viral response in persistent infections can ameliorate viral immunopathology (Fig. 7). This can be achieved with blocking peptides or, more recently, using "killer" dendritic cells that both abrogate anti-viral lymphocytes. It is important that viral replication is curtailed at the same time using drug-therapy.

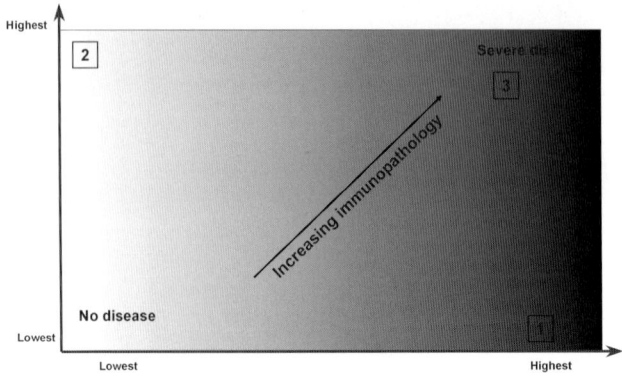

Fig. 7. Regulation of virally induced immunopathology as a function of viral load and the magnitude of the immune response. Similar to the situation in autoimmunity, augmenting or decreasing the anti-viral response during an ongoing infection can be beneficial or detrimental. Many viral infections will not fall under the extremes 1–3 indicated above, but in the "middle section." Since the viral load is a function of the efficacy of the immune response and concomitant anti-viral drug therapy, the prediction of the outcome of immune dampening or enhancing interventions is complex. It depends on the replication rate of the virus, amount of antigen-presenting cells infected, lytic damage of the viral infection to the host cell and the precise kinetics and effector molecules of the anti-viral response (cytokines, perforin, FAS etc.). Many of these factors have been characterized in experimental models and could form the basis for designing appropriate predictive model systems

Current studies are dissecting the precise mechanism(s) of action of regulatory cells (modulation of antigen-presenting cells, cytokines/chemokines, cell contact inhibition) as well as the requirements for their induction (endogenous autoreactive regulatory T-cell repertoire; route and dose of external antigen administration; expression level and involvement of the endogenous self-antigen in the autoimmune process). Paradigms developed from these studies will not only be useful in suppressing autoimmune diseases very selectively, but are likely to be useful in lowering the immunopathology that accompanies viral infections as well.

The molecular understanding of immune responses progresses at a vary rapid pace. Therefore, it is frequently impossible to make simple predictions, because the number of molecules and cells involved is too high and their interactive network is too complex. Furthermore, the relative contribution of the different 'players' has to be taken into account as a major factor and this is probably at least one of the reasons, why different research teams are frequently reporting seemingly opposing or conflicting results. Such issues might profit from appropriate mathematical or other computer-based modeling systems, which would ultimately allow us to more reliably predict the outcome of interventions for viral infections or autoimmune syndromes.

Acknowledgements. This is Publication No. 13038-NP from the Department of Neuropharmacology, The Scripps Research Institute, La Jolla, California. This work was supported by NIH grants R01AI4415 and R29 DK51091 to M.G.V. M.G.V. is also supported by a Juvenile Diabetes Foundation International Career Development Award No. 296120. We thank and Diana Frye for wonderful assistance with the manuscript preparation.

References

Ahmed R, Salmi A, Butler LD, Chiller JM, Oldstone MB (1984) Selection of genetic variants of lymphocytic choriomeningitis virus in spleens of persistently infected mice. Role in suppression of cytotoxic T lymphocyte response and viral persistence. J Exp Med 160:521–540

Akhtar I, Gold JP, Pan LY, Ferrara JL, Yang XD, Kim JI, Tan KN (1995) CD4+ beta islet cell-reactive T cell clones that suppress autoimmune diabetes in nonobese diabetic mice. J Exp Med 182:87–97

Alexander J, Fikes J, Hoffman S, Franke E, Sacci J, Appella E, Chicari FV, Guidotti LG, Chestnut RW, Livingston B, Sette A (1998) The optimization of helper T lymphocyte (HTL) function in vaccine development. Immunol Res 18:79–92

Amrani A, Verdaguer J, Anderson B, Utsugi T, Bou S, Santamaria P (1999) Perforin-independent beta-cell destruction by diabetogenic CD8+ T lymphocytes in transgenic nonobese diabetic mice. J Clin Invest 103:1201–1209

Andersen O, Lygner PE, Bergstrom T, Anderson M, Vahlne A (1993) Viral infections trigger multiple sclerosis relapses: a prospective seroepidmiological study. J Neurol 240:417–422

Bach J-F (1994a) Insulin dependent diabetes mellitus as an autoimmune disease. Endocrine Reviews 15:516–542

Bach J-F (1994b) Predictive medicine in autoimmune diseases: From the identification of genetic pre-disposition and environmental influence to precocious immunotherapy. Clin Immunol Immunopath 72:156–161

Bach J-F, Mathis D (1997) The NOD Mouse. Res Immunol 148:281–370

Bennett SRM, Carbone FR, Karamalis F, Miller JFAP, Heath WR (1997) Induction of a CD8+ cytotoxic T lymphocyte response by cross-priming requires cognate CD4+ T cell help. J Exp Med 186:65–70

Bergerot I, Fabien N, Maguer V, Thivolet C (1994) Oral administration of human insulin to NOD mice generates CD4+ T cells that suppress adoptive transfer of diabetes. 7:655–663

Bieganowska K, Hollsberg P, Buckle GJ, Lim D-G, Greten TF, Schneck J, Altman JD, Jacobson S, Ledis SL, Hanchard B, Chin J, Morgan O, Roth PA, Hafler DA (1999) Direct analysis of viral-specific CD8+ T cells with soluble HLA-A2/Tax11–19 tetramer complexes in patients with human T cell lymphotropic virus-associated myelopathy. J Immunol 62:1765–1771

Binder D, van den Broek MF, Kagi D, Bluethmann H, Fehr J, Hengartner H, Zinkernagel RM (1998) Aplastic anemia rescued by exhaustion of cytokine-secreting CD8+ T cells in persistent infection with lymphocytic choriomeningitis virus. J Exp Med 187:1903–1920

Bona CA, Casares S, Brumeanu T-D (1998) Towards development of T-cell vaccines. Immunol Today 19:126–132

Borrow P, Evans CF, Oldstone MB (1995) Virus-induced immunosuppression: immune system-mediated destruction of virus-infected dendritic cells results in generalized immune suppression. J Virol 69:1059–1070

Borrow P, Lewicki H, Wei X, Horwitz MS, Peffer N, Meyers H, Nelson JA, Gairin JE, Hahn BH, Oldstone MBA, Shaw GM (1997) Antiviral pressure exerted by HIV-1-specific cytotoxic T lymphocytes (CTLs) during primary infection demonstrated by rapid selection of CTL escape virus. Nature Med 3:205–212

Borrow P, Oldstone MBA (1997) Lymphocytic choriomeningitis Virus. In: Nathanson N (ed) Viral pathogenesis. Lippincott-Raven, Philadelphia, pp 593–627

Bot A, Holz A, Christen U, Wolfe T, Temann A, Flavell R, von Herrath MG (2000) Local IL-4 expression in the lung reduces pulmonary influenza-virus-specific secondary cytotoxic T cell responses. Virology (in press)

Brander C, Walker BD (1999) T lymphocyte responses in HIV-1 infection: implications for vaccine development. Curr Opin Immunol 11:451–459

Bridoux F, Badou A, Saoudi A, Bernard I, Druet E, Pasquier R, Druet P, Pelletier L (1997) Transforming growth factor beta (TGF-beta) -dependent inhibition of T helper cell 2 (Th2)-induced autoimmunity by self-major histocompatibility complex (MHC) Class II-specific regulatory CD4$^+$ T cell lines. J Exp Med 185:1769

Campbell IL, Iscaro A, Harrison LC (1988a) IFN-gamma and tumor necrosis factor-alpha. Cytotoxicity to murine islets of Langerhans. J Immunol 141:2325–2329

Campbell IL, Oxbrow L, West J, Harrison LC (1988b) Regulation of MHC protein expression in pancreatic beta-cells by interferon-gamma and tumor necrosis factor-alpha. Mol Endocrinol 2: 101–107

Cao H, Kanki P, Sankale J-L, Dieng-Sarr A, Mazzara GP, Kalams SA, Korber B, Mboup S, Walker BD (1997) Cytotoxic T-lymphocyte cross-reactivity among different human immunodeficiency virus type 1 clades: Implications for vaccine development. J Virology 71:8615–8623

Cao W, Henry MD, Borrow P, Yamada H, Elder JH, Ravkov EV, Nichol ST, Compans RW, Campbell KP, Oldstone MBA (1998) Identification of alpha-dystroglycan as a receptor for lymphocytic choriomeningitis virus and Lassa fever virus. Science 282:2079–2081

Carbone FR, Kurts C, Bennett SRM, Miller JFAP, Heath WR (1998) Cross-presentation: a general mechanism for CTL immunity and tolerance. Immunol Today 19:368–373

Charlton B, Lafferty KJ (1995) The Th1/Th2 balance in autoimmunity. Curr Opin Immunol 7: 793–798

Chatenoud L, Thervet E, Primo J, Bach JF (1994) Anti-CD3 antibody induces long-term remission of overt autoimmunity in nonobese diabetic mice. Proc Natl Acad Sci USA 91:123–127

Chen Y, Inobe J, Weiner HL (1995) Induction of oral tolerance to myelin basic protein in CD8-depleted mice: both CD4+ and CD8+ cells mediate active suppression. J Immunol 155:910–916

Chen Y, Kuchroo VK, Inobe JI, Hafler DA, Weiner HL (1994) Regulatory T cell clones induced by oral tolerance: Suppression of autoimmune encephalomyelitis. Science 265:1237–1240

Chervonsky AV, Wang Y, Wong FS, Visintin I, Flavell RA, Janeway Jr CA, Matis LA (1997) The role of Fas in autoimmune diabetes. Cell 89:17–24

Coon B, An L-L, Whitton JL, von Herrath MG (1999) DNA immunization to prevent autoimmune diabetes. JCI 104:189–194

Cope AP, Liblau RS, Yang XD, Congia M, Laudanna C, Schreiber RD, Probert L, Killias G, McDevitt HO (1997) Chronic tumor necrosis factor alters T cell responses by attenuating T cell receptor signaling. J Exp Med 185:1573–1584

de la Torre JC, Borrow P, Oldstone MBA (1991) Viral persistence and disease: cytopathology in the absence of cytolysis. British Medical Bulletin 47:838–851

DiLorenzo TP, Graser RT, Ono T, Christianson GJ, Chapman HD, Roopenian DC, Nathenson SG, Serreze DV (1998) Major histocompatibility complex class I-restricted T cells are required for all but the end stages of diabetes development in nonobese diabetic mice and use a prevalent T cell receptor alpha chain gene rearrangement. Proc Natl Acad Sci USA 95:12538–12543

Finzi D, Blankson J, Siliciano JD, Margolick JB, Chadwick K, Pierson T, Smith K, Lisziewicz J, Lori F, Flexner C, Quinn TC, Chaisson RE, Rosenberg E, Walker B, Gange S, Gallant J, Siliciano RF (1999) Latent infection of CD4+ T cells provides a mechanism for lifelong persistence of HIV-1, even in patients on effective combination therapy. Nature Med 5:512–517

Forster I, Hirose R, Arbeit JM, Clausen BE, Hanahan D (1995) Limited capacity for tolerization of CD4+ T cells specific for a pancreatic beta cell neo-antigen. Immunity 2:573–585

Frueh K, Ahn K, Djaballah H, Sempe P, van Endert PM, Tampe R, Peterson PA, Yang Y (1995) A viral inhibitor of peptide transporters for antigen presentation. Nature 375:415–418

Fu T-M, Guan L, Friedman A, Schofield TL, Ulmer JB, Liu MA, Donnelly JJ (1999) Dose dependence of CTL precursor frequency induced by a DNA vaccine and correlation with protective immunity against influenza virus challenge. J Immunol 162:4163–4170

Gairin JE, Mazarguil H, Hudrisier D, Oldstone MBA (1995) Optimal lymphocytic choriomeningitis virus sequences restricted by H-2D^b major histocompatibility complex class I molecules and presented to cytotoxic T lymphocytes. J Virol 69:2297–2305

Gairin JE, Oldstone MBA (1992) Design of high-affinity major histocompatibility complex-specific antagonist peptides that inhibit cytotoxic T-lymphocyte activity: implications for the control of viral disease. J Virol 66:6755–6762

Gallimore A, Glithero A, Godkin A, Tissot AC, Pluckthun A, Elliott T, Hengartner H, Zinkernagel R (1998) Induction and exhaustion of lymphocytic choriomeningitis virus-specific cytotoxic T lymphocytes visualized using soluble tetrameric major histocompatibility complex class I-peptide complexes. J Exp Med 187:1383–1393

Gamble DR, Taylor KW, Cumming H (1973) Coxsackie viruses and diabetes mellitus. Br Med J 4:260–262

Gandhi RT, Chen BK, Straus SE, Dale JK, Lenardo MJ, Baltimore D (1998) HIV-1 directly kills CD4$^+$ T cells by a fas-independent mechanism. J Exp Med 187:1113–1122

Garside P, Mowat AM (1997) Mechanisms of oral tolerance. Crit Rev Immunol 17:119–137

Genain C, Lee-Parritz D, Nguyen M, Massacesi L, Joshi N, Ferrante R, Hoffman K, Moseley K, Letvin N, Hauser S (1994) In healthy primates, circulating autoreactive T-cells mediate autoimmune disease. J Clin Invest 94:1339–1345

Gladisch R, Hoffmann W, Waldherr R (1976) Myocarditis and insulitis in Coxsackie virus infection. Z Kardiol 65:873–881

Green EA, Eynon EE, Flavell RA (1998) Local expression of TNFα in neonatal NOD mice promotes diabetes by enhancing presentation of islet antigens. Immunity 9:733–743

Guidotti LG, Borrow P, Hobbs MV, Matzke B, Gresser I, Oldstone MB, Chisari FV (1996) Viral crosstalk: intracellular inactivation of the hepatitis B virus during an unrelated viral infection of the liver. Proc Natl Acad Sci USA 93:4589–4594

Hafler DA (1999) The distinction blurs between an autoimmune versus microbial hypothesis in multiple sclerosis. J Clin Invest 104:527–529

Harding FA, Allison JP (1993) CD28-B7 interactions allow the induction of CD8$^+$ cytotoxic T-lymphocytes in the absence of exogenous help. J Exp Med 177:1791–1796

Harrison LC, Dempsey-Collier M, Kramer DR, Takahashi K (1996) Aerosol insulin induces regulatory CD8 gamma delta T cells that prevent murine insulin-dependent diabetes. J Exp Med 184:2167–2174

Harrison LC, Honeyman MC (1999) Cow's milk and type 1 diabetes. Diabetes 48:1501–1507

Haskins K, Wegmann D (1996) Diabetogenic T-cell clones. Perspectives in Diabetes 45:1299–1305

Herbelin A, Gombert J-M, Lepault F, Bach J-F, Chatenoud L (1998) Mature mainstream TCRαβ$^+$CD4$^+$ thymocytes expressing L-selectin mediate "active tolerance" in the nonobese diabetic mouse. J Immunol 161:2620–2628

Herold KG, Lenschow DJ, Bluestone JA (1997) CD28/B7 regulation of autoimmune diabetes. Immunol Res 16:71–84

Homann D, Holz A, Bot A, Coon B, Wolfe T, Petersen J, Dyrberg TP, Grusby MJ, von Herrath MG (1999) Autoreactive CD4+ lymphocytes protect from autoimmune diabetes via bystander suppression using the IL-4/STAT6 pathway. Immunity (in press)

Horwitz MS, Bradley LM, Harbertson J, Krahl T, Lee J, Sarvetnick N (1998) Diabetes induced by Coxsackie virus: Iniation by bystander damage and not molecular mimicry. Nature Med 4:781–785

Huang H, Paul WE (1998) Impaired interleukin 4 signaling in T helper type 1 cells. J Exp Med 187: 1305–1313

Itoh N, Imagawa A, Hanafusa T, Waguri M, Yamamoto K, Iwahashi H, Moriwaki M, Nakajima H, Miyagawa J, Namba M, Makino S, Nagata S, Kono N, Matsuzawa Y (1997) Requirement of Fas for the development of autoimmune diabetes in nonobese diabetic mice. J Exp Med 186:613–618

Jenson AB, Rosenberg HS, Notkins AL (1980) Pancreatic islet cell damage in children with fatal viral infections. Lancet 2:354–358

Kagi D, Ledermann B, Burki K, Seiler P, Odermatt B, Olsen KJ, Podack ER, Zinkernagel RM, Hengartner H (1994) Cytotoxicity mediated by T cells and natural killer cells greatly impaired in perforin-deficient mice. Nature 369:1–7

Kagi D, Odermatt B, Ohashi PS, Zinkernagel RM, Hengartner H (1996) Development of insulitis without diabetes in transgenic mice lacking perforin-dependent cytotoxicity. J Exp Med 183:2143–2152

Kagi D, Odermatt B, Seiler P, Zinkernagel RM, Mak TW, Hengartner H (1997) Reduced incidence and delayed onset of diabetes in perforin-deficient nonobese diabetic mice. J Exp Med 7:989–997

Kagi D, Seiler P, Pavlovic J, Ledermann B, Burki K, Zinkernagel RM, Hengartner H (1995) The roles of perforin- and Fas-dependent cytotoxicity in protection against cytopathic and noncytopathic viruses. Eur J Immunol 25:3256–3262

Kang SM, Schneider DB, Lin Z, Hanahan D, Dichek DA, Stock PG, Baekkeskov S (1997) Fas ligand expression in islets of Langerhans does not confer immune privilege and instead targets them for rapid destruction. Nature Med 3:738–743

Kapp JA, Ke Y (1997) The role of gammadelta TCR-bearing T cells in oral tolerance. Res Immunol 148:561–567

Katz J, Benoist C, Mathis DX (1995) T helper cell subsets in IDDM. Science 268:1185–1188

Katz-Levy Y, Neville KL, Girvin AM, Vanderlugt CL, Pope JG, Tan LJ, Miller SD (1999) Endogenous presentation of self myelin epitopes by CNS-resident APCs in Theiler's virus-infected mice. J Clin Invest 104:599–610

Kim YH, Kim S, Kim KA, Yagita H, Kayagaki N, Kim KW, Lee MS (1999) Apoptosis of pancreatic beta-cells detected in accelerated diabetes of NOD mice: no role of Fas-Fas ligand interaction in autoimmune diabetes. Eur J Immunol 29:455–465

King C, Davies J, Mueller R, Lee MS, Krahl T, Yeung B, O'Connor E, Sarvetnick N (1998a) TGF-beta1 alters APC preference, polarizing islet antigen responses toward a Th2 phenotype. Immunity 5:601–603

King C, Davies J, Mueller R, Lee M-S, Krahl T, Yeung B, O'Connor E, Sarvetnick N (1998b) TGF-β1 alters APC preference, polarizing islet antigen responses toward a Th2 phenotype. Immunity 8:601–613

Kurtzke JF, Delasnerie-Laupretre N, Wallin MT (1998) Multiple sclerosis in North African migrants to France. Acta Neurol Scand 98:302–309

Kurtzke JF, Hyllested K, Arbuckle JD, Bronnum-Hansen H, Wallin MT, Heltberg A, Jacobsen H, Olsen A, Eriksen LS (1997) Multiple sclerosis in the Faroe Islands. 7. Results of a case control questionnaire with multiple controls. Acta Neurol Scand 96:149–157

Lee M-S, Mueller R, Wicker LS, Peterson LB, Sarvetnick N (1996) IL-10 is necessary and sufficient for autoimmune diabetes in conjunction with NOD MHC homozygosity. J Exp Med 183:2663–2668

Lee M-S, von Herrath MG, Reiser H, Oldstone MBA, Sarvetnick N (1995) Sensitization to self antigens by in situ expression of interferon-γ. J Clin Invest 95:486–492

Leiter EH, Hamaguchi K (1990) Viruses and diabetes: diabetogenic role for endogenous retroviruses in NOD mice? J Autoimmun 1:31–40

Lernmark A, Baekkeskov S, Gerling I, Kastern W, Knutson C, Michelsen B (1985) Immunological aspects of type 1 and 2 diabetes mellitus. Adv Exp Med Biol 189:107–127

Liblau RS, Singer SM, McDevitt H (1995) Th1 and Th2 $CD4^+$ T cells in the pathogenesis of organ-specific autoimmune diseases. Immunol Today 16:34–38

Lider O, Santos LM, Lee CS, Higgins PJ, Weiner HL (1989) Suppression of experimental autoimmune encephalomyelitis by oral administration of myelin basic protein. II. Suppression of disease and in vitro immune responses is mediated by antigen-specific CD8+ T lymphocytes. J Immunol 142:748–752

Ludewig B, Odermatt B, Landmann S, Hengartner H, Zinkernagel RM (1998) Dendritic cells induce autoimmune diabetes and maintain disease via De Novo formation of local lymphoid tissue. J Exp Med 188:1493–1501

Ludewig B, Oehen S, Barchiesi F, Schwendener RA, Hengartner H, Zinkernagel RM (1999) Protective antiviral cytotoxic T cell memory is most efficiently maintained by restimulation via dendritic cells. J Immunol 163:1839–1844

Maini RN, Elliott M, Brennan FM, Williams RO, Feldmann M (1997) TNF blockade in rheumatoid arthritis: implications for therapy and pathogenesis. APMIS 105:257–263

Martin R, Ruddle NH, Reingold S, Hafler DA (1998) T helper cell differentiation in multiple sclerosis and autoimmunity. Immunol Today 19:495–498

Matsue H, Matsue K, Walters M, Okumura K, Yagita H, Takashima A (1999) Induction of antigen-specific immunosuppression by CD95L cDNA-transfected 'killer' dendritic cells. Nature Med 5:930–937

Miller A, al-Sabbagh A, Santos LM, Das MP, Weiner HL (1993) Epitopes of myelin basic protein that trigger TGF-beta release after oral tolerization are distinct from encephalitogenic epitopes and mediate epitope-driven bystander suppression. J Immunol 151:7307–7315

Miller A, Lider O, Weiner HL (1991) Antigen-driven bystander suppression after oral administration of antigens. J Exp Med 174:791–798

Miller JF, Heath WR, Allison J, Morahan G, Hoffmann M, Kurts C, Kosaka H (1997) T cell tolerance and autoimmunity. Ciba Found Symp 204:159–168

Morgan DJ, Kreuwel HTC, Fleck S, Levitsky HI, Pardoll DM, Sherman LA (1998) Activation of low avidity CTL specific for a self epitope results in tumor rejection but not autoimmunity[1]. J Immunol 160:643–651

Morgan DJ, Liblau R, Scott B, Fleck S, McDevitt HO, Sarvetnick N, Lo, D, Sherman LA (1996) CD8(+) T cell-mediated spontaneous diabetes in neonatal mice. J Immunol 157:978–983

Moskophidis D, Battegay M, van den Broek M, Laine E, Hoffmann-Rohrer U, Zinkernagel RM (1995) Role of virus and host variables in virus persistence or immunopathological disease caused by non-cytolytic virus. J Gen Virol 76:381–391

Mosmann TR, Coffman RL (1989) TH1 and TH2 cells: different patterns of lymphokine secretion lead to different functional properties. Ann Rev Immunol 7:145–173

Mueller RT, Sarvetnick N (1996) Pancreatic expression of IL-4 abrogates insulitis and diabetes in NOD mice. J Exp Med 184:1093–1099

Murali-Krishna K, Altman JD, Suresh M, Sourdive DJD, Zajac AJ, Miller JD, Slansky J, Ahmed R (1998) Counting antigen-specific CD8 T cells: A reevaluation of bystander activation during viral infection. Immunity 8:177–187

Notkins AL, Yoon J-W, Onodera T, Toniolo A, Jenson AB (1981) Virus-induced diabetes mellitus. Perspect Virol 11:141–162

Nussenblatt RB, Caspi RR, Mahdi R, Chan CC, Roberge F, Lider O, Weiner HL (1990) Inhibition of S-antigen induced experimental autoimmune uveoretinitis by oral induction of tolerance with S-antigen. J Immunol 144:1689–1695

Ohashi P, Oehen S, Buerki K, Pircher H, Ohashi C, Odermatt B, Malissen B, Zinkernagel R, Hengartner H (1991) Ablation of "tolerance" and induction of diabetes by virus infection in viral antigen transgenic mice. Cell 65:305–317

Oldstone MBA (1987) Molecular mimicry and autoimmune disease. Cell 50:819–820

Oldstone MBA (1989a) Infectious agents as etiologic triggers of autoimmune disease. In: Oldstone MBA (ed) CTMI, vol. 145. Springer, Heidelberg Berlin New York, pp 1–4

Oldstone MBA (1989b) Molecular mimicry as a mechanism for the cause and as a probe uncovering etiologic agent(s) of autoimmune disease. Curr Top Microbiol Immunol 145:127–135

Oldstone MBA (1990) Viral persistence and immune dysfunction. Hospital Practice 25:81–85

Oldstone MBA, Nerenberg M, Southern P, Price J, Lewicki H (1991) Virus infection triggers insulin-dependent diabetes mellitus in a transgenic model: Role of anti-self (virus) immune response. Cell 65:319–331

Oldstone MBA, von Herrath M, Lewicki H, Hudrisier D, Whitton JL, Gairin E (1999) Use of a high-affinity peptide that aborts MHC-restricted cytotoxic T lymphocyte activity against multiple viruses in vitro and virus-induced immunopathologic disease in vivo. Virology 256:246–257

Oldstone MBA, von Herrath MG, Evans CF, Horwitz MS (1996) Virus-induced autoimmune disease: transgenic approach to mimic insulin-dependent diabetes mellitus and multiple sclerosis. Curr Top Microbiol Immunol 206:67–83

Oxenius A, Bachmann MF, Ashton-Rickardt PG, Tonegawa S, Zinkernagel RM, Hengartner H (1995) Presentation of endogenous viral proteins in association with major histocompatibility complex class II: on the role of intracellular compartmentalization, invariant chain and the TAP transporter system. Eur J Immunol 25:3402–3411

Pakala SV, Chivetta M, Kelly CB, Katz JD (1999) In autoimmune diabetes the transition from benign to pernicious insulitis requires an islet cell response to tumor necrosis factor alpha. J Exp Med 189:1053–1062

Prabhaker BS, Haspel MV, McClintock PR, Notkins AL (1982) High frequency of antigenic variants among naturally occurring human Coxsackie B4 virus isolates identified by monoclonal antibodies. Nature 300:374–376

Prince G, Jenson AB, Billups L, Notkins AL (1978) Infection of human pancreatic beta cell cultures with mumps virus. Nature 27:158–161

Probert L, Akassoglou K, Kassiotis G, Pasparakis M, Alexopoulou L, Kollias G (1997) TNF-alpha transgenic and knockout models of CNS inflammation and degeneration. J Neuroimmunol 72:137–141

Rabinovitch A, Suarez-Pinzon W, Strynadka K, Ju, Q, Edelstein D, Brownlee M, Korbutt GS, Rajotte RV (1999) Transfection of human pancreatic islets with an anti-apoptotic gene (bcl-2) protects beta-cells from cytokine-induced destruction. Diabetes 48:1223–1229

Redondo MJ, Rewers M, Yu, L, Garg S, Pilcher CC, Elliott RB, Eisenbarth GS (1999) Genetic determination of islet cell autoimmunity in monozygotic twin, dizygotic twin, non-twin siblings of patients with type 1 diabetes: prospective twin study. BMJ 318:698–702

Rocken M, Racke M, Shevach EM (1996) IL-4-induced immune deviation as antigen-specific therapy for inflammatory autoimmune disease. Immunol Today 17:225–231

Rodriguez F, Zhang J, Whitton JL (1997) DNA immunization: Ubiquitination of a viral protein enhances cytotoxic T-lymphocyte induction and antiviral protection but abrogates antibody induction. J Virol 71:8497–8503

Sallusto F, Lanzavecchia A, Mackay CR (1998) Chemokines and chemokine receptors in T-cell priming and Th1/Th2-mediated responses. Immunol Today 19:568–574

Salojin KV, Zhang J, Madrenas J, Delovitch TL (1998) T-cell anergy and altered T-cell receptor signaling: effects on autoimmune disease. Immunol Today 19:468–473

Scholz C, Patton KT, Anderson DE, Freeman GJ, Hafler DA (1998) Expansion of autoreactive T cells in multiple sclerosis is independent of exogenous B7 costimulation[1]. J Immunol 160:1532–1538

Scott B, Liblau R, Degermann S, Marconi LA, Ogata L, Caton AJ, McDevitt HO, Lo D (1994) A role for non-MHC genetic polymorphism in susceptibility to spontaneous autoimmunity. Immunity 1:73–82

Sempe P, Richard MF, Bach JF, Boitard C (1994) Evidence of CD4+ regulatory T cells in the non-obese diabetic male mouse. Diabetologia. 37:337–343

Sevilla N, Homann D, von Herrath MG, Rodriguez F, Harkins S, Whitton JL, Oldstone MBA (1999) Virus-induced diabetes in a transgenic model: Quantitation of effector T cells needed to cause disease

Sigal LJ, Reiser H, Rock KL (1998) The role of B7-1 and B7-2 costimulation for the generation of CTL responses in vivo. J Immunol 161:2740–2745

Slifka MK, Rodriguez F, Whitton JL (1999) Rapid on/off cycling of cytokine production by virus-specific CD8+ T cells. Nature 401:76–79

Southern P, Oldstone MBA (1986) Medical consequences of persistent viral infection. N Eng J Med 314:359–367

Steinman L (1996) A few autoreactive cells in an autoimmune infiltrate control a vast population of nonspecific cells: A tale of smart bombs and the infantry. Proc Nat Acad Sci USA 93:2253–2256

Steinman L (1999) Absence of "original antigenic sin" in autoimmunity provides an unforeseen platform for immune therapy. J Exp Med 189:1021–1024

Tada Y, Ho, A, Koh DR, Mak TW (1996) Collagen-induced arthritis in CD4- or CD8-deficient mice: CD8+ T cells play a role in initiation and regulate recovery phase of collagen-induced arthritis. J Immunol 156:4520–4526

Thomas HE, Darwiche R, Corbett JA, Kay TWH (1999) Evidence that beta cell death in the nonobese diabetic mouse is Fas independent. J Immunol 163:1562–1569

Thomas HE, Parker JL, Schreiber RD, Kay TWH (1998) IFN-gamma action on pancreatic beta cells causes class I MHC upregulation but not diabetes. J Clin Invest 102:1249–1257

Tuohy VK, Yu M, Ling Y, Kawaczak JA, Kinkel RP (1999) Spontaneous regression of primary autoreactivity during chronic progression of experimental autoimmune encephalomyelitis and multiple sclerosis. J Exp Med 189:1033–1042

van der Most RG, Murali-Krishna K, Whitton JL, Oseroff C, Alexander J, Southwood S, Sidney J, Chesnut RW, Sette A, Ahmed R (1998) Identification of Db- and Kb-restricted subdominant cytotoxic T-cell responses in lymphocytic choriomeningitis virus-infected mice. Virology 240:158–167

Vella A, Teague TK, Ihle J, Kappler J, Marrack P (1997) Interleukin 4 (IL-4) or IL-7 prevents the death of resting T cells: Stat6 is probably not required for the effect of IL-4. J Exp Med 186:325–330

von Herrath M, Coon B, Homann D, Wolfe T, Guidotti LG (1999) Thymic tolerance to only one viral protein reduces lymphocytic choriomeningitis virus-induced immunopathology and increases survival in perforin-deficient mice. J Virol 73:5918–5925

von Herrath MG, Allison J, Miller JF, Oldstone MBA (1995a) Focal expression of interleukin-2 does not break unresponsiveness to "self" (viral) antigen expressed in beta cells but enhances development of autoimmune disease (diabetes) after initiation of an anti-self immune response. J Clin Invest 95:477–485

von Herrath MG, Coon B, Lewicki H, Mazarguil H, Gairin JE, Oldstone MBA (1998) In vivo treatment with a MHC Class I-Restricted blocking peptide can prevent virus-induced autoimmune diabetes. J Immunol 161:5087–5096

von Herrath MG, Coon B, Oldstone MBA (1997) Low-affinity cytotoxic T-lymphocytes requires IFN-gamma to clear an acute viral infection. Virology 229:349–359

von Herrath MG, Dockter J, Nerenberg M, Gairin JE, Oldstone MBA (1994a) Thymic selection and adaptability of cytotoxic T lymphocyte responses in transgenic mice expressing a viral protein in the thymus. J Exp Med 180:1901–1910

von Herrath MG, Dockter J, Oldstone MBA (1994b) How virus induces a rapid or slow-onset insulin-dependent diabetes mellitus in a transgenic model. Immunity 1:231–242

von Herrath MG, Dyrberg T, Oldstone MBA (1996a) Oral insulin treatment suppresses virus-induced antigen-specific destruction of beta cells and prevents autoimmune diabetes in transgenic mice. J Clin Invest 98:1324–1331

von Herrath MG, Evans CF, Horwitz MS, Oldstone MBA (1996b) Using transgenic mouse models to dissect the pathogenesis of virus-induced autoimmune disorders of the islets of Langerhans and the central nervous system. Immunol Rev 152:111–143 von Herrath MG, Guerder S, Lewicki H, Flavell R, Oldstone MBA (1995b) Coexpression of B7.1 and viral (self) transgenes in pancreatic β-cells can break peripheral ignorance and lead to spontaneous autoimmune diabetes. Immunity 3:727–738

von Herrath MG, Holz A (1997) Pathological changes in the islet milieu precede infiltration of islets and destruction of beta-cells by autoreactive lymphocytes in a transgenic model of virus-induced IDDM. J Autoimmun 10:231–238

von Herrath MG, Oldstone MB (1996) Virus-induced autoimmune disease. Curr Opin Immunol 8: 878–885

von Herrath MG, Oldstone MBA (1997) IFN-gamma is essential for beta-cell destruction by CTL. J Exp Med 185:531–539

Weiner HL (1997) Oral tolerance: immune mechanisms and treatment of autoimmune diseases. Immunol Today 18:335–343

Whitmire JK, Asano MS, Murali-Krishna K, Suresh M, Ahmed R (1998) Long-term CD4 Th1 and Th2 memory following acute lymphocytic choriomeningitis virus infection. J Virol 72:8281–8288

Whitton JL, Southern PJ, Oldstone MBA (1993) Analysis of the CTL responses to glycoprotein and nucleoprotein components of LCMV. J Virology 67:2903–2907

Wicker LS, Todd JA, Peterson LB (1995) Genetic control of autoimmune diabetes in the NOD mouse. Annu Rev Immunol 13:179–200

Wildner G, Hunig T, Thurau SR (1996) Orally induced, peptide-specific gamma/delta TCR+ cells suppress experimental autoimmune uveitis. Eur J Immmunol 9:2140–2148

Wong FS, Karttunen J, Dumont C, Wen L, Visintin I, Pilip IM, Shastri N, Pamer EG, Janeway Jr CA (1999) Identification of an MHC class I-restricted autoantigen in type 1 diabetes by screening an organ-specific cDNA library. Nature Med 5:1026–1031

Wong FS, Visintin I, Wen L, Flavell RA, Janeway JCA (1996) CD8 T cell clones from young nonobese diabetic (NOD) islets can transfer rapid onset of diabetes in NOD mice in the absence of CD4 cells. J Exp Med 183:67–76

Yang Y, Charlton B, Shimada A, Dal Canto R, Fathman CG (1996) Monoclonal T cells identified in early NOD islet infiltrates. Immunity 4:189–194

Yokoyama M, Zhang J, Whitton JL (1995) DNA Immunization confers protection against lethal LCMV infection. J Virol 69:2684–2688

Yoon JW, Austin M, Onodera T, Notkins AL (1979) Virus-induced diabetes mellitus: isolation of a virus from the pancreas of a child with diabetic ketoacidosis. New Engl J Med 300:1173–1179

Yoon J-W, Morishima T, McClintock PR, Austin M, Notkins AL (1984) Virus-induced diabetes mellitus: mengovirus infects pancreatic beta cells in strains of mice resistant to the diabetogenic effects of encephalomyocarditis virus. J Virol 50:684–690

Yoon J-W, Yoon C-S, Lim H-W, Huang QQ, Kang Y, Pyun KH, Hirasawa K, Sherwin RS, Jun H-S (1999) Control of autoimmune diabetes in NOD mice by GAD expression or suppression in beta cells. Science 284:1183–1187

Yoshino S (1996) Suppression of adjuvant arthritis in rats by oral administration of type II collagen in combination with type I interferon. J Pharm Pharmacol 48:702–705

Zajac AJ, Blattman JN, Murali-Krishna K, Sourdive DJD, Suresh M, Altman JD, Ahmed R (1998) Viral immune evasion due to persistence of activated T cells without effector function. J Exp Med 188:2205–2213

Zheng P, Liu Y (1997) Costimulation by B7 modulates specificity of cytotoxic T lymphocytes: A missing link that explains some bystander T cell Activation. J Exp Med 186:1787–1791

Zinkernagel RM, Hengartner H (1994) T-cell-mediated immunopathology *versus* direct cytolysis by virus: implications for HIV and AIDS. Immunol Today 15:262–268

LCMV and the Central Nervous System: Uncovering Basic Principles of CNS Physiology and Virus-Induced Disease

C.F. Evans[1,3], J.M. Redwine[1,4], C.E. Patterson[2], S. Askovic[2], and G.F. Rall[2]

1	Introduction	177
1.1	Factors that Govern Virus and Immune Cell Entry into the CNS	178
1.2	LCMV Infection of the Mouse Brain	179
2	Consequences of Viral Persistence in Neurons	179
2.1	Viral Perturbation of Differentiated "Luxury" Functions	180
2.2	Neuronal Loss During Persistence/Developmental Abnormalities	181
2.3	Behavioral Effects of Persistent LCMV Infection	182
2.4	Selection of Virus Quasispecies	182
3	Consequences of LCMV-Specific Immune Responses in the CNS	184
3.1	Acute Infection	184
3.1.1	Historical Description of LCMV Disease	184
3.1.2	Crucial Role of $CD8^+$ T Cells in Clearance	184
3.2	Clearance of Persistent Infection Following Adoptive Transfer	185
3.3	Absence of Class I MHC on Neurons	186
3.4	MHC Expression on Other Resident Brain Cells	187
3.5	Cytokines and Chemokines	187
3.6	A New Model of LCMV-Induced Hydrocephalus	189
4	The Role of Host Immunity in Neuroinvasion	190
5	Using LCMV to Model Human Autoimmune Demyelinating Disease	191
References		192

1 Introduction

Our understanding of the normal functions of the central nervous system (CNS) and of the mechanisms underlying neuroimmunological responses have greatly benefited from the use of lymphocytic choriomeningitis virus (LCMV) infection of its natural host, the mouse. One of the strengths of the LCMV system is its flexibility: infection can result in dramatically distinct outcomes in mice depending on variables such as host age, immunological competence, host genetic background,

---

[1] Department of Neuropharmacology, The Scripps Research Institute, 10550 N. Torrey Pines Road, La Jolla, CA 92037, USA
[2] Division of Basic Science, The Fox Chase Cancer Center, 7701 Burholme Avenue, Philadelphia, PA 19111, USA
[3] *Current address*: Digital Gene Technologies, 11149 N. Torrey Pines Road, La Jolla, CA 92037, USA
[4] *Current address*: Neurome, 11149 N. Torrey Pines Road, La Jolla, CA 92037, USA

virus dosage, virus strain and route of inoculation (reviewed in BORROW 1997; BUCHMEIER and ZAJAC 1999). Depending on how these variables are combined, the consequences of infection range from rapid onset, immune-mediated mortality to lifelong persistent infection in the absence of overt illness. While all of these outcomes can be induced in laboratory mice, mother-to-offspring transmission resulting in persistent LCMV infection predominates in the wild.

In this chapter, we will discuss much of the work done using LCMV infection of the mouse CNS, and will argue that the remarkable variability in the host response to LCMV has provided important information about basic brain physiology and the consequences of CNS virus infection in general. In particular, we will show that LCMV has been fundamental to our present understanding of how persistent viral infections can alter CNS function, what host factors regulate or restrict virus spread within the CNS, and what roles host immunity play in both the resolution of infection and the induction of CNS disease.

1.1 Factors that Govern Virus and Immune Cell Entry into the CNS

The complexity of understanding LCMV infection in the CNS is in great part due to the complexity of the CNS itself. Thus, in order to appreciate how LCMV can result in multiple effects within the mouse CNS, one must also take into consideration the physiology of the CNS. Because neurons are essential and generally nondividing cells, multiple barriers exist to protect this crucial, yet fragile, cell population from potentially dangerous factors that are carried in the blood. The barriers that contribute to the "immune privilege" of the CNS have been described in detail elsewhere (RALL 1998). In short, these barriers function to physically and biochemically restrict the access of immune cells or mediators into the CNS, or to limit their activity once in the brain parenchyma. For example, the blood-brain barrier, which is comprised of endothelial cells, basal membranes and astrocyte endfeet, controls the exchange of cells and metabolites between the circulation and the brain, predominately due to tight junctions between the capillary endothelium. Water and some lipid soluble molecules freely cross the barrier, and the vascular endothelial cells have protein transporters that actively transport nutrients from the blood into the brain. No antibodies or cells of the immune system are found within the normal brain (WEKERLE et al. 1986; HICKEY et al. 1991). However, during infection or traumatic injury to the brain, multiple immune components may cross into the brain parenchyma, including activated T cells (HICKEY et al. 1991), although their activity is further restricted by other factors that govern immune privilege. One of the best described of these restrictions is the absence or profoundly reduced expression of class I major histocompatibility complex (MHC) molecules on resident brain cells, especially neurons. Our present understanding of the basis for such restriction has been a subject of intense study (RALL 1998). With respect to neurotropic virus infections, the implications of these studies are important, since the absence of MHC expression on neurons would obviously limit the activity of class I-restricted cytotoxic T lymphocytes (CTL), which predominately mediate their antiviral effects by lysis of infected cells.

Perhaps because of the restricted surveillance of the CNS by the host immune response, many viruses have evolved to infect cells within the brain (RALL and OLDSTONE 1995). Viruses can gain access to the brain parenchyma by multiple routes, including interneuronal transfer which can bring viruses from the peripheral nervous system into the CNS, infection of capillary endothelial cells that comprise the blood-brain barrier, or infection of cells that routinely traffic through the parenchyma (the so-called Trojan horse approach). Understanding how viruses gain entry into the CNS, and how immune responses are generated and control such neurotropic viruses is of paramount importance in studying many virus-mediated human CNS diseases, including herpes encephalitis, poliomyelitis, subacute sclerosing panencephalitis and AIDS dementia. In addition, these studies are of relevance in determining what role, if any, viruses play in human CNS diseases of unknown etiology, including multiple sclerosis, autism, schizophrenia and amyotrophic lateral sclerosis.

1.2 LCMV Infection of the Mouse Brain

As described in detail elsewhere in this book, mice can be infected with LCMV to result in multiple outcomes, including: (1) a nonlethal acute infection that is cleared by the host immune response and that results in lifelong immunity, (2) an acute infection that results in immune-mediated lethality, and (3) a persistent infection of either neonatal mice or immunocompromised adults in which the virus persists throughout the lifetime of the host, usually in the absence of overt disease. Infection of the CNS occurs in both the lethal acute infection and the persistent infection, although the infected cells within the brain differ. In the lethal acute infection, LCMV is restricted to the cells that comprise the linings of the brain, including the meninges, the leptomeninges, the ependyma and the choroid plexus. In this infection, it is the host immune response, not the virus infection per se, that mediates disease (see Sect. 3.1). In the persistent infection, best studied in mice that were infected with LCMV at birth, viral antigens are restricted to neurons within the brain parenchyma. In these mice, no antiviral T cell response is generated, and disease is virus-mediated (see Sect. 2). What specific effects LCMV has on the mouse CNS, and what these studies have taught us about virus-CNS-immune response interactions in general, will be the focus of the remainder of this chapter.

2 Consequences of Viral Persistence in Neurons

Infection of neonatal mice (less than 24h old) with LCMV results in a lifelong persistent infection. In persistently infected mice, virtually all tissues are infected and within the CNS, viral replication is restricted to neurons (FAZAKERLEY et al. 1991). Because these mice appear healthy and have normal lifespans, it was initially thought that persistent infection occurred in the absence of disease. However, more careful scrutiny has revealed that persistent infection with LCMV can result in neurological impairment, including the perturbation of neuronally differentiated functions, altered CNS development, and behavioral effects.

2.1 Viral Perturbation of Differentiated "Luxury" Functions

In the 1970s, two studies implicated persistent LCMV infection in causing metabolic changes in neurons without overt neuropathology. First, in vitro studies of neuroblastoma cells persistently infected with LCMV showed altered activities of enzymes involved in acetylcholine metabolism in the absence of cell death (OLDSTONE et al. 1977). Second, mice persistently infected with LCMV had behavioral alterations including increased latency in open-field tests and decreased locomotor activity (HOTCHIN and SEEGAL 1977). Since these mice appeared normal in other respects, these studies were among the first to suggest that viral infections could cause disease in the absence of immunopathology or massive tissue destruction.

Further studies confirmed and extended these early observations, noting that persistent infections such as LCMV can perturb differentiated functions of cells (so-called luxury functions) without actually killing the infected cell. The most well-characterized example is the infection of anterior pituitary growth hormone-producing cells in C3H/ST mice. Following neonatal infection by LCMV Armstrong (LCMV-Arm), C3H/ST mice become persistently infected and develop a growth hormone (GH) deficiency syndrome (GHDS). This disease is characterized by retarded growth and hypoglycemia, and usually results in death by the age of 30 days (OLDSTONE et al. 1982). Despite the presence of replicating virus in many GH-producing cells, there is no evidence of damage to these cells, and no inflammation is detected. By 16 days post-infection, the amount of GH mRNA and protein is significantly reduced in infected versus uninfected controls (OLDSTONE et al. 1982; VALSAMAKIS et al. 1987). Infection of a pituitary cell line with LCMV-Arm resulted in decreased GH production at both the protein and mRNA levels (DE LA TORRE and OLDSTONE 1992). Further studies demonstrated that the viral infection interfered with the action of the GH transcriptional transactivator, Pit1, thus reducing the level of GH mRNA produced in virus-infected cells (DE LA TORRE and OLDSTONE 1992). Studies with reassorted chimeric viruses mapped the GHDS to the S RNA segment of LCMV (RIVIERE et al. 1985). In nonsusceptible strains (BALB/WEHI and SWR/J), normal levels of GH are found, and few GH-producing cells are infected. The basis of this apparent strain-specific infection remains enigmatic.

Subsequent studies of persistent LCMV infection revealed other examples of such "luxury function" effects within the CNS. LIPKIN et al. (1988) found that LCMV-Arm replicates in neurons that produce the neurotransmitter somatostatin.

Although the infected neurons had a normal morphology, reduced levels of somatostatin mRNA were detected. This was not due to a generalized reduction in brain levels of neurotransmitter mRNAs since mRNA levels of another neurotransmitter, cholecystokinin, were normal. More recent studies have demonstrated that persistent LCMV infection can also influence the expression of the phosphoprotein, neural growth-associated protein, or GAP43, which is associated with the membrane of presynaptic terminals and appears to play a role in learning and memory. Expression of GAP43 mRNA and protein were significantly decreased in the hippocampus of persistently infected mice (DE LA TORRE et al. 1996), and correlated with virus-induced behavioral effects, including a decreased ability to learn tasks. In vitro studies in a neuronal cell line demonstrated that LCMV infection alters both the rate of transcription of the GAP43 gene and the stability of GAP43 mRNA (CAO et al. 1997). Taken together, these findings suggest that persistent infection with a noncytolytic virus can alter the differentiated functions of neurons resulting in abnormalities in learning and memory, without overt immunopathology or neuronal loss.

2.2 Neuronal Loss During Persistence/Developmental Abnormalities

While LCMV infection of neurons in mice does not result in cell death, perinatal infection of rats with LCMV leads to neuronal loss in the developing cerebellum and hippocampus, although apparently via different mechanisms. Importantly, since LCMV is a noncytolytic virus, neuronal loss in either instance is not likely to be due to direct viral lysis. In infected newborn rats, virus is initially found in several areas of the brain, including the cerebellum, hippocampal dentate gyrus and the olfactory bulb, but is cleared from these regions by several months postinfection. Loss of neurons from the cerebellum occurs during the second week postinfection, during which time $CD8^+$ T cells can be detected in the brain. Cerebellar hypoplasia could be prevented by depleting T cells, indicating that damage to the cerebellar neurons during the acute infection was immune-mediated (MONJAN et al. 1974; PEARCE et al. 1999). In contrast, hippocampal neurons are lost over a period of months, even after infectious virus is cleared and lymphocytes are no longer detected in the CNS (MONJAN et al. 1974, 1975). These findings are consistent with a recent study which showed that loss of neurons from the dentate gyrus could not be prevented by depletion of T cells (PEARCE et al. 1999). Studies of dentate granule cell function indicated that LCMV infection disrupted inhibitory neurotransmission, and abnormalities in synaptic function persisted even after infectious virus was cleared from the CNS (PEARCE et al. 1996, 2000). These studies show that LCMV infection can trigger immune-mediated loss of neurons during the acute phase of infection, although the basis of such loss is not well understood. Moreover, LCMV results in a gradual loss of neurons by a mechanism that is not dependent on T cells or the presence of infectious virus. Such virus-induced, yet indirect neuronal death may be mechanistically relevant to human neurodegenerative diseases of unknown etiology.

2.3 Behavioral Effects of Persistent LCMV Infection

Many viruses that can infect the human CNS cause impairment in cognitive and psychomotor function. Thus, while mice that are persistently infected with LCMV do not show evidence of overt disease (as determined by gross assessment parameters such as weight loss, fur ruffling, seizures or morbidity), the possibility that cognitive impairment occurs in these mice cannot be excluded. The advent of sensitive and reproducible behavioral assays to determine acquisition of learned behaviors in mice has allowed for the testing of this hypothesis. Indeed, these assays have confirmed that, while appearing normal, mice persistently infected with LCMV display a marked behavioral impairment.

The consequences of viral infection on learning and memory in mice can be tested using relatively simple tasks that assess both spontaneous, unconditioned behaviors, as well as conditioned learned behaviors. In one behavioral study (GOLD et al. 1994), the short-term memory of persistently infected and uninfected mice was tested by a mild footshock test in a Y-maze. A deficit in the acquisition of footshock avoidance was noted in the infected mice, although with additional training sessions, the performance of virus-infected mice eventually reached that of vehicle-inoculated animals. A related study assessed the same skill acquisition in mice that were persistently infected at birth, but were then cleared of infection by an adoptive transfer of antiviral T cells from a haplotype-matched, immunized mouse. In such adoptive transfers, viral antigens are cleared in all tissues (including the CNS) by 120 days post-transfer, and mice survive. Following this 120 day clearance phase, "cured" mice were subjected to the same footshock avoidance assessments. Surprisingly, despite the absence of the offending infection, the learning deficits remained (BROT et al. 1997). Thus, similar to the hippocampal neuronal dropout that continues in the absence of virus infection, LCMV need not be present to exert a permanent deleterious effect.

In the search for viruses in human CNS diseases of unknown etiology, the "gold standard" of association is the detection of viral nucleic acids or proteins in specific CNS lesions. The behavioral defects that persist in mice in the absence of LCMV suggest that viruses may initiate a cascade of events leading to CNS disease that can continue even once the pathogen is eliminated.

2.4 Selection of Virus Quasispecies

RNA viruses have high mutation rates which result in the generation of heterogeneous viral populations called quasispecies (STEINHAUER and HOLLAND 1987). The selection of viral variants occurs in vivo due to multiple selective pressures, including the antiviral immune response. Many LCMV variants have been isolated from persistently infected mice. Interestingly, plaque-purified viral clones from the brain and lymphoid tissues of mice infected with LCMV-Arm as neonates showed striking clinical differences when injected intravenously into adult mice (AHMED et al. 1984; AHMED and OLDSTONE 1988). Isolates from the brain were similar to

LCMV-Arm and were cleared within 2 weeks post-infection, and in vitro killing assays were used to show that such isolates stimulated vigorous anti-LCMV CTL responses. In contrast, isolates from the spleen caused persistent infections and induced poor CTL responses. The most well characterized spleen variant, Clone 13, differs from LCMV-Arm by only five nucleotides that result in two amino acid changes: amino acid 260 in the viral GP (phenylalanine to leucine) and amino acid 1079 in the polymerase (lysine to glutamate) (SALVATO et al. 1988, 1991; MATLOUBIAN et al. 1990). The GP amino acid 260 change has been identified in more than 20 other viral isolates that cause persistent infection, and is therefore a good marker for viruses with the persistence phenotype.

To examine the kinetics of the selection of the amino acid 260 change, an RT-PCR assay was developed to distinguish between the Armstrong and Clone 13 sequences at amino acid 260 of the GP (EVANS et al. 1994). Neonatal mice were infected with LCMV-Arm, and RNA was prepared from various organs and cells at different times post-infection. The amino acid 260 phe to leu change was found in serum and lymphoid tissues by age 3–5 weeks, and in most other organs by 3 months, but was not detected in RNA from infected brains (EVANS et al. 1994). Since neurons are the primary cells infected in the CNS of mice persistently infected with LCMV from birth (RODRIGUEZ et al. 1983), these data suggested that neurons select Armstrong over variants with the phe to leu amino acid 260 change. This was confirmed when in vitro competition studies were done and it was found that Armstrong replicated more efficiently in two neuronal cell lines than did Clone 13, and that Armstrong outcompeted Clone 13 during serial passages in neuronal cells (EVANS et al. 1994). In agreement with these results, studies of several LCMV variants by Villarete et al. demonstrated that a phenylalanine at amino acid 260 correlated with the ability to grow in neuronal cells, with much lower viral replication occurring with variants containing a leu at position 260 (VILLARETE et al. 1994). When neonatal mice were infected with Clone 13, the CNS became persistently infected with virus with the Clone 13 genotype (DOCKTER et al. 1996). However, when mixtures of Armstrong and Clone 13 were injected into neonatal mice, Armstrong outcompeted Clone 13 for replication in the brain even when ten times more Clone 13 was initially used (DOCKTER et al. 1996). These experiments indicated that Clone 13 could infect the CNS in the absence of the parental LCMV-Arm, but could not outcompete LCMV-Arm for CNS infection. It is not known why replication of LCMV-Arm is favored over Clone 13 in neurons, nor is it known in which cell type Clone 13 is initially selected following neonatal infection with LCMV-Arm. Although there are only two amino acid changes between LCMV-Arm and Clone 13, it is clear that these changes lead to profound differences in viral pathogenesis during infection of adult mice.

These studies of the selection of LCMV variants demonstrated that organ-specific selection of viral variants occurs during persistent LCMV infection, and that brain and lymphoid organs strongly selected for different variants. In humans, brain specific selection of viral variants has been observed in HIV infection (CHENG-MAYER and LEVY 1988; EPSTEIN et al. 1991). Generation of viral variants probably contributes to the evolution of viruses (STEINHAUER and HOLLAND 1987), and to the ability of viruses to cause persistent infections. Identifying the basis for organ specific selection during viral infection will lead to a better understanding of how viruses interact with different cell types to establish persistent infection.

3 Consequences of LCMV-Specific Immune Responses in the CNS

3.1 Acute Infection

3.1.1 Historical Description of LCMV Disease

The name "lymphocytic choriomeningitis" is attributed to the fatal CNS disease that results from intracerebral inoculation of mice with as little as ten plaque forming units. Infected mice die within a remarkably narrow window, usually 6–9 days post-infection, and surprisingly, increasing the virus dose does not appear to hasten the disease process. The lethal disease that occurs in adult, immunocompetent mice was originally described by Lillie and Armstrong in 1945, and summarized by Hotchin in 1962 as follows:

> In the adult mouse, after intracerebral inoculation of 10 to 100 lethal doses, signs of disease appear after the fifth day with ruffling of the fur, a hunched posture and blepharitis. The signs become more severe, with weight loss, relative immobility and a 'jumpy' reaction to loud noises; by the eighth day the animal is likely to go into fatal convulsions... The usual postmortem position of the animals is in the convulsion position, with rear limbs extended, forelimbs flexed, neck extended, and thoracic spine flexed (HOTCHIN 1962).

Histologically, brains of moribund mice display a massive inflammation of the meninges and choroid plexus (hence, the virus name), with marked mononuclear cell infiltration in all the membranes, and in the absence of discrete neuronal pathology. Indeed, with the development of reagents for immunohistological detection of viral antigens, it was shown that virus replication is restricted to the meninges, with little or no virus appearing in CNS neurons (COLE et al. 1971). However, despite major advances in our understanding of the role of the host response in LCMV pathogenesis, and in the discovery of some of the lesions that accompany lethal infection, we still do not yet fully understand how LCMV kills its host, and why there is an apparent lack of correlation between viral load and mortality.

3.1.2 Crucial Role of $CD8^+$ T Cells in Clearance

Lethal choriomeningitis is mediated by cytotoxic T cells, since depletion of CTL by either neutralizing antibody or genetic deletion prevents mice from succumbing to the lethal disease. In 1972, Cole et al., employed a variety of immunosuppressive treatments to LCMV-infected mice to determine the basis of disease and the role of the inflammatory response. Inhibition of the host response ablated the CNS disease, even when administered 3 days post-infection (COLE et al. 1972). This work was subsequently confirmed and refined to show that $CD8^+$ T lymphocytes mediated the pathogenesis of the lethal acute disease, and that specific ablation of this cell population could convert the lethal disease into a persistent infection. Moreover, adoptive transfer of splenocytes from infected mice could cause death in immunosuppressed recipients previously inoculated with LCMV (ALLAN and DOHERTY 1985).

In one study, the question was asked: how do CTL cause mortality in normal mice following intracerebral challenge? Using electron microscopy, the observation was made that infected leptomeningeal and choroid plexus cells were destroyed by CTL, although the authors of this paper concluded that the degree of cell death seen in the CNS of infected mice did not satisfactorily explain the inevitably lethal outcome (SCHWENDEMANN et al. 1983). The lysis of leptomeningeal and choroid plexus cells by CTL is consistent with the observation that LCMV infection induces the upregulation of class I MHC on these cells (MUCKE and OLDSTONE 1992).

Other studies of LCMV infection have implicated the breakdown of the blood-brain barrier in the development of CNS disease. Horseradish peroxidase (HRP), which is normally excluded from the CNS of normal mice, was inoculated into mice with acute LCMV-induced meningitis (MARKER et al. 1984). HRP was able to leak into the extracellular spaces of meningeal vessel walls in mice with meningitis, and a clear correlation was found between blood-brain barrier dysfunction and mortality. The impairment in barrier function was found to be dependent on $CD8^+$ T cells, since antibody depletion of CTL prevented the passage of radiolabeled markers across endothelial junctions (ANDERSEN et al. 1991).

The importance of $CD8^+$ cytotoxicity was later verified using knockout mice with a specific deletion in perforin (KAGI et al. 1994). Mice lacking the pore-forming protein perforin have normal numbers of $CD8^+$ CTL, and generate an antiviral response, although such animals do not succumb to intracerebral challenge. Thus, $CD8^+$ CTL, through their ability to lyse infected target cells, are required for LCMV lethality, although whether death occurs due to cellular lysis, breakdown of the blood-brain barrier, edema, some other cause, or combinations of these factors, remains unknown.

3.2 Clearance of Persistent Infection Following Adoptive Transfer

Mice persistently infected mice from birth continue to shed virus throughout their lives, yet display few overt signs of illness. Virus can be cleared from such mice by an adoptive transfer of splenocytes, obtained from syngeneic mice previously challenged with LCMV by the nonlethal, intraperitoneal route. Interestingly, the kinetics of viral clearance following adoptive transfer were found to differ in various tissues. While clearance occurred within 7–15 days post-transfer from most tissues including the liver, pancreas and spleen, clearance from the brain and kidney was significantly delayed, and required as long as 120 days post-transfer (OLDSTONE et al. 1986). Moreover, clearance of virus from liver and pancreas was associated with T cell infiltration and tissue damage; clearance from the CNS was not associated with massive T cell infiltration, and occurred in the absence of detectable neuropathology (OLDSTONE et al. 1986). The kinetic and histological differences in these tissues suggested that clearance may be occurring by multiple mechanisms. Two theories have been advanced to explain the nonpathological clearance from the CNS. In one scenario, clearance is direct, but noncytolytic, and mediated by T cell-derived cytokines which are produced within the brain parenchyma. Indeed, IFN-γ is needed for the resolution of a persistent infection by adoptive transfer (TISHON et al. 1995). The ability of cytokines to mediate viral clearance may be true for multiple neurotropic infections, as recently shown for both vesicular stomatitis virus and mouse hepatitis virus (KOMATSU et al. 1996; PARRA et al. 1999). Another hypothesis that may explain the delayed kinetics of clearance from the CNS suggests that viral resolution occurs indirectly. This model assumes that the CNS is constantly being re-infected from the periphery in an LCMV persistent infection. As virus is cleared from the periphery, less virus is available to enter the CNS, and clearance occurs due to a depletion of the peripheral pools.

3.3 Absence of Class I MHC on Neurons

Regardless of the basis for such clearance, a noncytolytic strategy of virus clearance from neurons would be of obvious benefit to the host, given that most CNS neurons are nonrenewable (RALL 1998). One subject that has received intense scrutiny in neurobiology – and to which studies with LCMV have contributed (MUCKE and OLDSTONE 1992) – has been the low to absent expression of class I MHC determinants on CNS neurons. Many studies that have evaluated neuronal antigen presenting capacity in vitro found that MHC expression was undetectable in quiescent neurons, but could be induced by such stimuli as cytokines, viral infections, or direct injury (TING et al. 1987; GOPAS et al. 1992; NEUMANN et al. 1995, 1997; RALL 1998). These reports collectively argue that the normal low to negligible expression of class I molecules reflects a stringent regulatory control in CNS neurons (LAMPSON 1990).

Joly and Oldstone used LCMV infection of a neuroblastoma cell line, OBL-21, to show that the absence of functional class I MHC expression on the cell surface could be attributed to a lack of transcription of the MHC class I heavy chain and the TAP molecules, which are intracellular proteins required for shuttling cytosolic epitopes into the endoplasmic reticulum (JOLY et al. 1991; JOLY and OLDSTONE 1992). In the absence of MHC expression, such LCMV-infected neurons could not be recognized by antiviral CTL.

To determine what effect constitutive MHC expression would have on neurons in vivo, transgenic mice were established that expressed a class I MHC gene (D^b), driven by a constitutive, neuron-restricted promoter (neuron-specific enolase; NSE) (RALL et al. 1995). When NSE-D^b transgenic mice were infected with LCMV from birth, neurons became persistently infected as they do in nontransgenic mice. Upon adoptive transfer of LCMV-specific CTL into persistently infected D^b transgenic mice, large numbers of infiltrating $CD8^+$ T cells were found in the CNS, and virus was cleared more quickly than in nontransgenic mice, although no evidence for neuronal dropout was noted (RALL et al. 1995). When the permeability of the blood-brain barrier was examined by injecting mice intravenously with Evans blue, a dye which binds to serum albumin and is normally excluded from the CNS, the dye leaked into the parenchyma of the transgenic mice whereas it was restricted from nontransgenic mice. These studies suggested that even in the presence of "forced" MHC expression, neurons are still refractory to CTL-mediated lysis.

3.4 MHC Expression on Other Resident Brain Cells

To investigate MHC expression on other resident CNS cells during chronic inflammation, double transgenic mice were studied that expressed the LCMV nucleoprotein (NP) and the costimulatory protein, B7-1, in oligodendrocytes (C.F. Evans et al., manuscript in preparation). Peripheral infection of these mice with LCMV led to a chronic inflammation of the CNS, with large numbers of cells expressing MHC class I and class II, especially in white matter regions. This inflammation persisted despite the clearance of LCMV within 10 days post-infection, without the virus ever infecting the CNS. To identify resident CNS cells expressing MHC class I or class II, double immunofluorescence labeling with antibodies to MHC molecules and cell type specific markers followed by confocal microscopic analyses were done. MHC class I was expressed by oligodendrocytes, microglia and vessel endothelial cells, and MHC class II was expressed only by microglia (Fig. 1; J.M. Redwine et al., manuscript in preparation). Although in tissue culture astrocytes can be induced to express MHC class I and class II, no co-localization of either MHC molecule with astrocytes was observed in these in vivo experiments. These results suggested that in vivo, oligodendrocytes, microglia and endothelial cells may be targets of MHC class I restricted $CD8^+$ T cells, and this finding has important implications for understanding immunopathogenic mechanisms within the brain. Although infiltrating $CD8^+$ T cells are found within the CNS during many diseases, it is not known whether they can lyse cells in an antigen-specific manner. For example, $CD8^+$ T cells are found within the CNS of patients with the demyelinating disease multiple sclerosis (MS), however, whether these T cells cause damage is not known, nor is it known which antigens they recognize.

3.5 Cytokines and Chemokines

The unique, immune-privileged nature of the brain and spinal cord has enabled studies to determine the basis of T cell recruitment and function within the CNS. As

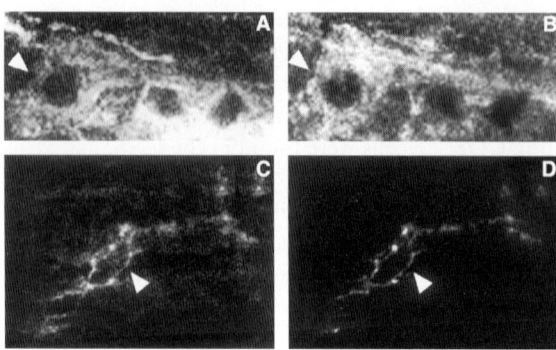

Fig. 1. Expression of MHC class I by oligodendrocytes and microglia. Transgenic mice expressing the LCMV-NP and B7-1 in oligodendrocytes were infected intraperitoneally with LCMV. Although the virus was cleared by 7 days post-infection, and the virus did not infect the CNS, a chronic inflammation of the CNS occurred. At 2 months post-infection, double label immunofluorescence imaged with confocal microscopy was used to show MHC class I expression on oligodendrocytes (*Panels A* and *B*) and microglia (*Panels C* and *D*). Oligodendrocytes were labeled with antibodies to MHC class I (Bachem, Torrance, CA) (*Panel A*) and 2',3'-cyclic nucleotide-3'-phosphodiesterase (CNP) (Sigma, St. Louis, MO) (*Panel B*). Microglia were labeled with antibodies to MHC class I (*Panel C*) and CD11b (Pharmingen, San Diego, CA) (*Panel D*)

described above, cytokines appear to play an important role in immune cell recruitment to, and viral clearance from the CNS. IFN-γ was first noted in the cerebrospinal fluid of mice with acute LCMV, coincident with overt disease (FREI et al. 1988). Subsequently, interleukin-6 (IL-6) was found in mice acutely infected with both LCMV and vesicular stomatitis virus (VSV). Interestingly, while T cell depletion reduced IL-6 levels in LCMV-infected mice, IL-6 was still detectable in VSV-infected mice, suggesting that resident brain cells (astrocytes and microglia) can synthesize cytokines, and do so in an apparently virus-specific manner (FREI et al. 1989). With the advent of RNase protection assays that afford the opportunity to monitor the RNA expression of multiple related genes, it was determined that many cytokines, including TNF-alpha, IL-1 alpha and IL-1 beta, along with IL-6 and IFN-γ, were detectable soon after IC challenge with LCMV (CAMPBELL et al. 1994). Interestingly, even in athymic mice (that do not have any mature T cells), expression of these cytokines was detectable, again indicating that cells within the CNS are a rich source of antiviral cytokines. Such cytokines may function not only to recruit and activate the host response, but may contribute to CNS disease as well, since depletion of interferon alpha and beta prevented the development of lymphocytic choriomeningitis, favoring the establishment of a persistent infection (SANDBERG et al. 1994).

Finally, LCMV infection has been of importance in revealing the role of chemokines, small, chemotactic proteins that can be synthesized by virtually all cells of the host. Lethal challenge with LCMV induces the expression of two T cell promoting chemokines, IP-10 and RANTES, by 3 days post-infection, prior to the entry of T cells, suggesting that intrathecally derived chemokines may influence the entry of antiviral T cells into the LCMV-infected brain (ASENSIO and CAMPBELL 1997).

3.6 A New Model of LCMV-Induced Hydrocephalus

While uncommon, LCMV can infect humans. Given the natural route of transmission of LCMV in mice from infected mother to progeny, it was of interest to determine whether this route of infection may also occur in humans. A number of reports have linked LCMV infection of pregnant humans to intrauterine LCMV infection of offspring (BARTON et al. 1995; BECHTEL et al. 1997; WRIGHT et al. 1997; ENDERS 1999). Congenital infection of humans can result in the death of the fetus or severe birth defects in surviving infants. Birth defects that have been associated with LCMV infection include hydrocephalus, microcephalus, chorioretinitis, mental retardation and psychomotor retardation (WRIGHT et al. 1997). In one study, nine out of 28 children (32%) with congenital hydrocephalus were serologically positive for LCMV infection, suggesting that LCMV may be a major cause of congenital hydrocephalus (SHEINBERGAS 1975). In recent studies, intraperitoneal inoculation of high doses ($>2 \times 10^5$ PFU) of LCMV-Arm into adult C57Bl/6 mice deficient in IFN-γ expression (IFN-$\gamma^{-/-}$) resulted in the development of hydrocephalus at 6–8 weeks post-infection in about 50% of the mice (L.P. Shriver and C.F. Evans, manuscript in preparation). Infectious virus could be isolated from the brains of mice with hydrocephalus, and viral antigen was detected within the CNS, particularly in cells with neuronal morphology. Massive periventricular infiltration by $CD8^+$ and $CD4^+$ T cells was observed around the enlarged ventricles (Fig. 2). Although neutrophils are not normally major components of LCMV-induced inflammation, large numbers of infiltrating neutrophils were also detected near the enlarged ventricles. IFN-γ wild-type mice infected with the same dose of LCMV by the same route cleared the infection by 2 weeks in the absence of pathology. TISHON et al. (1995) demonstrated that adoptive transfer of LCMV memory CTL from IFN-$\gamma^{-/-}$ mice into wild-type mice

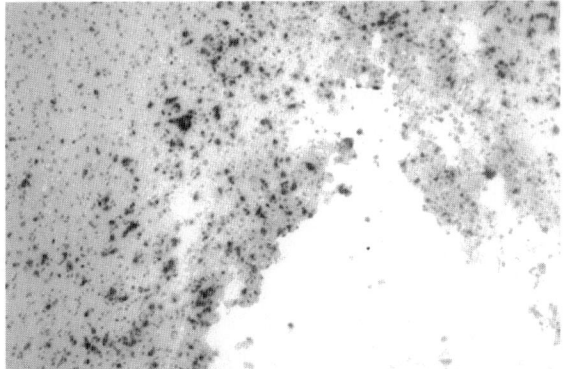

Fig. 2. $CD8^+$ T cell infiltration associated with hydrocephalus in an LCMV-infected IFN-$\gamma^{-/-}$ mouse. A monoclonal antibody to CD8 (Pharmingen, San Diego, CA) was used to identify $CD8^+$ T cells surrounding an enlarged ventricle in an IFN-$\gamma^{-/-}$ mouse that had received an intraperitoneal infection with LCMV 6 weeks previously. The chromagen used in the immunostaining procedure was diaminobenzidine, which gives a brown product where primary antibody bound. ×40

persistently infected with LCMV did not result in viral clearance, indicating that IFN-γ is required for clearance of a persistent infection. Thus, it appears that the expression of IFN-γ in wild-type mice helps to control viremia and prevent CNS inflammation and the development of hydrocephalus. Future studies in this new mouse model will address the mechanism by which hydrocephalus is induced, and may be of value in devising strategies to cure or prevent LCMV-induced hydrocephalus in human infants.

4 The Role of Host Immunity in Neuroinvasion

In order for a virus to establish a neuronal infection, it must spread from its primary site of infection to the central nervous system (CNS) before immune-mediated clearance occurs. Thus, the kinetics of neuroinvasion and the efficiency of immune induction play critical roles in the pathogenesis of neurotropic viral infections. To determine the neuroinvasive capacity of LCMV in the presence of an existent, but compromised, CTL response, the course of LCMV infection was examined in mice that possess only 10% of the normal complement of T lymphocytes, due to the lack of the CD3 δ subunit of the T cell receptor complex (CD3 δ KO mice) (DAVE et al. 1997). The CD3 complex is comprised of four subunits (gamma, delta, epsilon and zeta), each with extensive intracellular domains. Engagement of TCRs by peptide/MHC complexes on the surface of antigen presenting cells (APCs) results in CD3 complex-mediated signal transduction across the T cell membrane, culminating in the activation or anergy of engaged T cells (KAPPES et al. 1995). Mice which have a homozygous deletion of the CD3 delta subunit are impaired in positive T cell selection as evidenced by a 90% reduction in the number of single positive $CD4^+$ or $CD8^+$ T cells (DAVE et al. 1997). Moreover, mature peripheral T cells exhibit significantly lower surface TCR levels, thereby effectively reducing the strength of the signal generated upon TCR-peptide/MHC engagement. Thus, these mice represented a unique model for studying compromised host immune system-pathogen interactions.

Unlike immunocompetent mice which produced a massive immune response that caused death by 6–7 days post-infection, CD3 δ KO mice infected intracranially with LCMV mounted a weak response and survived (KAPPES et al. 2000). The presence of viral antigen gradually shifted from the class I MHC-positive meninges and ependyma to class I MHC-deficient CNS neurons 10–30 days post-inoculation. The infected CD3 δ KO mice developed a delayed T cell response which suppressed virus replication in peripheral tissues but not in the CNS. Subsequent adoptive transfer experiments supported the hypothesis that the lack of clearance from neurons was due to sequestration of LCMV in an immune-privileged cell type. Based on these results, it was proposed that a critical parameter in the pathogenesis of neurotropic viruses is the rate of immune activation; individuals with impaired T cell responses may be more vulnerable to persisting CNS infections.

5 Using LCMV to Model Human Autoimmune Demyelinating Disease

Multiple sclerosis (MS) is the most common human demyelinating disease. Although the causative agents of MS are not known, autoimmunity to myelin antigens has been implicated from studies of human MS patients and of various animal models. There have been many studies of T cell recognition of viral epitopes that have homology with "self" epitopes from various myelin proteins. As a result, many examples of cross-reactivity of T cells to viral and human peptides have been demonstrated, a concept termed molecular mimicry (OLDSTONE 1987; WUCHERPFENNIG and STOMINGER 1995; OLDSTONE 1998). A transgenic mouse model was developed using LCMV as a tool to investigate whether a virus could stimulate CNS autoimmune, demyelinating disease via molecular mimicry. In this model, the nucleoprotein (NP) of LCMV was expressed in oligodendrocytes of transgenic mice under the control of the myelin basic protein (MBP) promoter (MBP-NP mice). Intraperitoneal infection of adult mice with LCMV-Arm leads to infection of many tissues in the periphery but not the CNS, and the virus is cleared within 7–10 days by a vigorous anti-viral $CD8^+$ cytotoxic T cell response (WHITTON and OLDSTONE 1988; BORROW 1997). When transgenic MBP-NP mice were infected with LCMV, the virus was cleared with typical kinetics, and the mice developed anti-viral CTL and antibody responses equivalent to nontransgenic mice. Immunohistochemical analyses of the brains and spinal cords of transgenic mice infected with LCMV showed focal areas of infiltrating $CD8^+$ and $CD4^+$ T cells in white matter, and these cells persisted for the lifetime of the mice (Fig. 3). These infiltrates were accompanied by the upregulation of MHC class I and II molecules on a number of cell types, and local activation of microglia and astrocytes. When lymphocytes were isolated from the CNS of the MBP-NP mice and stimulated in vitro with LCMV, they were found to effectively lyse MHC-matched target cells infected with a recombinant vaccinia virus expressing the NP transgene. These results indicated that self-reactive T cells (since the LCMV-NP was a "self" protein in the transgenic mice) were present in the inflammatory infiltrates. Despite the chronic presence of autoreactive T cells in the CNS, no significant damage to oligodendrocytes was detected in this model.

These results are in striking contrast to an analogous model of autoimmunity in the pancreas. When the LCMV NP or GP were expressed as transgenes in the β cells of the pancreas, LCMV infection triggered infiltration of autoreactive T cells that destroyed the β cells and resulted in diabetes (OLDSTONE et al. 1991). It is not clear whether the immune responses are different in the CNS versus the pancreatic models, or whether the nature of the target organs influences the disease outcomes. Since the CNS has many features that suppress or restrict immune responses, T cells that can cause tissue damage in a peripheral organ may be inhibited within the CNS. For example, CNS-derived gangliosides can suppress T cell effector functions, and TGF-β, which is constitutively expressed in the CNS, has been shown to

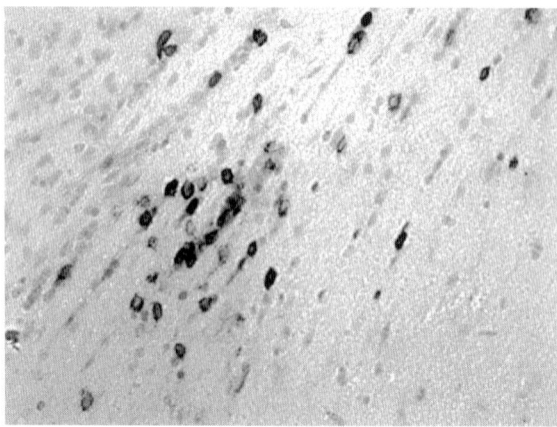

Fig. 3. Identification of chronic inflammation in a transgenic mouse model of virus-induced CNS disease. A monoclonal antibody to CD8 (Pharmingen, San Diego, CA) was used to detect CD8$^+$ T cells in CNS white matter. The tissue is from a mouse that expressed the LCMV NP as a transgene in oligodendrocytes. The mouse had received a peripheral infection with LCMV 12 months before sacrifice, and this staining identifies infiltrating CD8$^+$ T cells that are representative of focal areas of chronic inflammation observed in such mice(Evans et al. 1996). The chromagen used in the immunostaining procedure was diaminobenzidine, which gives a brown product where primary antibody bound. ×200

inhibit CTL activity (GORDON et al. 1998; IRANI 1998). It is likely that the generation of autoimmune-mediated pathology in the CNS occurs by different mechanisms than in the periphery, and dissecting out the differences will lead to a better understanding of immune and autoimmune responses within the brain.

Acknowledgements. We thank L.P. Shriver for excellent technical assistance and scientific suggestions, and S.B. Ugs and E. Baxter Kelby for helpful comments. This work was supported in part by a grant to C.F.E. from the National Multiple Sclerosis Society, and by National Institutes of Health grants NS37135 (to C.F.E), and MH56951 (to G.F.R.). J.M.R. was supported by a NIH Training Grant (5 T32 AG00080-21) and a postdoctoral fellowship from the National Multiple Sclerosis Society. C.E.P. was supported by NIH Training Grant CA06927. This is manuscript #13740-NP from The Scripps Research Institute.

References

Ahmed R, Oldstone MBA (1988) Organ-specific selection of viral variants during chronic infection. J Exp Med 167:1719–1724
Ahmed R, Salmi A, Butler LF, Chiller JM, Oldstone MBA (1984) Selection of genetic variants of lymphocytic choriomeningitis virus in spleens of persistently infected mice. J Exp Med 60:521–540
Allan JE, Doherty PC (1985) Immune T cells can protect or induce fatal neurological disease in murine LCMV. Cellular Immunology 90:401–407
Andersen IH, Marker O, Thomsen AR (1991) Breakdown of blood-brain barrier function in the murine lymphocytic choriomeningitis virus infection mediated by virus-specific CD8+ T cells. J Neuroimmunol 31:155–163

Asensio VC, Campbell IL (1997) Chemokine gene expression in the brains of mice with lymphocytic choriomeningitis. J Virol 71:7832–7840

Barton LL, Peters CJ, Ksiazek TG (1995) Lymphocytic choriomeningitis virus: an unrecognized teratogenic pathogen. Emerg Inf Dis 4:152–153

Bechtel RT, Haught KA, Mets MB (1997) Lymphocytic choriomeningitis virus: a new addition to the TORCH evaluation. Arch Ophthalmol 115:680–681

Borrow P (1997) Mechanisms of viral clearance and persistence. J Viral Hepat 4:16–24

Brot MD, Rall GF, Oldstone MBA, Koob GF, Gold LH (1997) Cognitive deficit remains following clearance of persistent viral infection in mice. J Neurovirology 3:265–273

Buchmeier MJ, Zajac AJ (1999) Lymphocytic choriomeningitis virus. In: Ahmed R, Chen IC (eds) Persistent viral infections. John Wiley, Chichester, pp 575–605

Campbell IL, Hobbs MV, Kemper P, Oldstone MB (1994) Cerebral expression of multiple cytokine genes in mice with LCMV. J Immunol 152:716–723

Cao W, Oldstone MB, De La Torre JC (1997) Viral persistent infection affects both transcriptional and posttranscriptional regulation of neuron-specific molecule GAP43. Virology 230:147–154

Cheng-Mayer C, Levy JA (1988) Distinct biological and serological properties of human immunodeficiency viruses from the brain. Ann Neurol 23:S58–S61

Cole GA, Gilden D, Monjan A, Nathanson N (1971) LCMV: pathogenesis of acute CNS disease. Federation Proceedings 30:1831–1841

Cole GA, Nathanson N, Prendergast RA (1972) Requirement for theta-bearing cells in LCMV-induced CNS disease. Nature 238:335–338

Dave VP, Cao Z, Browne C, Alarcon B, Fernandez-Miguel G, Lafille J, Hera Adl, Tonegawa S, Kappes DJ (1997) CD3 delta deficiency arrests development of the alpha-beta but not the gamma-delta cell lineage. EMBO J 16:1360–1370

de la Torre JC, Mallory M, Brot M, Gold L, Koob G, Oldstone MBA, Masliah E (1996) Viral persistence in neurons alters synaptic plasticity and cognitive functions without destruction of brain cells. Virology 220:508–515

de la Torre JC, Oldstone MB (1992) Selective disruption of growth hormone transcription machinery by viral infection. Proc Natl Acad Sci USA 20:9939–9943

Dockter J, Evans CF, Tishon A, Oldstone MBA (1996) Competitive selection in vivo by a cell for one variant over another: implications for RNA virus quasispecies in vivo. J Virol 70:1799–1803

Enders G (1999) Congenital lymphocytic choriomeningitis virus infection: An underdiagnosed disease. Pediatr Infect Dis J 18:652–655

Epstein LG, Kuiken C, Blumberg BM, Hartman S, Sharer LR, Clement M, Goudsmit J (1991) HIV-1 V3 domain variation in brain and spleen of children with AIDS: tissue-specific evolution within host-determined quasispecies. Virology 180:583–590

Evans CF, Borrow P, Torre JCdl, Oldstone MBA (1994) Virus-induced immunosuppression: kinetic analysis of the selection of a mutation associated with viral persistence. J Virol 68:7367–7373

Evans CF, Horwitz MS, Hobbs MV, Oldstone MBA (1996) Viral infection of transgenic mice expressing a viral protein in oligodendrocytes leads to chronic central nervous system autoimmune disease. J Exp Med 184:2371–2384

Fazakerley JK, Southern P, Bloom F, Buchmeier MJ (1991) High resolution in situ hybridization to determine the cellular distribution of LCMV RNA in the tissues of persistently infected mice: relevance to arenavirus disease and mechanisms of viral persistence. J Gen Virol 72:1611–1625

Frei K, Leist TP, Meager A, Gallo P, Leppert D, Zinkernagel RM, Fontana A (1988) Production of B cell stimulatory factor-2 and interferon gamma in the central nervous system during viral meningitis and encephalitis. Evaluation in a murine model infection and in patients. J Exp Med 168:449–453

Frei K, Malipiero UV, Leist TP, Zinkernagel RM, Schwab ME, Fontana A (1989) On the cellular source and function of interleukin 6 produced in the central nervous system in viral disease. Europ J Immunol 19:689–694

Gold LH, Brot MD, Polis I, Schroeder R, Tishon A, Torre JCdl, Oldstone MBA, Koob GF (1994) Behavioral effects of persistent LCMV infection in mice. Behav Neural Biol 62:100–109

Gopas J, Itzhaky D, Segev Y, Salzberg S, Trink B, Isakov N, Rager-Zisman B (1992) Persistent measles virus infection enhances class I MHC expression and immunogenicity of murine neuroblastoma cells. Cancer Immunol Immunother 34:313–320

Gordon LB, Nolan SC, Ksander BR, Knopf PM, Harling-Berg CJ (1998) Normal CSF suppresses the in vitro development of cytotoxic T cells: role of the brain microenvironment in CNS immune regulation. J Neuroimmunol 88:77–84

Hickey WF, Hsu BL, Kimura H (1991) T-lymphocyte entry into the CNS. J Neuroscience Res 28:254–260

Hotchin J (1962) The biology of LCMV infection: virus-induced immune disease. Cold Spring Harbor Symposium Quantitative Biology 27:479–499
Hotchin J, Seegal R (1977) Virus-induced behavioral alteration of mice. Science 196:671–674
Irani DN (1998) The susceptibility of mice to immune-mediated neurologic disease correlates with the degree to which their lymphocytes resist the effects of brain-derived gangliosides. J Immunol 161:2746–2752
Joly E, Mucke L, Oldstone MBA (1991) Viral persistence in neurons explained by a lack of MHC class I expression. Science 253:1283–1285
Joly E, Oldstone MBA (1992) Neuronal cells are deficient in loading peptides onto MHC class I molecules. Neuron
Kagi D, Liedermann B, Burki K, Seller P, Odermatt B, Olsen KJ, Podack E, Zinkernagel R, Hengartner H (1994) Cytotoxicity mediated by T cells and NK cells is greatly impaired in perforin-deficient mice. Nature 369:31–37
Kappes DJ, Alarcon B, Reguiero J (1995) T lymphocyte receptor deficiencies. Curr Opin Immunol 7: 441–447
Kappes DJ, Lawrence DM, Vaughn MM, Dave VP, Belman AR, Rall GF (2000) Protection of CD3 delta knockout mice from lymphocytic choriomeningitis virus-induced immunopathology: implications for viral neuroinvasion. Virology 269:248–256
Komatsu T, Bi Z, Reiss CS (1996) Interferon gamma-induced type I nitric oxide sythase activity inhibits viral replication in neurons. J Neuroimmunol 68:101–108
Lampson LA (1990) MHC regulation in neural cells: distribution of peripheral and internal β 2 microglobulin and class I molecules in human neuroblastoma cell lines. J Immunol 144:512–520
Lillie RD, Armstrong C (1945) Pathology of lymphocytic choriomeningitis virus. Arch Pathol 40:141–152
Lipkin WI, Battenberg ELF, Bloom FE, Oldstone MBA (1988) Viral infection of neurons can depress neurotransmitter mRNA levels without histologic injury. Brain Res 451:333–339
Marker O, Nielsen MH, Diemer NH (1984) The permeability of the blood-brain barrier in mice suffering from fatal lymphocytic choriomeningitis virus infection. Acta Neuropathol (Berl) 63:229–239
Matloubian M, Somasundaram T, Kolhekar SR, Selvakumar R, Ahmed R (1990) Genetic basis of viral persistence: single amino acid change in the viral glycoprotein affects ability of lymphocytic choriomeningitis virus to persist in adult mice. J Exp Med 172:1043–1048
Monjan AA, Bohl LS, Hudgens GA (1975) Neurobiology of LCM virus infection in rodents. Bull World Health Organ 52:487–492
Monjan AA, Cole GA, Nathanson N (1974) Pathogenesis of cerebellar hypoplasia produced by lymphocytic choriomeningitis virus infection of neonatal rats: protective effect of immunosuppression with anti-lymphoid serum. Infect Immun 10:499–502
Mucke L, Oldstone MBA (1992) The expression of MHC class I antigens in the brain differs markedly in acute and persistent infections with LCMV. J Neuroimmunol 36:193–198
Neumann H, Cavalie A, Jenne DE, Wekerle H (1995) Induction of MHC class I genes in neurons. Science 269:549–552
Neumann H, Schmidt H, Cavalie A, Jenne D, Wekerle H (1997) MHC class I gene expression in single neurons of the central nervous system: differential regulation by interferon gamma and tumor necrosis factor alpha. J Exp Med 185:305–316
Oldstone MB, Homstoen J, Welsh RM (1977) Alterations of acetylcholine enzymes in neuroblastoma cells persistently infected with lymphocytic choriomeningitis virus. J Cell Physiol 91:459–472
Oldstone MBA (1987) Molecular mimicry and autoimmune disease. Cell 50:819–820
Oldstone MBA (1998) Molecular mimicry and immune-mediated disease. FASEB J 12:1255–1265
Oldstone MBA, Blount P, Southern PJ, Lampert PW (1986) Cytoimmunotherapy for persistent virus infection reveals a unique clearance pattern from the central nervous system. Nature 321:239–243
Oldstone MBA, Nerenberg M, Southern P, Price J, Lewicki H (1991) Virus infection triggers insulin-dependent diabetes mellitus in a transgenic model: role of anti self (virus) immune response. Cell 65:319–331
Oldstone MBA, Sinha YN, Blout P, Tishon A, Rodriguez M, von Wendel R, Lampert PW (1982) Virus-induced alterations in homeostasis: alterations in differentiated functions of infected cells in vivo. Science 218:1125–1127
Parra B, Hinton DR, Marten NW, Bergmann CC, Lin MT, Yang CS, Stohlman SA (1999) IFN-gamma is required for viral clearance from central nervous system oligodendroglia. J Immunol 162:1641–1647
Pearce BD, Po CL, Pisell TL, Miller AH (1999) Lymphocytic responses and the gradual hippocampal neuron loss following infection with lymphocytic choriomeningitis virus (LCMV). J Neuroimmunol 101:137–147

Pearce BD, Steffensen SC, Paoletti AD, Henriksen SJ, Buchmeier MJ (1996) Persistent dentate granule cell hyperexcitability after neonatal infection with LCMV. J Neurosci 16:220–228

Pearce BD, Valadi NM, Po CL, Miller AH (2000) Viral infection of developing GABAergic neurons in a model of hippocampal disinhibition [In Process Citation]. Neuroreport 11:2433–2438

Rall GF (1998) CNS neurons: the basis and benefits of low MHC expression. Curr Top Microbiol Immunol 232:115–134

Rall GF, Mucke L, Oldstone MBA (1995) Consequences of cytotoxic T lymphocyte interaction with major histocompatibility complex-expressing neurons in vivo. J Exp Med 182:1201–1212

Rall GF, Oldstone MBA (1995) Virus-neuron-cytotoxic T lymphocyte interactions. Curr Topics Microbiol Immun 202:261–273

Riviere Y, Ahmed R, Southern PJ, Buchmeier MJ, Dutko FJ, Oldstone MB (1985) The S RNA segment of lymphocytic choriomeningitis virus codes for the nucleoprotein and glycoproteins 1 and 2. J Virol 53:966–968

Rodriguez M, Buchmeier MJ, Oldstone MBA, Lampert PW (1983) Ultrastructural localization of viral antigens in the CNS of mice persistently infected with LCMV. Amer J Pathology 110:95–100

Salvato M, Borrow P, Shimomaye E, Oldstone MBA (1991) Molecular basis of viral persistence: a single amino acid change in the glycoprotein of lymphocytic choriomeningitis virus is associated with suppression of the antiviral cytotoxic T-lymphocyte response and establishment of persistence. J Virol 65:1863–1869

Salvato M, E.Shimomaye, Southern P, Oldstone MBA (1988) Virus-lymphocyte interactions: IV. Molecular characterization of LCMV Armstrong (CTL+) small genomic segment and that of its variant clone 13 (CTL$^-$). Virology 164:517–522

Sandberg K, Kemper P, Stalder A, Zhang J, Hobbs MV, Whitton JL, Campbell IL (1994) Altered tissue distribution of viral replication and T cell spreading is pivotal in the protection against fatal lymphocytic choriomeningitis in mice after neutralization of IFN-alph/beta. J Immunol 153:220–231

Schwendemann G, Lohler J, Lehmann-Grube F (1983) Evidence for CTL-target cell interaction in brains of mice infected intracerebrally with LCMV. Acta Neuropathol 61:183–195

Sheinbergas MM (1975) Antibody to lyjphocytic choriomeningitis virus in children with congenital hydrocephalus. Acta Virol 19:165–166

Steinhauer DA, Holland JJ (1987) Rapid evolution of RNA viruses. Annu Rev Microbiol 41:409–433

Ting JP-Y, Takaguchi M, Macchi M, Frelinger JA (1987) The expression and detection of MHC class I antigens on murine neuroblastoma and ependymoblastoma lines. J Neuroimm 14:87–98

Tishon A, Lewicki H, Rall G, Herrath Mv, Oldstone MBA (1995) An essential role for type 1 interferon gamma in terminating persistent viral infection. Virology 212:244–250

Valsamakis A, Riviere Y, Oldstone MBA (1987) Perturbation of differentiated functions in vivo during persistent viral infection. III. Decreased growth hormone mRNA. Virol 156:214–220

Villarete L, somasundarum T, Ahmed R (1994) Tissue-mediated selection of viral variants: correlation between glycoprotein mutation and growth in neuronal cells. J Virology 68:7490–7496

Wekerle M, Linington C, Lassmann H, Meyermann R (1986) Cellular immune reactivity in the CNS. TINS June:271–277

Whitton JL, Oldstone MBA (1988) The recognition of virus-infected cells by cytotoxic T lymphocytes. Int Pediatrics 3:16–21

Wright R, Johnson D, Neumann M, Ksiazek TG, Rollin P, Keech RV, Bonthius DJ, Hitchon P, Grouse CF, Bell WE, Bale JF (1997) Congenital LCMV syndrome: a disease that mimics congenital toxoplasmosis or cytomegalovirus infection. Pediatrics 100:

Wucherpfennig KW, Stominger JL (1995) Molecular mimicry in T cell-mediated autoimmunity: viral peptides activate human T cell clones specific for myelin basic protein. Cell 80:695–705

Contribution of LCMV Towards Deciphering Biology of Quasispecies In Vivo

N. Sevilla[1], E. Domingo[2], and J.C. de la Torre[1]

1 Overview of RNA Virus Evolution: Relevance to Arenaviruses	197
1.1 Error-Prone Replication: Mutation, Recombination, and Reassortment	198
1.2 Quasispecies Swarms	199
1.3 Adaptive Value of Mutant Spectra	201
2 Mutation Frequencies in Arenavirus Populations	202
3 Contribution of LCMV Genetic Variability to Persistence and Disease	202
3.1 Overview of LCMV Biology	202
3.2 Selection of Immunosuppressive Variants During LCMV Persistence	204
3.3 Contribution of LCMV Variants to Virally Induced Growth Hormone Deficiency Syndrome	209
4 Perspectives	215
References	216

1 Overview of RNA Virus Evolution: Relevance to Arenaviruses

Arenaviruses have often been viewed as relatively stable genetically with amino acid sequence homologies of 90%–95% among different strains of lymphocytic choriomeningitis virus (LCMV) and of 44%–63% for homologous proteins of different arenaviruses: LCMV, Pichinde, Junin, Machupo and Lassa (Southern and Bishop 1987; Southern and Oldstone 1988). Yet considerable variation in biological properties among LCMV strains soon became apparent (reviewed in Dutko and Oldstone 1983; Southern and Oldstone 1988). Hotchin already recognized the importance of passage history in determining the biological properties of LCMV (Hotchin 1962). He showed that early mouse brain passages of LCMV-induced immunologic tolerance in newborn mice and mortality was low. In contrast, late mouse brain passages of LCMV lost the tolerance-inducing capacity, and mortality was high. Neonatal infection of certain mouse strains with LCMV strains Armstrong (ARM) and E-350-induced growth hormone deficiency and severe

---

[1] Department of Neuropharmacology, The Scripps Research Institute, 10550 North Torrey Pines Road, IMM-6, La Jolla, CA 92037, USA
[2] Centro de Biologia Molecular Severo Ochoa, Universidad Autonoma de Madrid, Cantoblanco, 28049 Madrid, Spain

hypoglycemia, which frequently resulted in the death of the infected mice, while strains WE and Traub did not cause this syndrome. This difference was associated with the ability of LCMV ARM and E-350, but not WE and Traub, to replicate at high levels in the GH-producing cells in the anterior pituitary (OLDSTONE et al. 1985). Other biological differences among LCMV strains include the capacity of the virus to invade β-cells in the islets of Langerhans, to cause alterations in glucose tolerance, or differences in the formation of immune complexes and lethality for adult guinea pigs (SOUTHERN and OLDSTONE 1988), as well as to induce generalized immunosuppression in adult mice. Selection of LCMV variants with distinguishable phenotypes occurs in different organs of infected mice. Most isolates from the central nervous system of mice persistently infected since birth with ARM tend to produce acute infection in adult mice, whereas isolates from the spleen of the same mice tend to persist (AHMED and OLDSTONE 1988). These dramatic phenotypic differences among genetically closely related LCMV isolates led to the suggestion that a few amino acid replacements in LCMV proteins could suffice to produce important alterations in its biological properties (SOUTHERN and OLDSTONE 1988). This opens the important question of to what extent a quasispecies dynamics may play a role in arenavirus adaptability and pathogenesis. To approach this question we will first review briefly the main steps involved in the generation of RNA virus diversity and the adaptive manifestations of quasispecies swarms, with reference to arenaviruses when appropriate. Then we will assess the current evidence of quasispecies involvement in arenavirus adaptability and pathogenesis.

1.1 Error-Prone Replication: Mutation, Recombination, and Reassortment

Mutation rates during RNA genome replication are in the range of 10^{-3}–10^{-5} substitutions per nucleotide copied (BATSCHELET et al. 1976; HOLLAND et al. 1992; DRAKE and HOLLAND 1999). This means that RNA genomes of 3,000–32,000 nucleotides in length (which is the range of the non-defective viral RNAs characterized to date) are highly unlikely to undergo two rounds of template copying without introducing an incorrect nucleotide into the product. Insertions and deletions (*indels*) are generally less frequent but they may also contribute to RNA genome variation. High mutation rates are expected from the lack of proofreading-repair activities in viral RNA-dependent RNA polymerases (RdRp), as evidenced by structural and functional studies (STEINHAUER et al. 1992; FRIEDBERG et al. 1995; SOUSA 1996; HANSEN et al. 1997; MEYERHANS and VARTANIAN 1999). Sequence alignment of the L gene product of LCMV with other viral RdRp does not provide any evidence of the presence of a proofreading exonuclease domain, although detailed enzyme activity studies have not been performed with any of the arenavirus L polymerases. Also, it cannot be excluded that the intracellular replicase complex could include some cellular subunit with a capacity for error correction, perhaps of the type shown for HIV-1 reverse transcriptase when excising some chain terminating nucleotides (MEYER et al. 1998, 1999).

Many viruses, particularly some positive-strand RNA viruses such as picornaviruses and coronaviruses, show active recombination in vivo (NAGY and SIMON 1987; LAI 1992; GROMEIER et al. 1999). Recombination has been viewed as having potentially two opposite effects: generation of divergent genome combinations to produce large evolutionary jumps, or, conversely, the possibility to rescue fit genomes from debilitated, mutated parents. Recombination also happens in negative-strand RNA viruses, like LCMV; however, it occurs several orders of magnitude less frequency than in most positive-strand RNA viruses. Moreover, there is no experimental evidence for recombination between two negative-strand ribonucleoproteins (RNPs), suggesting that perhaps only intramolecular recombination plays a role in negative-strand RNA viruses. Arenaviruses, however, as with other segmented RNA viruses, can undergo segment reassortment as shown by the analysis of the progeny of coinfection of the same cell by two distinguishable parental viruses. The formation of reassortants can be, at least in cell culture, very efficient. In some cases their frequency is close to the theoretical predicted maximum of 50%. However, even among genetically closely related viruses, some particular combinations appear to be excluded. Hence, some specific reassortants are never recovered (TENG et al. 1996a). Arenavirus reassortants can display pathogenic potential, which is not manifested by any of the two parental viruses (RIVIERE and OLDSTONE 1986). It has been suggested that reassortment may have contributed to the emergence of new arenaviruses with new disease associations (SOUTHERN and OLDSTONE 1988).

1.2 Quasispecies Swarms

High mutation rates, and also in some systems, active recombination and genome segment reassortments, as in the case of arenaviruses, constitute a triggering incentive for a complex chain of events. Pressure for variation entails ranking for performance. Replicative ensembles of RNA viruses are organized as dynamic mutant distributions termed viral quasispecies (EIGEN 1971, 1996; EIGEN and BIEBRICHER 1988; HOLLAND et al. 1992; DOMINGO 1999) (Fig. 1). In each infected cell, replicating genomes compete for whatever resources might be limiting (nucleotide substrates, ribosomes, membrane sites, etc.), and needed to complete the virus life cycle. RNAs replicating in two separate cells do not enter into competition at least immediately and, therefore, must be regarded as separate quasispecies in separate compartments. However, the progeny from different cells (or even tissues or organs) may meet in a common compartment (a new organ such as the brain, or the blood stream) and compete for binding to cell receptors and to replicate in new subsets of cells and tissues. The process of replication with errors lends itself to spatial and temporal heterogeneities in RNA genome populations, as is indeed observed experimentally among many other examples of human, animal and plant RNA genetic elements (MEYERHANS et al. 1989; MARTELL et al. 1992; SALA et al. 1994; BORROW et al. 1997; KARLSSON et al. 1999; KIMATA et al. 1999). Selection of organ-specific LCMV variants has been extensively documented (AHMED and OLDSTONE 1988).

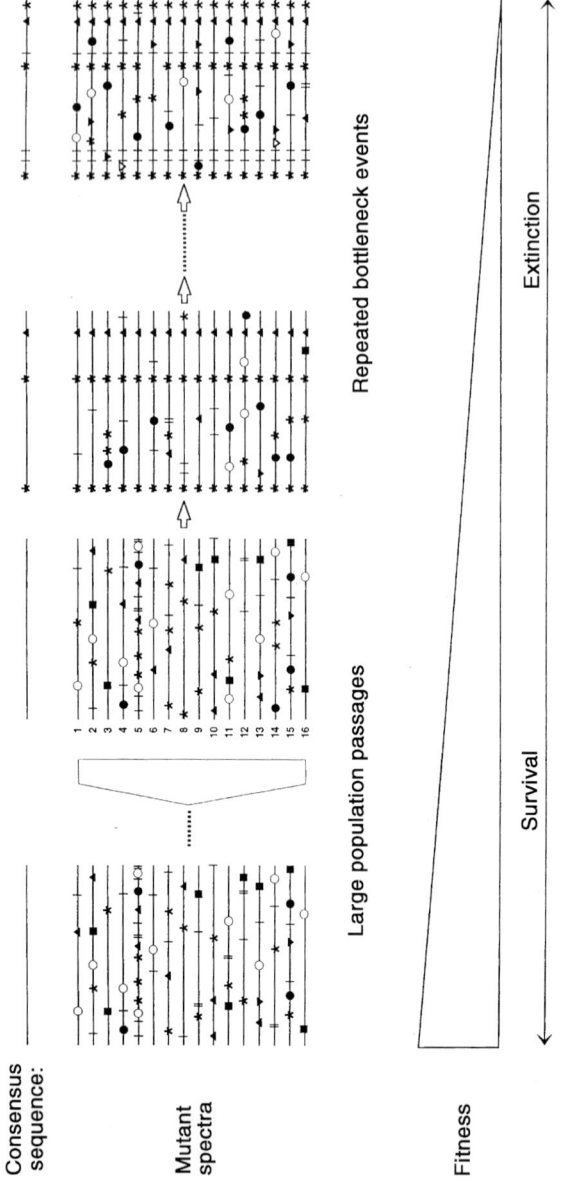

Fig. 1. Schematic representation of possible outcomes in the evolution of viral quasispecies. Individual RNA viral genomes are depicted as *horizontal lines* and mutations as *symbols on the lines*. Starting with the second mutant distribution from the *left* (with numbered genomes), large population passages (*big arrow, left*) generally lead to fitness gain and virus survival. In contrast, repeated bottlenecks (*small arrows, right*) lead to fitness decrease, and the virus approaches extinction. Fitness gain or loss is determined by the initial viral fitness and by the size of genetic bottleneck (NOVELLA et al. 1995). For an introduction to viral quasispecies and fitness variations see EIGEN and BIEBRICHER (1988), DOMINGO and HOLLAND (1997) and DOMINGO (1999). Virologists use a definition of quasispecies which is an extension of the original theoretical concept developed by Eigen, Schuster and their colleagues: viral quasispecies are dynamic distributions of non-identical but closely related mutant and recombinant viral genomes subjected to a continuous process of genetic variation, competition and selection, and which act as a unit of selection. Mutant distributions are often perturbed by stochastic effects, such as bottleneck events, as depicted in the scheme. (Figure reproduced from DOMINGO 1999, with permission from Academic Press)

1.3 Adaptive Value of Mutant Spectra

Viral quasispecies contain multitudes of variants that can manifest alternative phenotypes in response to environmental demands. Current evidence that mutant distributions affect the biology of RNA viruses can be divided in two main lines: (a) demonstration of the presence and selection of variants in quasispecies evolving in vivo, and (b) recent evidence that mutant spectrum complexity may have a predictive value with regard to the response to antiviral treatments and to shifts from acute into chronic phases of viral disease.

Regarding the presence of biologically significant variants in quasispecies, a well documented example is the presence, already, of HIV variants resistant to viral inhibitors in patients who have not been subjected to therapy with the inhibitors (NAJERA et al. 1995; HAVLIR et al. 1996; LECH et al. 1996; RIBEIRO et al. 1998). Likewise, selection of CTL-escape HIV-1 and SIV mutants is increasingly regarded as a likely contributor to viral persistence and pathogenesis (BORROW et al. 1997; GOULDER et al. 1997; MCMICHAEL and PHILLIPS 1997; BORROW and SHAW 1998; QUINONES-MATEU et al. 1998; EVANS et al. 1999; KIMATA et al. 1999). Therefore, although most mutations in mutant swarms of RNA viruses may not be of short-term or even long-term selective value for the virus (SALA and WAIN-HOBSON 1999), the evolution of viral quasispecies can be adaptive, and may exert an influence in viral pathogenesis (FORNS et al. 1999; DOMINGO et al. 2000; FLINT et al. 2000). Additionally, fitness variations among components of mutant spectra have been documented in vivo in animals infected with FMDV but not subjected to any therapy (CARRILLO et al. 1998).

A second aspect that has produced increasing evidence is the effect of the complexity of mutant spectra (quantified by mutation frequency or by the number of genomes with a different sequence – the Shannon entropy) in pathogenic potential, response to treatments and the establishment of chronicity. ROWE et al. (1997) provided evidence that the mutant spectrum complexity of the coronavirus murine hepatitis virus, which contains multiple point mutants and recombinants, influenced the pathogenic potential of this virus in mice. Since the demonstration that the important human pathogen hepatitis C virus (HCV) followed a quasispecies dynamics in vivo (MARTELL et al. 1992) there has been great interest in elucidating whether parameters that can be quantified in HCV quasispecies could anticipate responses to treatment (frequently a combination of interferon and ribavirin) or to the chronicity of infection. Although the outcome of a viral infection is necessarily the result of a complex set of host and viral influences, recent results with HCV of genotype 1B suggest that sustained viral clearance as a result of treatment was associated with low viral load and with low mutant spectrum complexity (PAWLOTSKY et al. 1998). More recently, the establishment of chronicity in HCV as the outcome of an acute infection has been correlated with the evolution of quasispecies from infected patients (FARCI et al. 2000). Thus the trend in the last few years has been in revealing a number of the implications of quasispecies complexity for disease progression and response to treatment. What is the current evidence for arenaviruses?

2 Mutation Frequencies in Arenavirus Populations

No systematic measurements of mutant spectrum complexity in arenavirus populations have been reported. Yet early studies suggested high mutation rates during LCMV replication (HOTCHIN 1962). CTL-escape mutants were selected in T-cell receptor (TCR) transgenic mice infected with high doses (10^6PFU) of LCMV (PIRCHER et al. 1990). Escape mutants included single amino acid replacements in T cell epitopes. CTLs from infected, transgenic mice lysed target cells coated with peptides representing the wild-type T cell epitopes, but not cells coated with variant peptides (PIRCHER et al. 1990). Observations in cell culture also suggest high frequencies of CTL-escape mutants. The latter were readily selected when 3×10^5 B6-SV40 primary fibroblasts, under CTL selective pressure, were infected at a multiplicity of infection (m.o.i.) of 1 or 10^{-3}, but not 10^{-5}PFU/cell, suggesting that escape mutants either preexisted in the LCMV population or occurred early in the course of one replication round in the infected cells (AEBISCHER et al. 1991). In another study, escape mutants were obtained by infecting 6×10^2PFU of LCMV ARM (LEWICKI et al. 1995). Although frequencies of escape variants could not be estimated from these studies, the fact that they were obtained after 1h infection with 6×10^2PFU suggests that their frequency should be at least in the range of 10^{-4}–10^{-5}.

In spite of a strong CTL response involved in the control of viremia in LCMV infections in mice, a neutralizing antibody response may be involved in the control of viremia and viral clearance in the absence of $CD8^+$ T cells (PLANZ et al. 1997; CIUREA et al. 2000). High doses (2×10^6PFU) of LCMV-WE led to an enhanced antibody response, and infection of $CD8^{-/-}$ mice resulted in early clearance of virus. Yet neutralization-resistant LCMV mutants emerged after weeks of CTL absence. The mutants included amino acid substitutions within GP1, and manifested a strong tendency to persist in mice (CIUREA et al. 2000). In these analyses, in addition to amino acid replacements, silent mutations (synonymous, not leading to amino acid substitutions) were found at frequencies of 3×10^{-4} substitutions per nucleotide, which would argue for considerable complexity of LCMV in these infected mice (CIUREA et al. 2000).

3 Contribution of LCMV Genetic Variability to Persistence and Disease

3.1 Overview of LCMV Biology

LCMV, the prototypic member of the family Arenaviridae, has been used extensively as a model system for the study of viral persistence and pathogenesis. LCMV can infect its natural host, the mouse, either acutely or persistently. The outcome of the infection depends on the genetics, age and immune status of the host, as well as on viral genetic determinants, route of inoculation and amount of virus inoculum (BUCHMEIER et al. 1980; LEHMANN-GRUBE 1984; BORROW and OLDSTONE 1997). Investigations using this viral model have been central to defining the basic virologic and immunologic concepts including the major histocompatibility complex (MHC) restriction of T-cell recognition (ZINKERNAGEL and DOHERTY 1979), tolerance, immunological memory, immune mediated pathology, mechanisms of viral clearance, as well as the strategies by which viruses evade the host immune responses (OLDSTONE 1989, 1991, 1998). Studies of LCMV virus-host interaction have also uncovered the ability of non-cytolytic persistent viruses to induce disease by interfering with specialized functions of infected cells, revealing a new way by which viruses do harm in the absence of the classic hallmarks of cytolysis and inflammation (OLDSTONE and DE LA TORRE 1993; OLDSTONE 1984; DE LA TORRE and OLDSTONE 1996).

LCMV is an enveloped virus with a bisegmented negative single-stranded RNA genome. The genomic L (ca 7.3kb) and S (ca 3.5kb) RNA segments contain non-overlapping coding information. Each RNA segment has an ambisense coding strategy, encoding two polypeptides in opposite orientation, separated by an intergenic region. The S RNA encodes the viral glycoprotein precursor, GPC, (ca 75kDa) and the nucleoprotein, NP, (ca 63kDa) (AUPERIN et al. 1984; BUCHMEIER et al. 1987; FRANZE-FERNANDEZ et al. 1987; SOUTHERN et al. 1987), whereas the L RNA encodes the putative viral L polymerase (ca 200kDa) and a small polypeptide Z (ca 11kDa) that contains a RING finger-domain (SINGH et al. 1987; IAPALUCCI et al. 1989a,b; SALVATO et al. 1989; SALVATO and SHIMOMAYE 1989), whose function is only poorly understood (SALVATO et al. 1989, 1992; GARCIN et al. 1993; SALVATO 1993). LCMV RNA synthesis is confined to the cytoplasm of infected cells, although whether a nuclear component is involved is still controversial. A detailed description of the molecular biology of arenaviruses is presented in the chapter by Southern et al. The viral glycoprotein precursor GPC is post-translationally cleaved to yield the two mature virion glycoproteins GP-1 (40–46kDa) and GP-2 (35kDa) (WRIGHT et al. 1990). GP-1 is located at the top of the spike, away from the membrane, and is held in place by ionic interactions with the N-terminus of the transmembrane GP-2, that forms the stalk of the spike (BURNS and BUCHMEIER 1993). Neutralizing antibodies predominantly recognize conformational epitopes within GP-1, which is the virion attachment protein that mediates virus interaction with host cell surface receptors. Alpha-dystroglycan (α-DG), has been recently identified as the receptor for LCMV (CAO et al. 1998), but the nature of the interactions between α-DG and GP-1 remain to be determined.

Closely related variants of the same LCMV strain can display remarkable phenotypic differences in vivo, including those related to disease manifestations. We will next discuss two different LCMV-host interactions that illustrate how the generation and selection of specific variants during the natural course of LCMV infection can significantly influence the outcome of the infection.

3.2 Selection of Immunosuppressive Variants During LCMV Persistence

Mice infected at birth or in utero with LCMV become persistently infected for life because they are unable to mount an effective antiviral CTL response. This is due to virus replication in the thymus which leads to a specific deletion of LCMV-reactive T cells by a negative selection process similar to that involved in the elimination of high affinity autoreactive T cells (MATLOUBIAN et al. 1990; SALVATO et al. 1991). This long-term persistence with continuous virus replication permits the emergence of variants that have a growth advantage in certain cell types (AHMED et al. 1984). These viral variants have different biological properties, which correlate with the type of tissue from which they are isolated. One type of variant predominates in the CNS (TISHON et al. 1993; EVANS et al. 1994), and another type predominates in lymphocytes and macrophages of the immune system (PAREKH and BUCHMEIER 1986; AHMED et al. 1991; BORROW and OLDSTONE 1992; TISHON et al. 1993). Most of the CNS isolates, exhibit the same phenotype as the parental ARM strain used to infect mice. These viruses cause an acute infection when injected intravenously (i.v.) into immunocompetent adult mice. This infection is efficiently terminated within seven to ten days by the host anti-LCMV $CD8^+$ cytotoxic T lymphocytes (CTL). These viruses are referred to as CTL^+P^-. In contrast, the vast majority of variants isolated from lymphocytes and macrophages, cause a generalized immunosuppression as judged by the absence of a LCMV-specific CTL response, absence of CTL response to other viruses and absence to antibody responses to soluble and particulated foreign antigens (BARANOWSKI et al. 2000). These viruses are referred to as CTL^-P^+.

Viral determinants involved in this immunosuppressive phenotype have been mapped. Clone 13 (Cl 13), the prototypic immunosuppressive variant, was derived from the non-immunosuppressive parental ARM. Mice infected i.v. with reassortant viruses between Cl 13 and ARM showed that only reassortants containing the S segment from Cl 13 caused immunosuppression, whereas mice infected with the reassortants whose S fragment came from ARM exhibited high levels of LCMV-specific CTL and the infection was cleared in 2 weeks (MATLOUBIAN et al. 1993). Comparisons of the complete genome sequence between ARM and Cl13 showed just five nucleotide changes. Only two of these changes caused amino acid substitutions: one in position 260 of the glycoprotein GP-1 [F(ARM) $\rightarrow$ L(Cl 13)] and another in the amino acid 1079 of the viral polymerase [K(ARM) $\rightarrow$ Q(Cl 13)] (PAREKH and BUCHMEIER 1986; SALVATO et al. 1988; MATLOUBIAN et al. 1993; SEVILLA et al. 2000). These findings indicate that the capability of LCMV to cause immunosuppression maps to the S genomic RNA, also providing strong evidence for an association of amino acid substitution F260L in GP1 with the CTL^-P^+ phenotype.

The contribution of immune-mediated selective pressures to the selection of Cl 13-like variants has been examined recently (SEVILLA et al. 2000). The majority (23 out of 36) of virus isolates from lymphocytes and macrophages of CD4 knock out (ko) mice, as well as perforin ko and TNF-α ko mice persistently infected with

ARM, exhibited a CTL^-P^+ phenotype when inoculated i.v. into adult immunocompetent mice (Table 1). In contrast, all the CNS isolates analyzed from the same mice exhibited a CTL^+P^- phenotype; they were able to elicit a strong anti-LCMV response, which cleared the virus within 2 weeks. Selection of CTL^-P^+ variants occurred at similar frequencies in normal or immunocompromised neonatally infected mice. Hence, it is highly unlikely that the host immune response represents a main driving force in this selection process. Detailed molecular characterization of these variants has provided strong additional evidence for an association between the nature of the amino acid at position 260 in GP-1 and the virus' ability to cause immunosuppression. The complete sequence of GP-1 showed that all of the CTL^-P^+ variants had a L or I in position 260. Some isolates had additional changes in GP-1 but they showed the same CTL^-P^+ phenotype as those with the single amino acid substitution F260L or F260I (Fig. 2A). Thus, it is highly unlikely that these additional amino acid substitutions within GP-1 contributed to the immunosuppressive phenotype. The change K (ARM) to Q (Cl13) at amino acid position 1079 in the LCMV polymerase had been previously also associated with the immunosuppressive phenotype (Fig. 2B). The sequence of selective regions of the polymerase ORF showed that only some CTL^-P^+ variants had the change K(1079)Q, suggesting that any role played by the polymerase in the immunosuppressive phenotype does not map to amino acid 1079.

Arenaviruses are considered to be genetically rather stable. Nevertheless, in isolates from immunocompromised mice accumulated mutations at a rate of 2×10^{-3} and 4×10^{-3} substitutions per nucleotide for GP-1 and L polymerase, respectively (SEVILLA et al. 2000). The selection of variants with the change F260L

Table 1. Phenotypic characterization of LCMV variants[a]

Virus isolate[b]	Carrier mouse[c]	LCMV-specific CTLs in spleen (E/T ratio of 0:1)[d]	LCMV titer (PFU) per ml of serum (D 30 p.i.)
ARM 53b		80	<50
Cl 13		2	3×10^4
PBL: 2 isolates	CD4 ko	0	6×10^5
Mac: 2 isolates	5 months after infection	0	1×10^5
PBL: 11 isolates	Perforin ko 7 months after infection	1.2 ± 2.1	$2 \pm 0.5 \times 10^4$
CD8: 5 isolates	Perforin ko 24 days after infection	75 ± 8	<50
CD4: 3 isolates	24 days after infection		
CD4: 2 isolates	TNF-α ko	0	1×10^4
CD8: 2 isolates	3 months after infection	0	1×10^4
PBL: 4 isolates		3.6 ± 6	$1 \pm 1 \times 10^2$
Brain: 4 isolates		66 ± 15	<50

[a] BALB/c ByJ mice were infected i.v. with 2×10^6 PFU of isolated virus. CTL response was checked at day 7 post-infection (p.i.) and viral titers in sera were measured at day 30 p.i.. The data shown are averages of values from two to four mice per group.
[b] Virus isolates were obtained by picking plaques formed by macrophages (Mac), CD4+ cells (Cd4), CD8+ cells (CD8), peripheral blood lymphocytes (PBL) and brain homogenate.
[c] These virus isolates had been selected in immunocompromised mice [CD4 knock out (ko), perforin ko and TNF-α ko] at different times after inoculation.
[d] These values are percentages of 51Cr release from BALB/c C17 (H-2d) targets at an E/T ratio of 50:1.

in GP-1 in mice persistently infected with LCMV appears to be organ-specific. A rapid selection of variants with the F260L change in GP1 occurs in the spleen and liver, but this selection is relatively slower in the kidney, and has never been observed in the brain where the parental genotype (with F at position 260 in GP1) is maintained (TISHON et al. 1993). Virus transmission from host to host or from one tissue to another within an infected host, often involves small virus populations or even a single, or very few, virus particles (inter-host or intra-host genetic bottleneck transmissions). These founder effects might facilitate that RNA species present at low levels in the original viral quasispecies, can be randomly sampled and colonize a new host environment (e.g., tissue or cell type), where they might become the dominant species. All virus isolates recovered from the lymphocytes of immunocompromised mice at 24 days of persistent infection with ARM, showed the same CTL^+P^- phenotype as ARM. However, by 2 weeks after infection, ARM had already reached high levels of replication in most main organs, including the spleen. Therefore, founder effects do not appear to contribute to the initial selection of Cl 13-like variants in the spleen of mice persistently infected since birth.

Both in vivo and tissue culture competition assays between ARM and Cl 13 showed that ARM dominates over Cl 13 in neurons, but Cl 13 has a higher fitness than ARM during replication in liver and spleen (KING et al. 1992). The preferential growth of one virus over other in certain cell types has also been associated with amino acid replacements at position 260 in GP-1 (KING et al. 1992). However, the specific selective pressures and molecular interactions involved in this selection remain to be elucidated.

Variants with a CTL^-P^+ phenotype, like Cl 13, also have a preferential tropism for dendritic cells ($CD11c^+$ and $DEC-205^+$ cells) of the white pulp of the spleen (Fig. 3) (BORROW et al. 1995; SEVILLA et al. 2000). In contrast, CTL^+P^- variants, like ARM, replicate preferentially in macrophage ($F4/80^+$) cells of the red pulp of the spleen. The preferential tropism of Cl 13-like variants for dendritic cells has been associated with a generalized suppression of the host immune system. This is thought to occur by a mechanism in which the infected dendritic cells are targets for destruction by the antiviral immune response (ODERMATT et al. 1991; BORROW et al. 1995). The mechanism by which the mutation F260L in GP-1 contributes to the specific targeting of dendritic cells by Cl 13-like variants is likely related to changes in the affinity of the interaction between α-DG and GP-1 (SEVILLA et al. 2000). Evidence indicates that GP-1 is the LCMV protein that binds to the cellular receptor α-DG (PIRCHER et al. 1989; DOCKTER et al. 1996; CAO et al. 1998). Immobilized α-DG was shown to bind with more affinity to Cl 13 than ARM (CAO et al. 1998). Moreover, all the immunosuppressive isolates analyzed that were recovered from immunocompromised mice and had L or I at position 260 of GP-1, also showed about 2 logs higher affinity for α-DG than those viruses with F instead of L or I. These findings indicate that a small aliphatic (L or I), but not a bulky aromatic (F), amino acid in position 260 of GP-1 is associated with high affinity binding to α-DG. It is then plausible that within the original ARM quasispecies, Cl 13-like variants are selected because of their higher affinity binding to α-DG. It is worth noting how the selection of a viral variant with higher receptor binding

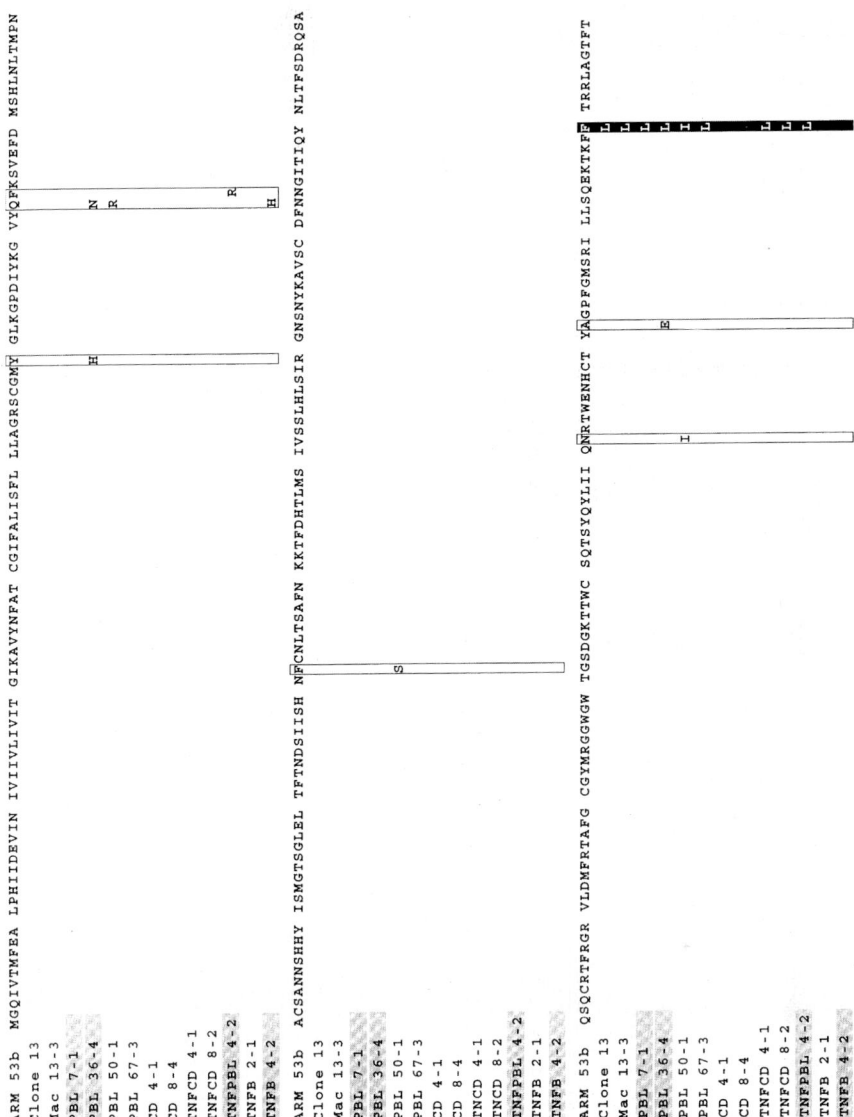

Fig. 2A.

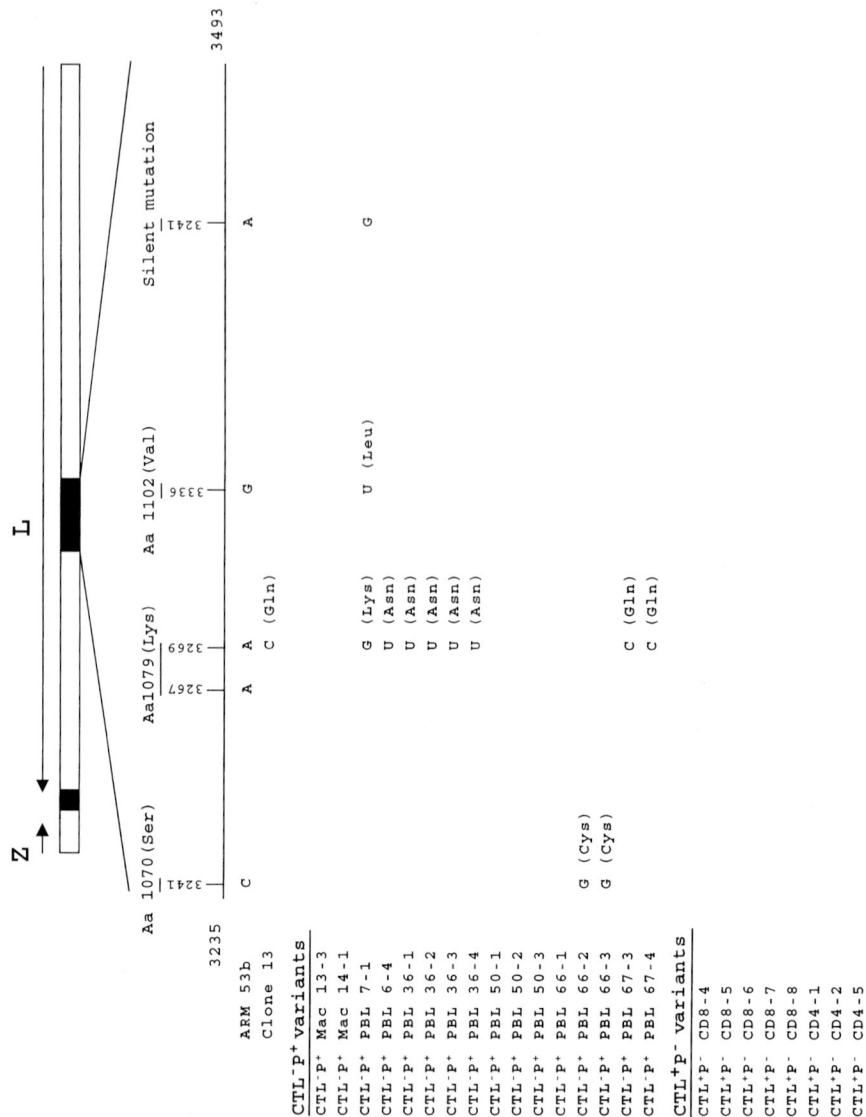

Fig. 2B.

Fig. 2. A Comparison of amino acid sequences of GP-1 of LCMV ARM 53b, Clone 13 and several viral isolates from immunocompromised mice. The origin of these isolates is described in Table 1. Only those amino acids that differ from the parental sequence are indicated. Variants with additional mutations to the one at amino acid (aa) position 260 are marked with a *box*. The *black box* indicates aa position 260. Most of the immunosuppressive isolates have only the change Phe 260 Leu or Ile. Isolates with additional changes were: PBL 36-4: Tyr 60 His, Gln 73 Asn and Ala 242 Glu; PBL 50-1: Gln 73 Arg, Phe 122 Ser and Asn 232 Ile; TNFPBL 4-2: Phe 74 Arg. Non-immunosuppressive isolates did not have aa changes in GP-1, except TNFB 4-2, with a change Gln 73 His. **B** Alignment of nucleotide sequences of a fragment of the L protein for several of the LCMV isolates described. The *upper part* of the figure shows the fragment of the L protein amplified by RT-PCR (from nucleotide 3235 to 3493). Only those nucleotides that differed from the parental sequence are indicated. The aa sequence is shown in *parentheses*, and the phenotype of each virus is indicated on the *left side*. No changes were found in the non-immunosuppressive variants

affinity, may also result in a virus with a very different effect on the host biology, namely a CTL^-P^+ phenotype, as compared to the parental CTL^+P^- virus. These findings illustrate the relationship between virus-host molecular interactions and the selection of biologically relevant viral variants.

3.3 Contribution of LCMV Variants to Virally Induced Growth Hormone Deficiency Syndrome

Neonatal infection of C3H/St mice with certain LCMV strains leads to a persistent infection that is associated with a growth hormone (GH) deficiency syndrome (GHDS) (Fig. 4). This disorder is manifested by marked growth retardation and the development of severe hypoglycemia, which frequently leads to the death of the infected mice (OLDSTONE et al. 1982, 1984). Both viral and host genetic determinants contribute to GHDS (OLDSTONE et al. 1984; TISHON and OLDSTONE 1990). The development of GHDS is associated with high levels of virus load in the GH-producing cells of the anterior pituitary. However, the pituitary remains free from any apparent structural damage or inflammation, yet levels of GH, both mRNA and protein are significantly diminished (OLDSTONE et al. 1985; VALSAMAKIS et al. 1987; KLAVINSKIS and OLDSTONE 1989). Molecular studies using PC cells, a pituitary derived somatotroph cell line, indicated that LCMV-induced reduction in GH synthesis is likely due to the altered activity of the GH transcriptional transactivator GHF1 (Pit-1) in the infected cells (DE LA TORRE and OLDSTONE 1992). This, in turn, results in lower levels of GH promoter activity and the subsequent decrease in mRNA synthesis and protein production.

Genetic studies using reassortant viruses between strains of LCMV, which do (ARM), or do not (WE) cause GHSD, mapped the ability to cause this disorder to the S RNA (RIVIERE et al. 1985). However, the numerous amino acid differences within the S RNA between these two LCMV strains impeded a more detailed analysis aimed at precisely defining the viral determinants responsible for the development of GHDS (RIVIERE et al. 1985; TENG et al. 1996a). As with all other known RdRp, the LCMV polymerase is expected to be error prone. This hypothesis would predict that the GHDS-nil WE parental population is a quasispecies containing a mixture of genetically closely related genomes in which variants with

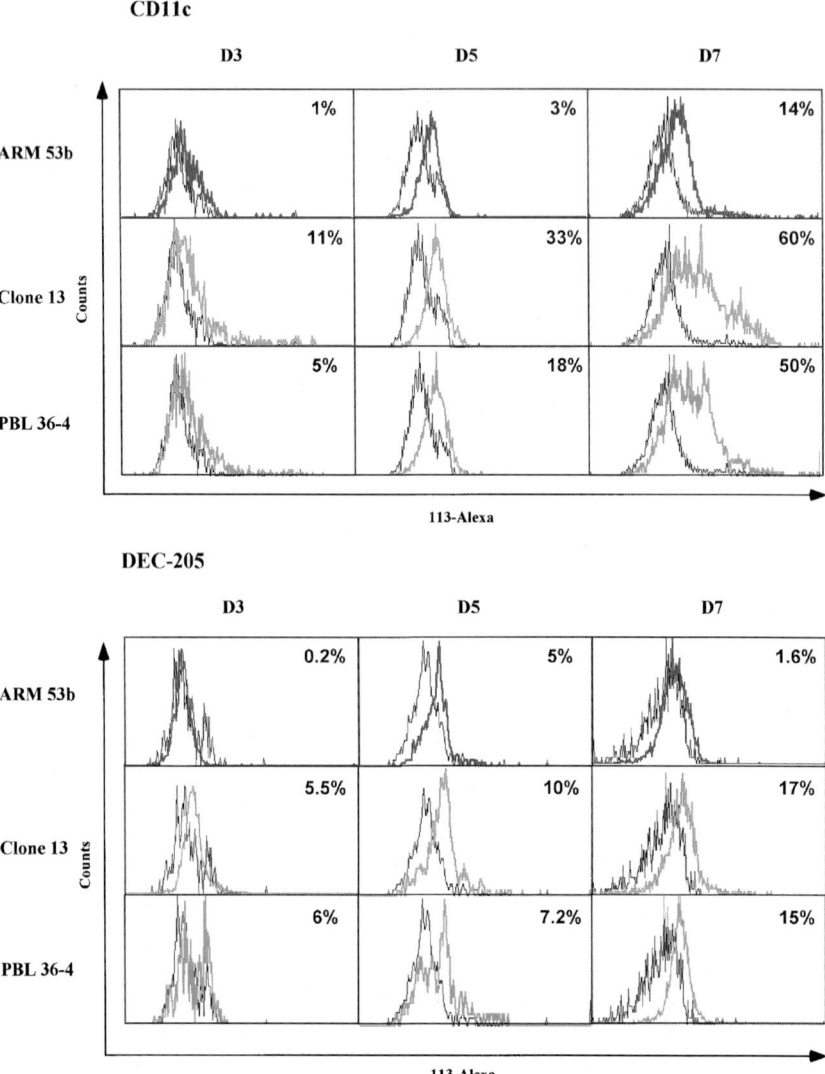

Fig. 3. Replication of immunosuppressive and non-immunosuppressive variants in dendritic cells. Four adult BALB ByJ mice were infected i.v. with 2×10^6 PFUs of either ARM 53b, Cl 13 or PBL 36-4. $CD11c^+$ and $DEC-205^+$ cells were labeled with specific antibodies from single cell splenocytes preparations from pooled spleens. Intracellular virus antigen was detected using a mouse monoclonal antibody (1.1.3) to the LCMV NP. The 1.1.3 antibody was labeled by conjugation to Alexa. Histograms show 113-Alexa positive cells that were previously gated for $CD11c^+$ (*upper panel*) or $DEC-205^+$ (*lower panel*) cells. The *black lines* represent the staining in a naive mouse. Numbers in each histogram correspond to the percentage of 113-Alexa positive cells over the background in each cell population

the potential to cause GHDS may be already present at low levels. Clonal analysis of the parental WE population showed that the majority (58/61) of the clones characterized (e.g., WEc54) behaved as the parental WE clonal population and did not cause GHDS (BUESA-GOMEZ et al. 1996). However, three of the clones isolated from the WE parental population (e.g., WEc2.5), did cause marked growth retardation, severe hypoglycemia and high mortality, upon intracerebral inoculation into C3H/St neonates (BUESA-GOMEZ et al. 1996) (Fig. 5). These results provided evidence that variants with the ability to cause GHDS (GHDS+) may constitute up to 5% of the RNA species within a GHDS-nil replicating WE parental population. WE variants that caused GHDS replicated at high levels in the GH-producing cells of the anterior pituitary, and caused reduced levels of GH synthesis in the infected mice. In contrast, the majority of the WE clonal isolates had the same GHDS-nil phenotype as the parental WE, and replicated poorly in the GH-producing cells (BUESA-GOMEZ et al. 1996).

Clonal isolates from the same parental virus population are expected to be genetically very closely related. Therefore, the molecular characterization of GHDS-nil (e.g., Wec54) and GHDS+ (e.g., Wec2.5) WE variants provides a useful experimental avenue to investigate the precise viral determinants required for the development of GHDS. The biological characterization of reassortants between WEc54 or WEc2.5 variants and ARM (GHDS+), confirmed that the virus' ability to cause GHDS is associated with the S RNA (TENG et al. 1996a). Sequence analysis of the S RNA of WEc54 and WEc2.5 showed that the genetic basis for their phenotypic differences is a single amino acid substitution S to F at position 153 in the GP1 (TENG et al. 1996a). With the exception of their different tropism for GH-producing cells, both WEc54 and WEc2.5 have the same tissue and cell distribution in GHDS susceptible infected mice. Therefore, the single amino acid substitution S153F in GP-1 appears to confer WEc2.5-like variants with the ability to recognize a virus receptor that is specifically expressed in the GH-producing cells. PC cells did not exhibit the restriction for LCMV entry observed in vivo. Both GHDS (+) and (−) LCMV variants were able to replicate in PC cells and both caused a similar decrease in GH mRNA steady-state levels (TENG et al. 1996a). These findings suggest that in addition to GP-1, involved in receptor recognition and virus entry into the GH target cells, other viral determinants also contribute to LCMV-mediated impairment of GH synthesis (Fig. 6). Coimmunoprecipitation studies have revealed a specific interaction between LCMV NP and GHFI. Whether this interaction is responsible for the LCMV mediated effect on GH promoter activity remains to be determined. Moreover, it cannot be ruled out that Z and L could also contribute to this virally induced disturbance in GH synthesis.

Interestingly, WEc54 and WEc2.5 also exhibited distinct CTL^-P^+ and CTL^+P^- phenotypes, respectively. As with Cl 13, the CTL^-P^+ phenotype displayed by WEc54 correlated with the virus' ability to target dendritic cells in the spleen. It is then intriguing that the same mutation S153F in GP-1 extends the tropism of WE to the GH-producing cells, but abrogates its capability to target dendritic cells. The mechanisms underlying the association of a single amino acid at position 153 in GP-1 with two different and seemingly unrelated phenotypes, namely, the ability to cause GHDS and immunosuppression, remain to be determined. Nevertheless, these findings illustrate how only minor changes in the virus genome may have a great impact in the biology of the infected host.

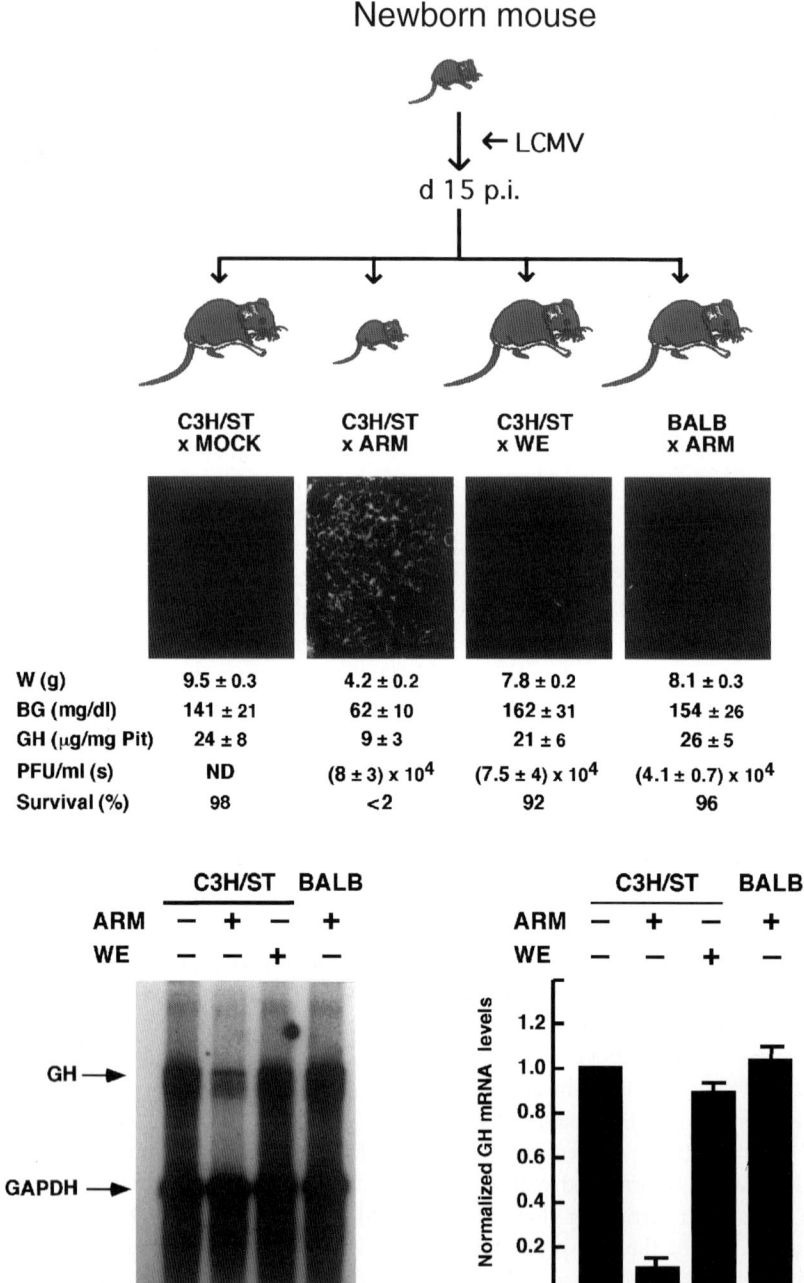

Another intriguing question relates to the mechanisms by which GHDS+ WE variants are stably maintained, although phenotypically silent, within the GHDS-nil WE population, and how changes in the virus population dynamics within the

Fig. 4. The main features of LCMV-induced growth hormone deficiency syndrome (GHDS). Neonatal infection of C3H/St mice with LCMV strain ARM leads to a persistent infection that is associated with an GHDS, manifested as marked growth retardation and development of severe hypoglycemia, which frequently leads to the death of infected mice. Both viral and host genetic determinants contribute to GHDS. Thus, the non-susceptible BALB mice do not develop GHDS after neonatal infection with strain ARM of LCMV. Likewise, some strains of LCMV like WE are unable to cause GHDS in the susceptible C3H/St mouse strain. Development of GHDS correlates with high levels of viral load in the GH-producing cells in the anterior pituitary, but absence of inflammatory changes and detectable structural damage within the pituitary. Infection of GH-producing cells is associated with decreased levels of GH for both mRNA and protein

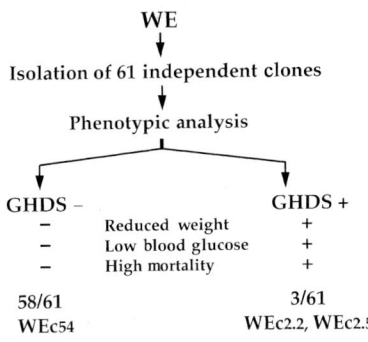

Fig. 5. Variants able to cause GHDS are present within the GHDS-nil WE parental population. Clones (total of 61) isolated from the parental GHDS-nil WE population, were phenotypically characterized with respect to their ability to induce GHDS in C3H/St mice. The majority of the clones (58/61) exhibited the GHDS-nil phenotype of the parental WE population. However, three of the 61 clones examined were able to cause GHDS upon inoculation into newborn C3H/St mice

infected host may influence disease progression. The single amino acid change in GP1 (S153F) responsible for the phenotypic difference between WEc54 (GHDS-nil) and Wec2.5 (GHDS+), was caused by a single nucleotide substitution G to A. As a consequence of this nucleotide change, cDNAs derived from WEc54 and WEc2.5 differ by the presence or absence, respectively, of a EcoRI site, which provided the basis for a sensitive RT-PCR and restriction enzyme cleavage assay to analyze the dynamics of these WE variants during the natural course of infection (TENG et al. 1996b) (Fig. 7). Results from these studies indicated that neonatal C3H/St mice infected with a mixture of Wec54 (GHDS-nil) and Wec2.5 (GHDS+) developed GHDS only when Wec2.5 represented a fraction higher than 0.1 in an inoculum of 1,000PFU (TENG et al. 1996b). It needs to be mentioned that mice infected with less than 10PFU of Wec2.5 alone developed GHDS, indicating that the amount of WEc2.5 (10PFU), corresponding to a fraction of 0.1 in the inoculum, should have been sufficient to induce GHDS. Interestingly, WEc2.5 was not entirely outcompeted by a large excess of Wec54 and was stably maintained at low levels, but phenotypically silent, in the infected mouse. As predicted, Wec54 and WEc2.5 also exhibited quasispecies structures but with sequence distributions that differed between them and with respect to the parental WE population (TENG et al. 1996b) (Fig. 8). About 50% (14/32) of the analyzed clones derived from the EcoRI-resistant PCR band obtained from the RNA of cells infected with the parental WE population, had the G to A change resulting in the S153F GP-1 mutation present in Wec2.5. In contrast, when the same analysis was applied to the WEc54 quasispecies, none of the 28 clones analyzed had the WEc2.5 genotype (TENG et al. 1996b).

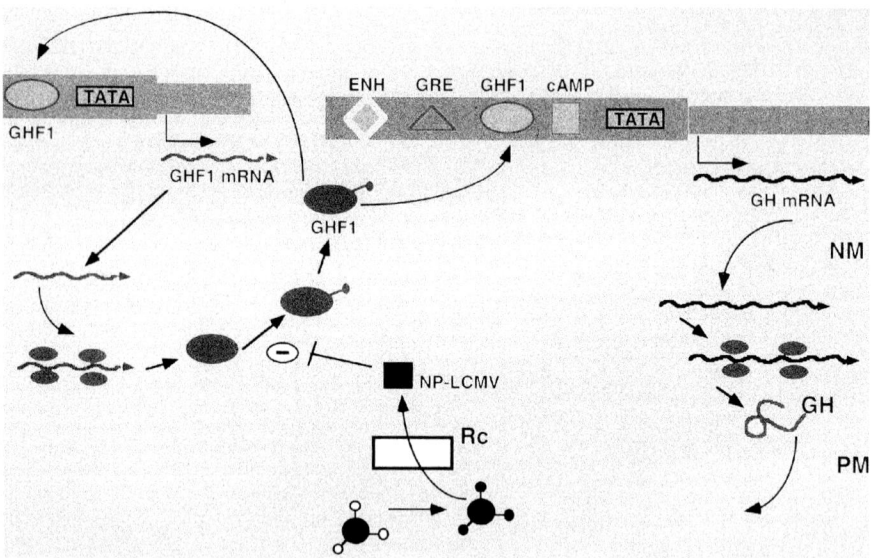

Fig. 6. Proposed model for virus-cell interactions leading to GHDS. Recognition and entry of LCMV into GH-producing cells is mediated by the interaction between the virus GP-1 ligand and a virus cellular receptor (*Rc*). GHDS-nil viruses like WEc54 have a GP-1 unable to use the Rc present in GH cells. Single amino acid substitutions within the GP-1 can affect its interaction with the Rc, and change the virus phenotype from GHDS-nil to GHDS+. Within the infected cells, other viral factors, like the NP, can interfere with the function of GHF1 and affect the GH promoter transcriptional activity. This, in turn, will cause reduced levels of GH synthesis

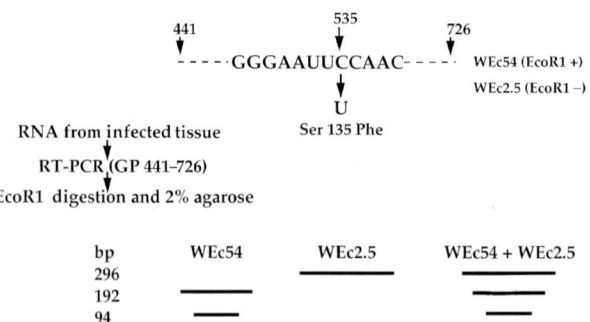

Fig. 7. RT-PCR/EcoR1 assay to distinguish between WEc54 (GHDS-nil) and WEc2.5 (GHDS+). Total RNA from mice infected with different proportions of WEc54 and WEc2.5 viruses, is isolated and analyzed by RT-PCR using specific primers to amplify nt 441 to 726. After digestion of PCR products with EcoR1, samples are analyzed by agarose gel electrophoresis. WEc54 GP-1 sequence contains a EcoR1 site around position 535, whereas this restriction site is absent in the WEc2.5 corresponding segment of GP-1. If the EcoR1 site is present, digestion with the restriction enzyme will result in two bands of 192 and 94bp, whereas undigested PCR product will migrate as a product of 296bp

	WEc54-like			WEc2.5-like			S135P		
Codon	CCC	TTA	AGG	CCC	TTA	AAG	CCC	TTA	GGG
Amino acid	G	N	S	G	N	P	G	N	P
	151	152	153	151	152	153	151	152	153
WEc54		12/19			0/19			7/19 (~0.2%)	
WEc54:WEc2.5 (100:1)		7/15			3/15 (~0.2%)			5/15 (~0.2%)	
WE		14/32			16/32 (~1.0%)			2/32 (~0.12%)	

Fig. 8. WEc2.5 (GHDS+) variants can be silently but stably maintained within a population largely composed of GHDS-nil viral species. RNA isolated from mice infected with Wec54, a mixture of WEc54 and WEc2.5 (100:1), or the WE parental clonal population, was subjected to the RT-PCR/EcoR1 assay described in Fig. 7. The different sequences derived from the EcoR1-resistant PCR bands are indicated. Percentages shown in parenthesis were estimated based on the observation of the EcoR1-resistant bands represented only 1%–2% of the total PCR product

These findings illustrate how differences in the sequence distributions between two quasispecies with very similar, or identical, consensus sequences can significantly influence the virus' impact on the infected host.

Direct fitness measurements of WEc54 and WEc2.5 variants have not been performed. However, one-step growth studies in cultured cells showed that WEc2.5 appears to replicate slightly faster and to higher titers than WEc54. The question then is why WEc2.5 species do not raise to dominance over WEc54 species within the WE population. In this respect, studies with VSV have shown that virus clones with relatively high fitness could be hidden and suppressed by viral quasispecies containing a large excess of variants with lower average fitness (DE LA TORRE and HOLLAND 1990). An important message emerging from these findings is that the phenotypic manifestations, including disease symptoms, of a virus infection can be significantly influenced by the specific threshold levels at which particular variants are present within the population. Hence, pathogenic variants, of high fitness, can be maintained within a non-pathogenic viral population. Low probability bottleneck transmissions in which the amount of a phenotypically suppressed variant transmitted might exceed the threshold limitation, will drastically alter the outcome of the infection in the newly infected host or tissue.

4 Perspectives

It is now well established that most, if not all, RNA viruses have the genetic structure of quasispecies. Increasing amounts of information obtained from different viral systems are contributing to the elucidation of the complex dynamics of viral quasispecies. This is important to understand not only virus evolution, but also viral pathogenesis. Such population dynamics may often confound efforts to correlate phenotypic manifestations of virus-induced disease with single genome variants of a defined sequence.

Viral systems, like LCMV, for which both cell culture systems and suitable laboratory animal models are available, provide amenable models for the investigation of the impact of viral quasispecies on the biology of the infected host. The findings on LCMV-host interactions discussed here also support the view that persistent viral infections can be viewed as a succession of invasions modulated by the response of the host. The manifestations of these chronic infections will differ not only because of the genetic, physiological and immunological differences among the infected hosts, but also because each host is likely to experience a unique array of quasispecies challenges during an infection. These features of viral infections are mostly overlooked because disease symptoms are often, but not always, similar for a given type of virus. Moreover, it is worth noting that viral variants with a very different impact on the biology of the infected host can, however, exhibit undistinguishable biochemical and immunological properties.

References

Aebischer T, Moskophidis D, Rohrer UH, Zinkernagel RM, Hengartner H (1991) In vitro selection of lymphocytic choriomeningitis virus escape mutants by cytotoxic T lymphocytes. Proc Natl Acad Sci USA 88:11047–11051

Ahmed R, Hahn CS, Somasundaram T, Villarete L, Matloubian M, Strauss JH (1991) Molecular basis of organ-specific selection of viral variants during chronic infection. J Virol 65:4242–4247

Ahmed R, Oldstone MB (1988) Organ-specific selection of viral variants during chronic infection [published erratum appears in J Exp Med 1988 Jul 1;168(1):457]. J Exp Med 167:1719–1724

Ahmed R, Salmi A, Butler LD, Chiller JM, Oldstone MB (1984) Selection of genetic variants of lymphocytic choriomeningitis virus in spleens of persistently infected mice. Role in suppression of cytotoxic T lymphocyte response and viral persistence. J Exp Med 160:521–540

Auperin DD, Romanowski V, Galinski M, Bishop DHL (1984) Sequencing studies of Pichinde arenavirus S RNA indicate a novel coding strategy, an ambisense viral S RNA. J Virol 52:897–904

Baranowski E, Ruiz-Jarabo CM, Sevilla N, Andreu D, Beck E, Domingo E (2000) Cell recognition by foot-and-mouth disease virus that lacks the RGD integrin-binding motif: flexibility in aphthovirus receptor usage. J Virol 74:1641–1647

Batschelet E, Domingo E, Weissmann C (1976) The proportion of revertant and mutant phage in a growing population, as a function of mutation and growth rate. Gene 1:27–32

Borrow P, Evans C, Oldstone MBA (1995) Virus-induced immune suppression: immune system-mediated destruction of virus-infected dendritic cells results in generalized immune suppression. J Virol 69:1059–1070

Borrow P, Lewicki H, Wei X, Horwitz MS, Peffer N, Meyers H, Nelson JA, Gairin JE, Hahn BH, Oldstone MB, Shaw GM (1997) Antiviral pressure exerted by HIV-1-specific cytotoxic T lymphocytes (CTLs) during primary infection demonstrated by rapid selection of CTL-escape virus [see comments]. Nat Med 3:205–211

Borrow P, Oldstone MB (1992) Characterization of lymphocytic choriomeningitis virus-binding protein(s): a candidate cellular receptor for the virus. J Virol 66:7270–7281

Borrow P, Oldstone MBA (1997) Lymphocytic Choriomeningitis Virus. In: Nathanson N, et al. (eds) Viral Pathogenesis. Lippincott-Raven, Philadelphia

Borrow P, Shaw GM (1998) Cytotoxic T-lymphocyte escape viral variants: how important are they in viral evasion of immune clearance in vivo? Immunol Rev 164:37–51

Buchmeier MJ, Southern PJ, Parekh BS, Wooddell MK, Oldstone MBA (1987) Site-specific antibodies define a cleavage site conserved among arenavirus GP-C glycoproteins. J Virol 61:982–985

Buchmeier MJ, Welsh RM, Dutko FJ, Oldstone MBA (1980) The virology and immunobiology of lymphocytic choriomeningitis virus infection. Adv Immunol 30:275–331

Buesa-Gomez J, Teng MN, Oldstone CE, Oldstone MB, de la Torre JC (1996) Variants able to cause growth hormone deficiency syndrome are present within the disease-nil WE strain of lymphocytic choriomeningitis virus. J Virol 70:8988–8992

Burns JW, Buchmeier MJ (1993) Glycoproteins of the arenaviruses. In: Salvato MS (ed) The arenaviridae. Plenum Press, New York

Cao W, Henry MD, Borrow P, Yamada H, Elder JH, Ravkov EV, Nichol ST, Compans RW, Campbell KP, Oldstone MBA (1998) Identification of alpha-dystroglycan as a receptor for lymphocytic choriomeningitis virus and Lassa fever virus. Science 282:2079–2081

Carrillo C, Borca M, Moore DM, Morgan DO, Sobrino F (1998) In vivo analysis of the stability and fitness of variants recovered from foot-and-mouth disease virus quasispecies. J Gen Virol 79:1699–1706

Ciurea A, Klenerman P, Hunziker L, Horvath E, Senn BM, Ochsenbein AF, Hengartner H, Zinkernagel RM (2000) Viral persistence in vivo through selection of neutralizing antibody-escape variants. Proc Natl Acad Sci USA 97:2749–2754

de la Torre JC, Holland JJ (1990) RNA virus quasispecies populations can suppress vastly superior mutant progeny. J Virol 90:6278–6281

de la Torre JC, Oldstone MB (1992) Selective disruption of growth hormone transcription machinery by viral infection. Proc Natl Acad Sci USA 92:9939–9943

de la Torre JC, Oldstone MBA (1996) The anatomy of viral persistence: Mechanisms of persistence and associated disease. Adv Virus Res 46:311–343

Dockter J, Evans CF, Tishon A, Oldstone MB (1996) Competitive selection in vivo by a cell for one variant over another: implications for RNA virus quasispecies in vivo. J Virol 70:1799–1803

Domingo E (1999) Quasispecies. In: Granoff A, Webster RG (eds) Encyclopedia of Virology. Academic Press, London

Domingo E, Holland JJ, Eigen M (2000) Quasispecies and RNA virus evolution: Principles and Consequences. Landes Bioscience, Austin

Domingo E, Holland JJ (1997) RNA virus mutations and fitness for survival. Annu Rev Microbiol 51:151–178

Drake JW, Holland JJ (1999) Mutation rates among RNA viruses. Proc Natl Acad Sci USA 96:13910–13913

Dutko FJ, Oldstone MBA (1983) Genomic and biological variation among commonly used lymphocytic choriomeningitis virus strains. J Gen Virol 64:1689–1698

Eigen M (1971) Selforganization of matter and the evolution of biological macromolecules. Naturwissenschaften 58:465–523

Eigen M (1996) On the nature of virus quasispecies [letter; comment]. Trends Microbiol 4:216–218

Eigen M, Biebricher CK (1988) Sequence space and quasispecies distribution. In: Domingo E, Ahlquist P, Holland JJ (eds) RNA Genetics. CRC Press, Boca Raton

Evans CF, Borrow P, de la Torre JC, Oldstone MB (1994) Virus-induced immunosuppression: kinetic analysis of the selection of a mutation associated with viral persistence. J Virol 94:7367–7373

Evans DT, O'Connor DH, Jing P, Dzuris JL, Sidney J, da Silva J, Allen TM, Horton H, Venham JE, Rudersdorf RA, Vogel T, Pauza CD, Bontrop RE, DeMars R, Sette A, Hughes AL, Watkins DI (1999) Virus-specific cytotoxic T-lymphocyte responses select for amino-acid variation in simian immunodeficiency virus Env and Nef. Nat Med 5:1270–1276

Farci P, Shimoda A, Coiana A, Diaz G, Peddis G, Melpolder JC, Strazzera A, Chien DY, Munoz SJ, Balestrieri A, Purcell RH, Alter HF (2000) The outcome of acute hepatitis C predicted by the evolution of the viral quasispecies. Science 288:339–344

Flint SJ, Enquist LW, Krug RM, Racaniello VR, Skalka AM (2000) Molecular biology, pathogenesis and control. In: Virology. ASM Press, Washington DC

Forns X, Purcell RH, Bukh J (1999) Quasispecies in viral persistence and pathogenesis of hepatitis C virus. Trends Microbiol 7:402–410

Franze-Fernandez M-T, Zetina C, Iapalucci S, Lucero MA, Bouissou C, Lopez R, Rey O, Deheli M, Cohen GN, Zakin MM (1987) Molecular structure and early events in the replication of Tacaribe arenavirus S RNA. Virus Res 7:309–324

Friedberg EC, Walker GC, Siede W (1995) DNA repair and mutagenesis. In: American Society for Microbiology. Washington DC

Garcin D, Rochat S, Kolakofsky D (1993) The Tacaribe arenavirus small zinc finger protein is required for both mRNA synthesis and genome replication. J Virol 67:807–812

Goulder P, Price D, Nowak M, Rowland-Jones S, Phillips R, McMichael A (1997) Co-evolution of human immunodeficiency virus and cytotoxic T-lymphocyte responses. Immunol Rev 159:17–29

Gromeier M, Wimmer E, Gorbalenya AE (1999) Genetics, pathogenesis and evolution of picornaviruses. In: Domingo E, Webster RG (eds) Origin and Evolution of Viruses. Academic Press, San Diego

Hansen JL, Long AM, Schultz SC (1997) Structure of the RNA-dependent RNA polymerase of poliovirus. Structure 5:1109–1122

Havlir DV, Eastman S, Gamst A, Richman DD (1996) Nevirapine-resistant human immunodeficiency virus: kinetics of replication and estimated prevalence in untreated patients. J Virol 70:7894–7899

Holland JJ, de la Torre JC, Steinhauer DA (1992) RNA virus populations as quasispecies. [Review]. Curr Top Microbiol Immunol 176:1–20

Hotchin J (1962) The biology of lymphocytic choriomeningitis infection: virus-induced immune disease. Cold Spring Harbor Symp Quant Biol 176:1–20

Iapalucci S, Lopez N, Rey O, Zakin MM, Cohen GN, Franze-Fernandez M-T (1989a) The 5' region of Tacaribe virus L RNA encodes a protein with a potential metal binding domain. Virology 173: 357–361

Iapalucci S, Lopez R, Rey O, Lopez N, Franze-Fernandez M-T (1989b) Tacaribe virus L gene encodes a protein of 2210 amino acid residues. Virology 170:40–47

Karlsson AC, Gaines H, Sallberg M, Lindback S, Sonnerborg A (1999) Reappearance of founder virus sequence in human immunodeficiency virus type 1-infected patients. J Virol 73:6191–6196

Kimata JT, Kuller L, Anderson DB, Dailey P, Overbaugh J (1999) Emerging cytopathic and antigenic simian immunodeficiency virus variants influence AIDS progression [see comments]. Nat Med 5: 535–541

King CC, Jamieson BD, Reddy K, Bali N, Concepcion RJ, Ahmed R (1992) Viral infection of the thymus. J Virol 66:3155–3160

Klavinskis LS, Oldstone MBA (1989) Lymphocytic choriomeningitis virus selectively alters differentiated but not housekeeping functions: block in expression of growth hormone gene is at the level of transcription initiation. Virology 168:232–235

Lai MM (1992) Genetic recombination in RNA viruses. Curr Top Microbiol Immunol 176:21–32

Lech WJ, Wang G, Yang YL, Chee Y, Dorman K, McCrae D, Lazzeroni LC, Erickson JW, Sinsheimer JS, Kaplan AH (1996) In vivo sequence diversity of the protease of human immunodeficiency virus type 1: presence of protease inhibitor-resistant variants in untreated subjects. J Virol 70:2038–2043

Lehmann-Grube F (1984) Portraits of viruses: arenaviruses. Inter-virology 22:121–145

Lewicki H, Tishon A, Borrow P, Evans C, Gairin JE, Hahn KM, Jewell DA, Wilson IA, Oldstone MBA (1995) CTL escape viral variants. I. Generation and molecular characterization. Virology 210:29–40

Martell M, Esteban JI, Quer J, Genesca J, Weiner A, Esteban R, Guardia J, Gomez J (1992) Hepatitis C virus (HCV) circulates as a population of different but closely related genomes: quasispecies nature of HCV genome distribution. J Virol 66:3225–3229

Matloubian M, Kolhekar SR, Somasundaram T, Ahmed R (1993) Molecular determinants of macrophage tropism and viral persistence: importance of single amino acid changes in the polymerase and glycoprotein of lymphocytic choriomeningitis virus. J Virol 67:7340–7349

Matloubian M, Somasundaram T, Kolhekar SR, Selvakumar R, Ahmed R (1990) Genetic basis of viral persistence: single amino acid change in the viral glycoprotein affects ability of lymphocytic choriomeningitis virus to persist in adult mice. J Exp Med 172:1043–1048

McMichael AJ, Phillips RE (1997) Escape of human immunodeficiency virus from immune control. Annu Rev Immunol 15:271–296

Meyer PR, Matsuura SE, Mian AM, So AG, Scott WA (1999) A mechanism of AZT resistance: an increase in nucleotide-dependent primer unblocking by mutant HIV-1 reverse transcriptase. Mol Cell 4:35–43

Meyer PR, Matsuura SE, So AG, Scott WA (1998) Unblocking of chain-terminated primer by HIV-1 reverse transcriptase through a nucleotide-dependent mechanism. Proc Natl Acad Sci USA 95: 13471–13476

Meyerhans A, Cheynier R, Albert J, Seth M, Kwok S, Sninsky J, Morfeldt-Manson L, Asjo B, Wain-Hobson S (1989) Temporal fluctuations in HIV quasispecies in vivo are not reflected by sequential HIV isolations. Cell 58:901–910

Meyerhans A, Vartanian JP (1999) The fidelity of cellular and viral polymerases and its manipulation for hypermutagenesis. In: Domingo E, Webster RG, Holland JJ (eds) Origin and Evolution of Viruses. Academic Press, San Diego

Nagy PD, Simon AE (1987) New insights into the mechanisms of RNA recombination. Virology 235:1–9
Najera I, Holguin A, Quinones-Mateu ME, Munoz-Fernandez MA, Najera R, Lopez-Galindez C, Domingo E (1995) Pol gene quasispecies of human immunodeficiency virus: mutations associated with drug resistance in virus from patients undergoing no drug therapy. J Virol 69:23–31
Novella IS, Elena SF, Moya A, Domingo E, Holland JJ (1995) Size of genetic bottlenecks leading to virus fitness loss is determined by mean initial population fitness. J Virol 69:2869–2872
Odermatt B, Eppler M, Leist TP, Hengartner H, Zinkernagel RM (1991) Virus-triggered acquired immunodeficiency by cytotoxic T-cell-dependent destruction of antigen-presenting cells and lymph follicle structure. Proc Natl Acad Sci USA 88:8252–8256
Oldstone MB, de la Torre JC (1993) Viral diseases of the next century. [Review]. Transactions of the American Clinical & Climatological Association 93:62–68
Oldstone MBA (1984) Virus can alter cell function without causing cell pathology: disordered function leads to imbalance of homeostasis and disease. In: Notkins AL, Oldstone MBA (eds) Concepts in viral pathogenesis. Springer, Heidelberg Berlin New York
Oldstone MBA (1989) Virla persistence. Cell 56:517–520
Oldstone MBA (1991) Molecular anatomy of viral persistence. J Virol 65:6381–6386
Oldstone MBA (1998) Viral persistence: mechanisms and consequences. Curr Opin Microbiol 1:436–441
Oldstone MBA, Ahmed R, Buchmeier MJ, Blout P, Tishon A (1985) Perturbation of differentiated functions during viral infection in vivo. Virology 142:158–174
Oldstone MBA, Rodriguez M, Daughaday WH, Lampert PW (1984) Viral perturbation of endocrine function: disordered cell function leads to disturbed homeostasis and disease. Nature 307:278–281
Oldstone MBA, Sinha YN, Blout P, Tishon A, Rodriguez M, von Wedel R, Lampert PW (1982) Virus-induced alterations in homeostasis: alterations in differentiated functions of infected cells in vivo. Science 218:1125–1127
Parekh BS, Buchmeier MJ (1986) Proteins of lymphocytic choriomeningitis virus: antigenic topography of the viral glycoproteins. Virology 153:168–178
Pawlotsky JM, Germanidis G, Neumann AU, Pellerin M, Frainais PO, Dhumeaux D (1998) Interferon resistance of hepatitis C virus genotype 1b: relationship to non-structural 5 A gene quasispecies mutations. J Virol 72:2795–2805
Pircher H, Burki K, Lang R, Hengartner H, Zinkernagel RM (1989) Tolerance induction in double specific T-cell receptor transgenic mice varies with antigen. Nature 342:559–561
Pircher H, Moskophidis D, Rohrer U, Burki K, Hengartner H, Zinkernagel RM (1990) Viral escape by selection of cytotoxic T cell-resistant virus variants in vivo. Nature 346:629–633
Planz O, Ehl S, Furrer E, Horvath E, Brundler MA, Hengartner H, Zinkernagel RM (1997) A critical role for neutralizing-antibody-producing B cells, CD4(+) T cells, and interferons in persistent and acute infections of mice with lymphocytic choriomeningitis virus: implications for adoptive immunotherapy of virus carriers. Proc Natl Acad Sci USA 94:6874–6879
Quinones-Mateu ME, Albright JL, Mas A, Soriano V, Arts EJ (1998) Analysis of pol gene heterogeneity, viral quasispecies, and drug resistance in individuals infected with group O strains of human immunodeficiency virus type 1. J Virol 72:9002–9015
Ribeiro RM, Bonhoeffer S, Nowak MA (1998) The frequency of resistant mutant virus before antiviral therapy. Aids 12:461–465
Riviere Y, Ahmed R, Southern P, Oldstone MBA (1985) Perturbation of differentiated functions during viral infection in vivo. II. Viral reassortants map growth hormone defect to the S RNA of the lymphocytic choriomeningitis virus genome. Virology 142:175–182
Riviere Y, Oldstone MB (1986) Genetic reassortants of lymphocytic choriomeningitis virus: unexpected disease and mechanism of pathogenesis. J Virol 59:363–368
Sala M, Wain-Hobson S (1999) Drift and conservation in RNA virus evolution: are they adapting or merely changing? In: Domingo E, Webster RG, Holland JJ (eds) Origin and Evolution of Viruses. Academic Press, San Diego
Sala M, Zambruno G, Vartanian JP, Marconi A, Bertazzoni U, Wain-Hobson S (1994) Spatial discontinuities in human immunodeficiency virus type 1 quasispecies derived from epidermal Langerhans cells of a patient with AIDS and evidence for double infection. J Virol 68:5280–5283
Salvato M, Borrow P, Shimomaye E, Oldstone MB (1991) Molecular basis of viral persistence: a single amino acid change in the glycoprotein of lymphocytic choriomeningitis virus is associated with suppression of the antiviral cytotoxic T-lymphocyte response and establishment of persistence. J Virol 65:1863–1869
Salvato M, Shimomaye EM, Oldstone MBA (1989) The primary structure of the lymphocytic choriomeningitis virus L gene encodes a putative RNA polymerase. Virology 169:377–384

Salvato M, Shimomaye EM, Southern P, Oldstone MBA (1988) Virus-lymphocyte interactions: IV. Molecular characterization of LCMV Armstrong (CTL+) small genomic segment and that of its variant, clone 13 (CTL⁻). Virology 164:517–522

Salvato MS (1993) Molecular biology of the prototype arenavirus, lymphocytic choriomeningitis virus. In: Salvato MS (ed) The Arenaviridae. Plenum, New York

Salvato MS, Schweighofer KJ, Burns J, Shimomaye EM (1992) Biochemical and immunological evidence that the 11-kDa zinc-binding protein of lymphocytic choriomeningitis virus is a structural component of the virus. Virus Res 22:185–198

Salvato MS, Shimomaye EM (1989) The completed sequence of lymphocytic choriomeningitis virus reveals a unique RNA structure and a gene for a zinc finger protein. Virology 173:1–10

Sevilla N, Kunz S, Holz A, Lewicki H, Homann D, Yamada H, Campbell KP, de La Torre JC, Oldstone MB (2000) Immunosuppression and resultant viral persistence by specific viral targeting of dendritic cells. J Exp Med 192:1249–1260

Singh MK, Fuller-Pace FV, Buchmeier MJ, Southern PJ (1987) Analysis of genomic L RNA segment of lymphocytic choriomeningitis virus. Virology 161:448–456

Sousa R (1996) Structural and mechanistic relationships between nucleic acid polymerases. Trends Biochem Sci 21:186–190

Southern PJ, Bishop DH (1987) Sequence comparison among arenaviruses. Curr Top Microbiol Immunol 133:19–39

Southern PJ, Singh MK, Riviere Y, Jacoby DR, Buchmeier MJ, Oldstone MBA (1987) Molecular characterization of the genomic S RNA segment from lymphocytic choriomeningitis virus. Virology 157:145–155

Steinhauer DA, Domingo E, Holland JJ (1992) Lack of evidence for proofreading mechanisms associated with an RNA virus polymerase. Gene 122:281–288

Teng MN, Borrow P, Oldstone MBA, de la Torre JC (1996a) A single amino acid change in the glycoprotein of lymphocytic choriomeningitis virus is associated with the ability to cause growth hormone deficiency syndrome. J Virol 70:8438–8443

Teng MN, Oldstone MB, de la Torre JC (1996b) Suppression of lymphocytic choriomeningitis virus–induced growth hormone deficiency syndrome by disease-negative virus variants. Virology 223:113–119

Tishon A, Borrow P, Evans C, Oldstone MB (1993) Virus-induced immunosuppression. 1. Age at infection relates to a selective or generalized defect. Virology 195:397–405

Tishon A, Oldstone MB (1990) Perturbation of differentiated functions during viral infection in vivo. In vivo relationship of host genes and lymphocytic choriomeningitis virus to growth hormone deficiency. Am J Pathol 137:965–969

Valsamakis A, Riviere Y, Oldstone MBA (1987) Perturbation of differentiated functions in vivo during persistent viral infection. III. Decreased growth hormone mRNA. Virology 156:214–220

Wright KE, Spiro RC, Burns JW, Buchmeier MJ (1990) Post-translational processing of the glycoproteins of lymphocytic choriomeningitis virus. Virology 177:175–183

Zinkernagel RM, Doherty PC (1979) MHC-restricted cytotoxic T cells: studies on the biological role of polymorphic major transplantation antigens determining T-cell restriction specificity, function, and responsiveness. Adv Immunol 27:51–177

Designing Arenaviral Vaccines

J.L. WHITTON

1	Introduction	221
2	The Viruses and the Diseases that They Cause	222
2.1	LCMV	222
2.2	Lassa Virus	223
2.3	Junin Virus	224
2.4	Machupo Virus	224
2.5	Guanarito Virus	224
2.6	Sabia Virus	225
3	Routes of Infection Used by Arenaviruses	225
4	The Host Immune Response to Arenaviral Infections	226
4.1	New World Viruses	226
4.2	Old World Viruses	227
5	Immunopathology During Arenaviral Infections and After Vaccination	228
5.1	Antibody-Mediated Immunopathology	228
5.2	T-Cell-Mediated Immunopathology	229
5.3	Vaccine-Primed Immunopathology	229
6	Maximizing Antiviral Efficacy While Minimizing Immunopathology	230
6.1	Most Virus-Specific $CD8^+$ T Cells Are "Silent" at the Peak of the Antiviral Response	230
6.2	$CD8^+$ T Cells Are Exquisitely Responsive to Antigen Contact	230
7	Designing an Arenaviral Vaccine	231
7.1	What Responses Would an Ideal Arenaviral Vaccine Induce?	231
7.2	Conventional and Recombinant Viral Vaccines	232
7.3	Should DNA Immunization Be Applied to Arenaviral Vaccines?	232
	References	233

1 Introduction

Several hemorrhagic fevers, including Lassa fever (LF; Lassa virus), Argentinian hemorrhagic fever (AHF; Junin virus), Bolivian hemorrhagic fever (BHF; Machupo virus) and Venezuelan hemorrhagic fever (VHF; Guanarito virus) are caused by arenaviruses. These diseases can be devastating, and often lethal. LCMV, the

---

Department of Neuropharmacology, CVN-9, The Scripps Research Institute, 10550 N. Torrey Pines Road, La Jolla, CA 92037, USA

prototype of the Arenaviridae, can cause aseptic meningitis, and is a known teratogen. Despite the pathogenicity of the arenaviruses, to date only one vaccine (against AHF) has been evaluated in humans. In this chapter we shall review the design of arenaviral vaccines. The rational design of a vaccine requires that we consider (a) the viruses themselves, (b) the route(s) by which they infect the unfortunate host, (c) the host immune response induced, (d) possible negative consequences of immunization, and (e) the vaccine approaches available. Some of these topics are discussed in detail in other chapters of this volume, and thus will be described quite briefly below.

2 The Viruses and the Diseases that They Cause

The arenaviruses can be classified phylogenetically into Old World (which includes LCMV and Lassa virus) and New World; this latter group has been further divided into three lineages, A–C. The four most pathogenic New World agents (Junin, Machupo, Guanarito and Sabia) all belong to lineage B, suggesting that the highly pathogenic phenotype may derive from a common ancestral virus (BOWEN et al. 1996).

2.1 LCMV

LCMV is the prototype of the family. Arenaviruses are pleomorphic enveloped viruses, with bi-segmented single-stranded RNA genomes; each segment is ambisense, containing two convergently arranged genes separated by an intergenic hairpin loop. The long segment (L) encodes the putative polymerase (the L protein), and a zinc finger protein (Z) (SALVATO and SHIMOMAYE 1989; SALVATO et al. 1989), while the short (S) segment encodes the nucleoprotein (NP) and glycoprotein (GP-C) (AUPERIN et al. 1984; RIVIERE et al. 1985). The latter protein undergoes post-translation processing to yield the two mature virion glycoproteins, GP1 and GP2 (BUCHMEIER and OLDSTONE 1979). LCMV is most often used as a model system in which to study immune responses and mechanisms of viral pathogenesis, and as such has been enormously valuable; it is less frequently studied as a human pathogen. Nevertheless, a pathogen it is; indeed there are several well documented incidences of infection in laboratory workers, most often working with hamsters (see VANZEE et al. 1975; HINMAN et al. 1975; BIGGAR et al. 1977), but in one case, with immunodeficient (nude) mice (DYKEWICZ et al. 1992). The natural host for the virus is the house mouse (*Mus musculus*), but outbreaks in humans appear to originate more readily from hamsters than from mice, and cases in the general population have been traced back to pet hamster breeding facilities (BIGGAR et al. 1975; MAETZ et al. 1976). LCMV can be lethal in primates. In one well documented case, eight cynomolgus monkeys were inadvertently exposed to LCMV; six were infected, all of which died (PETERS et al. 1987). Human infections are often subclinical, but can present as fever, myalgia and headaches, and sometimes as full-blown aseptic meningitis (VANZEE et al. 1975; DEKONENKO et al. 1985; ROEBROEK et al. 1994). Although horizontal human-to-human transmission of LCMV has not been documented, the virus can pass from mother to fetus, and LCMV is a proven teratogen (SHEINBERGAS 1976; SHEINBERGAS et al. 1981, 1984; BARTON et al. 1995; BARTON and METS 1999); in one study of 16 children with evidence of prenatal exposure, 14 were stricken with hydrocephalus (SHEINBERGAS 1976). As further evidence of the infectious potential of LCMV, studies within the past decade have shown a remarkably high human seropositivity ($\sim$5%) in American cities such as Baltimore Md. (CHILDS et al. 1991) and Birmingham Ala. (STEPHENSEN et al. 1992), and mouse trapping revealed that $\sim$9% of rodents were seropositive (CHILDS et al. 1992). However the risk of infection declined with improved socio-economic conditions, and there was a marked reduction in seropositivity in younger age groups (PARK et al. 1997).

2.2 Lassa Virus

Lassa fever is the topic of another chapter in these volumes, and so will be reviewed only briefly. Most of the highly pathogenic arenaviruses are named after a nearby geographical feature, and Lassa virus is named for the town of Lassa. The natural reservoir of this virus is *Mastomys natalensis*. Lassa virus infects hundreds of thousands of people each year in Africa, with an annual death toll of $\sim$5,000–10,000 (McCORMICK 1987); the ratio of fatality to infection is estimated at 1%–2% (McCORMICK et al. 1987). Symptoms of infection include fever, facial edema, and, in severe disease, hemorrhage and encephalopathy (CUMMINS et al. 1992). Platelet abnormalities have been observed (CUMMINS et al. 1989), and pathophysiology may include dysregulated production of various cytokines. Survivors frequently have nerve deafness, the precise etiology of which remains uncertain (CUMMINS et al. 1990). Virus is found in rodent urine (WALKER et al. 1975), and transmission from rodent to human is thought to be by inhalation. Better storage of food products, in areas less open to rodent infestation, appears to diminish the risk of infection. There is clear evidence of human-to-human transmission, usually in a hospital setting; the risk of nosocomial spread is reduced by good medical and nursing practices (FISHER-HOCH et al. 1995). Some cases of Lassa fever respond well to ribavirin, but transfer of convalescent human plasma is of little benefit (McCORMICK et al. 1986). A vaccine comprising inactivated Lassa virus failed to induce protective immunity in rhesus monkeys (McCORMICK et al. 1992), but this species was protected by the administration of a recombinant vaccinia virus encoding Lassa virus proteins (FISHER-HOCH et al. 1989); guinea pigs were also protected by a recombinant vaccine (MORRISON et al. 1989). To date no human vaccine trials have taken place.

2.3 Junin Virus

Junin virus has been identified in several rodent species, including *Calomys musculinus*. These rodents are largely rural, rarely infest populated areas, and the virus is endemic in a restricted area of the north-central Argentine pampas, where ~12% of humans are seropositive. Although infection may be subclinical, there are 100–800 cases of AHF each year (MAIZTEGUI et al. 1998). The incidence of infection peaks at the time of crop harvest, consistent with the inhalation of infectious rodent material during farming activities (MAIZTEGUI 1975). Men are therefore infected more commonly than women. After an incubation period of 1–2 weeks, non-specific symptoms appear, followed ~1 week later by more severe cardiovascular, renal and neurologic involvement. In most cases (~80%), remission occurs within 10 days, but mortality is approximately 16% in untreated cases. Treatment takes the form of administration of human convalescent plasma which, if given within 8 days of onset, reduces mortality to ~1% (MAIZTEGUI et al. 1979). Ribavirin may also be effective (McKEE et al. 1988; ENRIA and MAIZTEGUI 1994). A live attenuated Junin virus vaccine, Candid-1, has proven safe and effective in guinea pig and primate trials (McKEE et al. 1992, 1993), and a recent study demonstrated its safety and efficacy in men (MAIZTEGUI et al. 1998); women were excluded from this study because of their lower risk of contracting AHF, and the potential teratogenicity of a live vaccine given during pregnancy. The development of a vaccine against AHF is the topic of another chapter in this volume.

2.4 Machupo Virus

The natural rodent reservoir for this virus is *Calomys callosus* and, in contrast to Junin virus, most human exposures to Machupo virus occur within the home. Indeed, this is the only arenaviral infection for which rodent trapping has been shown to have a dramatic effect; about 3,000 mice were trapped in homes in San Joaquin, Bolivia, with a consequent massive reduction in the incidence of BHF (MACKENZIE 1965; MERCADO 1975). Preliminary data suggest that ribavirin may be effective in mitigating BHF (KILGORE et al. 1997), but a larger clinical study is required to reach statistical significance. The transfer of convalescent serum is effective in animal models (EDDY et al. 1975), and may be effective in ameliorating human disease.

2.5 Guanarito Virus

This virus, isolated in 1989 in the Guanarito municipality in Venezuela, was first thought to be carried by the cotton rat *Sigmodon alstoni*, but it is now clear that the virus identified in that rodent is new arenavirus named Pirital; Guanarito virus is instead tightly restricted to *Zygodontomys brevicauda* (FULHORST et al. 1997, 1999). The virus is endemic in a circumscribed area on the central Venezuelan plains, and the disease peaks each year between November and January, during the period of maximal agricultural activity. Intensive surveillance has failed to identify VHF outside of this region (WEAVER et al. 2000). Full-blown VHF has a mortality rate of ~33% (SALAS et al. 1991; DE MANZIONE et al. 1998).

2.6 Sabia Virus

This virus was first isolated in Brazil, from a fatal case of hemorrhagic fever (LISIEUX et al. 1994). The natural history of this agent remains obscure, and widespread outbreaks of disease have not been reported. Infection is not inevitably fatal since a laboratory technician working with the original isolate developed a prolonged but non-fatal flu-like syndrome (LISIEUX et al. 1994), and a scientist who became infected at Yale University while purifying large quantities of virus also survived (GANDSMAN et al. 1997). Despite the evident infectivity for humans, careful procedures can prevent human-to-human transmission; when enhanced precautions (mainly prevention of aerosols) were employed to manage a patient with Sabia infection, no nosocomial infections were found in 142 case contacts (ARMSTRONG et al. 1999).

3 Routes of Infection Used by Arenaviruses

As outlined above, arenaviral diseases are zoonoses, transmitted from rodents to man, and the rodent reservoir differs for each of the agents. The routes by which arenaviruses travel from rodent to human remain somewhat conjectural, but they represent an important issue. They will determine which public health measures should be taken to prevent infection, and may affect our choice of vaccine strategy. Most data point to aerosol transmission, most likely by inoculation or inhalation. Mouse-to-mouse transmission of LCMV is increased by fighting, suggesting that inoculation into an abrasion is an efficient way to transmit arenaviruses. However, direct contact is by no means an absolute requirement. The lethal infections of cynomolgus monkeys, described above, resulted from the monkeys being caged in the same room as LCMV-infected mice (PETERS et al. 1987). In one hospital outbreak of LCMV, medical personnel contracted the virus from infected hamsters; infection was acquired merely by being in the animal room, and direct physical contact with contaminated material was not required (HINMAN et al. 1975). Similar epidemiological findings were observed in other cases of hamster-to-human transmission (BIGGAR et al. 1975; MAETZ et al. 1976). Junin virus infection is most common during the harvest season, when agricultural machinery disperses aerosols containing infected rodent urine and blood; drivers and their assistants have a high incidence of disease (MAIZTEGUI 1975). It is thought that infection may occur through skin abrasions, or through oral, respiratory or conjunctival mucosa;

consistent with the latter, guinea pigs and marmosets are readily infected by the mucosal route (SAMOILOVICH et al. 1983, 1984). It has been suggested that LCMV – and possibly other arenaviruses – might be transmitted by ingestion. Viruses which spread by this route must resist the action of gastric acid, which has a pH of ~1.5. Some viruses are resistant to acid (e.g., picornaviruses), while others (e.g., reoviruses) carry acid-activated proteins by which they exploit the gastric environment. However, LCMV glycoproteins are relatively acid-sensitive, and the virus is irreversibly inactivated by a pH of <6.0 (DI SIMONE et al. 1994). Nevertheless, infection is possible if the virus is delivered in an appropriate diluent; e.g., the milk of persistently-infected dams spreads LCMV to suckling pups (SKINNER and KNIGHT 1973). Consistent with the possibility that buffering is required to protect the virus, administration by gavage of LCMV suspended in Hepes-buffered medium can result in infection (RAI et al. 1997). These data demonstrate that gastric acid is a major, albeit surmountable, obstacle to gastro-intestinal transmission of these agents. In summary, arenavirus infections are most likely transmitted by direct contact between infectious materials and abraded skin, or by mucosal exposure to aerosols. Particularly in the latter case, it is reasonable to suggest that mucosal immunity might be beneficial in preventing infection.

4 The Host Immune Response to Arenaviral Infections

Is there a common thread in the host immune effector mechanisms used to control arenaviral infections? There appears to be a relationship between the nature of the response to primary challenge, and the phylogeny of the target virus. In general, New World arenaviruses appear to induce strong antibody-mediated immune responses, while Old World viruses do so less effectively, instead being countered mainly by $CD8^+$ T cell immunity. However, as will be discussed below, this division of labor is not absolute.

4.1 New World Viruses

Neutralizing antibodies play an important role in protecting against New World arenaviruses (reviewed by HOWARD 1987). There is little available information regarding IgA responses in arenavirus infections, which is somewhat surprising given the importance of mucosal routes of infection. Antibody responses appear critical to recovery from Junin virus infection, and passive transfer of immune serum at an early stage of infection can significantly reduce the morbidity and mortality (MAIZTEGUI et al. 1979). Similarly, convalescent serum appears able to limit BHF in animal models (EDDY et al. 1975). The cellular immune responses have not been extensively characterized in any of these infections. However, Junin-specific lytic activity can be observed in experimental models (KENYON and PETERS 1986), and human studies with Candid-1 showed that Junin virus-specific T cell responses could be detected in vaccinees (cited in MAIZTEGUI et al. 1998), and that viremia often disappeared several weeks before the appearance of detectable antibody (McKEE et al. 1993), suggesting that virus-specific T cell responses may play a role in limiting viral replication.

4.2 Old World Viruses

As might be expected, the immune correlates of recovery and subsequent immunity have been very extensively studied in the murine model of LCMV infection. LCMV-specific neutralizing antibodies play little role in combating the acute primary infection, since they appear only many weeks later. However the antibody response can contribute to ultimate viral clearance; mice lacking B cells (and, therefore, lacking antibodies) were able to clear a low-dose LCMV challenge, but were unable to eradicate virus following high-dose challenge (BRUNDLER et al. 1996). Furthermore, neutralizing antibodies contribute to the eradication of persistent LCMV infection (PLANZ et al. 1997). The delayed appearance of neutralizing antibodies may be a consequence of an ingenious viral ploy. GP molecules in the virion envelope bind to B cells which express GP-specific antibodies, and the virus thereby enters these cells, using the antibodies as a receptor. Virus-specific B cells are thus preferentially infected, and are then destroyed by the antiviral cytotoxic T cell (CTL) response, preventing the host from producing GP-specific neutralizing antibodies (PLANZ et al. 1996), which may in turn facilitate the establishment of persistent infection. The production of NP-specific antibodies is not affected in this way, but such antibodies are of limited biological relevance.

The critical role played by $CD8^+$ T cells in anti-LCMV immunity has been extensively documented. If these cells are depleted, or if they lack perforin (the major cytolytic protein), primary infection is not eradicated (KAGI et al. 1994; WALSH et al. 1994), demonstrating that antibodies alone cannot counter this virus. Furthermore, successful vaccination does not require the induction of LCMV-specific antibodies; vaccines encoding isolated LCMV CTL epitopes can confer complete protection against virus challenge (KLAVINSKIS et al. 1989a; WHITTON 1990; WHITTON et al. 1993). Recently, the LCMV model has been used to reevaluate the extent and specificity of $CD8^+$ T cell responses to virus infection. Remarkably, at 7 days post-infection, the peak of the antiviral response, some 50% of all splenic $CD8^+$ T cells are virus specific (MURALI-KRISHNA et al. 1998); this corresponds to $\sim 10^7$ LCMV-specific cells in the mouse. After the virus is eradicated, the frequency of virus-specific cells falls, but remains a very substantial proportion ($\sim 10\%$) of the $CD8^+$ T cell pool (MURALI-KRISHNA et al. 1998).

Similarly to LCMV, but in contrast to its New World relatives, primary Lassa virus infection appears to be controlled mainly by cellular immune responses, and reasonable titers of neutralizing antibodies to Lassa virus appear only late in the disease process, usually long after the patient has clinically recovered (JAHRLING

1983). However, individuals who have recovered from Lassa fever may continue to shed virus for several weeks (LEIFER et al. 1970), and it is tempting to suggest that this prolonged shedding may result from the delay in neutralizing antibody production. The paucity of neutralizing antibody may explain the failure of Lassa-convalescent plasma to reduce mortality in humans (McCORMICK et al. 1986). A killed Lassa vaccine did not protect, despite inducing antibodies (McCORMICK et al. 1992), while a live recombinant vaccinia virus expressing Lassa virus glycoprotein protects monkeys against Lassa challenge (FISHER-HOCH et al. 1989). Although these and other data (S.P. Fisher-Hoch and J.B. McCormick, personal communication) suggest that a successful vaccine against Lassa fever should induce cellular immune responses, a role for vaccine-induced neutralizing antibody cannot be excluded, since some studies have indicated that serum transfer can confer protection in animal models (JAHRLING 1983; JAHRLING and PETERS 1984), and perhaps even in man (MONATH et al. 1974).

5 Immunopathology During Arenaviral Infections and After Vaccination

The beneficial effects of vigorous immunity are not always without cost because; in combating a microbial infection, the immune responses may harm the host. This issue must be considered when designing vaccines because, under some circumstances, vaccination can render an individual more susceptible to immunopathology following challenge by the pathogenic agent.

5.1 Antibody-Mediated Immunopathology

Arenaviruses often establish chronic or persistent infections in their hosts. Although best characterized in the LCMV model system, carriers states have also has been observed in animal models for Machupo (JOHNSON et al. 1965) and Junin (SABATTINI et al. 1977), and a patient who recovered from Lassa fever continued to shed virus for at least one month after infection (LEIFER et al. 1970). One consequence of chronic or persistent infection is the constant production of viral antigen, which may combine with antibody. These antigen-antibody complexes may be deposited in various tissues, including the basement membrane of the kidney, and in joints, causing glomerulonephritis and arthritis (OLDSTONE 1975). Several studies have suggested that the therapeutic administration of high-titer antibodies against Junin or Machupo virus may result in a late-onset neurological syndrome; the majority of cases are self-limiting, although fatal neuropathy has been observed (EDDY et al. 1975; MAIZTEGUI et al. 1979). It is not clear whether or not this syndrome is related to deposition of antigen-antibody complexes.

5.2 T-Cell-Mediated Immunopathology

In the process of clearing a virus, $CD8^+$ T cells almost inevitably harm the host. $CD8^+$ T cells employ two distinct effector mechanisms – lysis of target cells (often by the release of granules containing perforin) and secretion of antiviral cytokines such as IFNγ and TNFα – and both of these effector arms have been implicated in immunopathology during arenavirus infection.

Perforin-Mediated Immunopathology. Intracranial (i.c.) LCMV infection is lethal to a naive mouse, and death results not from the infection per se, but instead from the resulting primary $CD8^+$ T cell response. Ablation of $CD8^+$ T cells, or removal of perforin [using perforin "knockout" (PKO) mice], allows the animal to survive i.c. infection (KAGI et al. 1994; WALSH et al. 1994).

Cytokine-Mediated Immunopathology. Although it is easy to see how cell lysis might harm the host, the potential for cytokines to wreak havoc is, perhaps, less widely appreciated. However, cytokines are extremely toxic molecules; their release causes many of the signs and symptoms associated with virus infections (headache, fever, etc.). In high concentrations they can be lethal (SLIFKA and WHITTON 2000b). Others studies have reported that LCMV infection can induce IFNγ-mediated immunopathology (SELIN et al. 1998) and, under certain circumstances, cytokines produced by LCMV-specific $CD8^+$ T cells can cause lethal systemic shock (PUGLIELLI et al. 1999).

5.3 Vaccine-Primed Immunopathology

Given the topic of this chapter, it is particularly important to note that vaccination can prime an immune response which, upon subsequent exposure to the pathogen, can seriously harm the vaccinee. Thus, vaccination is not without risk. For example, in the 1960s, killed vaccines were developed against measles virus and respiratory syncytial virus. Children receiving these vaccines mounted strong immune responses but, upon encountering the related pathogen, became much more seriously ill than did unvaccinated children (FULGINITI et al. 1967; KAPIKIAN et al. 1969). Similar disease potentiation has also been demonstrated in the LCMV model system. As stated above, CTL are vital to the control of LCMV infection, but are also responsible for death following i.c. challenge. One might imagine that a vaccine which induced CTL would render the mouse more sensitive to subsequent i.c. challenge, but the opposite is true – under most circumstances, the vaccinated animal survives, and clears the virus. The explanation is straightforward. The vaccinated host can mount an accelerated CTL response, which rapidly limits virus replication in the CNS. As a result, viral dissemination throughout the choriomeninges is much reduced, and the animal survives. However, if one carefully titrates the vaccine-induced CTL response against the i.c. challenge dose of virus, one can observe disease potentiation – a vaccinated mouse succumbs to i.c. challenge more rapidly than does an unvaccinated animal (OEHEN et al. 1991; BATTEGAY et al. 1992). In this case, the CTL response is too weak to prevent virus dissemination throughout the CNS, but remains more robust than that present in an unvaccinated mouse; hence, the vaccinee dies sooner.

6 Maximizing Antiviral Efficacy While Minimizing Immunopathology

Cellular immune responses play a pivotal role in determining the outcome of many viral infections, and there is reason to believe that they may be the dominant mediator of protective immunity against Lassa virus. However, viruses do not go gently into the good night; their eradication involves the lysis of infected host cells and/or the release of potentially toxic immunomodulatory molecules. Intuitively, one might anticipate that evolutionary pressures would have generated a system which was sufficiently potent to combat the invading microbe, but sufficiently flexible to minimize the consequent damage to the host. Understanding these events may allow us to design improved vaccines, or to develop therapeutic measures which would diminish immunopathology without altering antiviral efficacy.

6.1 Most Virus-Specific $CD8^+$ T Cells Are "Silent" at the Peak of the Antiviral Response

Until recently, it was assumed that virus-specific T cell effector functions were constitutively expressed, and remained so until the virus infection had been eradicated. However, we found that IFNγ mRNA expression in the spleen was extremely low during the anti-LCMV immune response (RODRIGUEZ et al. 1998), leading us to suspect that cytokine production was in some way regulated by virus-specific $CD8^+$ T cells. We investigated this using the sensitive technique of intracellular cytokine staining (ICCS) to detect cytokine production by these cells. We found that approximately 50% of all $CD8^+$ T cells were virus-specific, a number consistent with earlier analyses using peptide/MHC "tetramer" techniques (MURALI-KRISHNA et al. 1998), but that – rather remarkably – even at the peak of the antiviral immune response, the majority of $CD8^+$ T cells did not constitutively produce cytokines (SLIFKA et al. 1999). Although similar data could be found in previous publications, the biological significance had not been appreciated; the mouse contains $\sim 10^7$ LCMV-specific $CD8^+$ T cells which can produce cytokines, and since cytokine release can be harmful or even lethal to the host, this effector function is very closely regulated.

6.2 $CD8^+$ T Cells Are Exquisitely Responsive to Antigen Contact

The absence of constitutive cytokine expression, even at the peak of the antiviral immune response, suggested that virus-specific T cells might be capable of rapidly down-regulating cytokine production. To test this hypothesis, we examined the effects of disrupting T cell:APC contact. We found that IFNγ production was terminated immediately upon disruption of contact between the CD8$^+$ T cell and its cognate antigen, and that T cells could re-initiate IFNγ production very quickly after antigen was restored (SLIFKA et al. 1999). These results showed that virus-specific CD8$^+$ T cells are exquisitely sensitive to antigen contact and are able to rapidly cycle cytokine production on, off, and on again. We carried out similar analyses of LCMV-specific memory cells, which are the cornerstone of vaccine-induced immunity in this model system, and found that they, too, are constitutively "silent", but quickly initiate cytokine production upon encountering antigen (SLIFKA and WHITTON 2000a).

7 Designing an Arenaviral Vaccine

Of the six viruses described, arguably the greatest single problem is posed by Lassa virus, which infects ~300,000 people annually, killing ~3,000. However, Junin virus infection is the direct result of human activity in erstwhilerural areas, and Guanarito and Sabia viruses enter the human population as a result of incursion into the rodent habitat. As we continue to encroach on rural and wild habitats, it is likely that these infections – and others as yet unknown – will increase. The relatively low risk of human-to-human transmission is encouraging, but we should not become sanguine; all of the arenaviruses cited in Section 2 cause significant morbidity and mortality, and their highly pathogenic nature suggests that vaccines should be made available. The obstacles to delivering new vaccines to the clinic are great, and to date there has been progress only for Junin virus. Those who have undertaken such a task are to be applauded. Below I ignore the formidable practical considerations, and address only the conceptual questions of arenaviral vaccine development.

7.1 What Responses Would an Ideal Arenaviral Vaccine Induce?

Although I argue above that antibodies and T cells contribute differently to the control of primary infection by New World and Old World arenaviruses, it would be foolish to restrict a vaccine to the induction of one or other aspect of immunity. We and others have shown definitively that a vaccine which induces only CD8$^+$ T cells can confer solid protection against LCMV (SCHULZ et al. 1989; HANY et al. 1989; KLAVINSKIS et al. 1989a,b, 1990; WHITTON 1990; AICHELE et al. 1990; WHITTON et al. 1993), but this is not to say that vaccine-induced neutralizing antibodies would be ineffective; indeed, monoclonal antibody can protect against LCMV challenge (BALDRIDGE and BUCHMEIER 1992). The reason that Old World arenaviruses fail to induce good neutralizing antibodies is not because the glycoproteins are intrinsically non-immunogenic; it is because, as outlined in section 4.2 above, GP-specific B cells are selectively infected, and killed by virus-specific CTL. One could argue that the evolution by these viruses of such an elegant mechanism of immune evasion, indicates that they might otherwise be rather susceptible to the effects of neutralizing antibody. Note that the molecular explanation for the failure of LCMV and Lassa virus to induce neutralizing antibodies suggests a possible loophole; perhaps neutralizing antibodies could be induced by GP delivered by some means other than the native virus. In summary, an anti-arenaviral vaccine should, ideally, induce neutralizing antibodies as well as long-lasting cellular responses.

7.2 Conventional and Recombinant Viral Vaccines

As stated above, a live attenuated vaccine, Candid-1, has been developed against AHF. Vaccine studies have been undertaken for Lassa virus, using recombinant vaccinia viruses, and have proven effective in guinea pigs and in primates. Mopeia virus, a close relative of Lassa, appears to be non-pathogenic in primates and, possibly, in humans. Although closely related to Lassa virus by gene sequence, Mopeia differs slightly in its intergenic structure (WILSON and CLEGG 1991), and this may be the factor dictating the virus' low pathogenicity. Monkeys immunized with Mopeia were solidly protected against Lassa challenge (FISHER-HOCH et al. 1989). One wonders if Mopeia should be considered for development as a candidate vaccine, despite its current categorization as a biosafety level 3 agent. Reassortants have been made between Lassa and Mopeia viruses (LUKASHEVICH 1992), but the pathogenicity of the reassortant viruses in monkeys – and in humans – is not known.

7.3 Should DNA Immunization Be Applied to Arenaviral Vaccines?

DNA immunization has been the subject of much study over the past 7 years, and several clinical trials are currently under way. Naysayers cite the possible risks of DNA vaccines, and claim that federal approval might be unlikely; in this author's opinion, DNA is safer than live attenuated viral vaccines, some of which can revert to virulence within days of administration. Such issues aside, what is the promise of DNA immunization? DNA vaccines have been the subject of several reviews (HASSETT and WHITTON 1996; DONNELLY et al. 1997; ROBINSON 1999; RODRIGUEZ and WHITTON 2000). In general, plasmid DNA induces better $CD8^+$ T cell responses than antibody responses, and this approach may, therefore, be better suited to immunizing against the Old World arenaviruses. We have recently used ICCS to track the immune responses induced by plasmid DNA encoding LCMV NP, and have found that a single inoculation of DNA induces $CD8^+$ T cells which, by 15 days post-injection, constitute 1%–2% of all $CD8^+$ T cells in the vaccinee (HASSETT et al. 2000a). Although lower than the response induced by LCMV, this figure is impressive; it indicates that the mouse has $\sim 2 \times 10^5$ virus-specific $CD8^+$ T cells, well above the number required to protect against virus challenge. Furthermore, the response is long-lasting, remaining at or near this level for at least one year post-inoculation (HASSETT et al. 2000b). Although DNA often induces good $CD8^+$ T cell responses, we and others have attempted to enhance the induction of this cell type by manipulating the encoded antigens to direct them into the MHC class I antigen presentation pathway; several of these studies have been successful (RODRIGUEZ et al. 1997, 1998; TOBERY and SILICIANO 1997, 1999). In some cases, antigen manipulation had a dramatic effect, transforming a non-protective vaccine into one which confers solid immunity (RODRIGUEZ et al. 1998; XIANG et al. 2000). In addition to $CD8^+$ T cells, DNA vaccines usually induce antibodies, but of lower titer than is seen following infection or live virus vaccination. We have carried out many experiments using plasmid DNA immunization to induce cellular and antibody responses to LCMV NP and GP (YOKOYAMA et al. 1995; HASSETT and WHITTON 1996; HASSETT et al. 1999, 2000b; WHITTON et al. 1999; AN et al. 2000; RODRIGUEZ and WHITTON 2000), and studies are under way to determine if plasmid-delivered GP can induce neutralizing antibodies. However, recent work indicates that some antigen modifications can greatly enhance antibody induction (BOYLE et al. 1998) and, if applicable to all antigens, this may be a useful adjunct for antiviral vaccination. In conclusion, while it is obviously premature to suggest that arenaviral DNA vaccines should be tested in humans, a strong case can be made – especially for Lassa virus – for testing these vaccines in primate models.

Acknowledgements. I thank the members of my lab. for their outstanding efforts, and Annette Lord for excellent secretarial support. This work was supported by NIH grant AI27028 and AI37186. This is manuscript number 12768-NP from the Scripps Research Institute.

References

Aichele P, Hengartner H, Zinkernagel RM, Schulz M (1990) Antiviral cytotoxic T cell response induced by in vivo priming with a free synthetic peptide. J Exp Med 171:1815–1820
An LL, Rodriguez F, Harkins S, Zhang J, Whitton JL (2000) Quantitative and qualitative analyses of the immune responses induced by a multivalent minigene DNA vaccine. Vaccine 18:2132–2141
Armstrong LR, Dembry LM, Rainey PM, Russi MB, Khan AS, Fischer SH, Edberg SC, Ksiazek TG, Rollin PE, Peters CJ (1999) Management of a Sabia virus-infected patients in a US hospital. Infect Control Hosp Epidemiol 20:176–182
Auperin DD, Romanowski V, Galinski M, Bishop DHL (1984) Sequencing studies of Pichinde arenavirus S RNA indicate a novel coding strategy an ambisense viral s RNA. J Virol 52:897–904
Baldridge JR, Buchmeier MJ (1992) Mechanisms of antibody-mediated protection against lymphocytic choriomeningitis virus infection: mother-to-baby transfer of humoral protection. J Virol 66: 4252–4257
Barton LL, Mets MB (1999) Lymphocytic choriomeningitis virus: pediatric pathogen and fetal teratogen. Pediatr Infect Dis J 18:540–541
Barton LL, Peters CJ, Ksiazek TG (1995) Lymphocytic choriomeningitis virus: an unrecognized teratogenic pathogen. Emerg Infect Dis 1:152–153

Battegay M, Oehen S, Schulz M, Hengartner H, Zinkernagel RM (1992) Vaccination with a synthetic peptide modulates lymphocytic choriomeningitis virus-mediated immunopathology. J Virol 66:1199–1201

Biggar RJ, Schmidt TJ, Woodall JP (1977) Lymphocytic choriomeningitis in laboratory personnel exposed to hamsters inadvertently infected with LCM virus. J Am Vet Med Assoc 171:829–832

Biggar RJ, Woodall JP, Walter PD, Haughie GE (1975) Lymphocytic choriomeningitis outbreak associated with pet hamsters. Fifty-seven cases from New York State. JAMA 232:494–500

Bowen MD, Peters CJ, Nichol ST (1996) The phylogeny of New World (Tacaribe complex) arenaviruses. Virol 219:285–290

Boyle JS, Brady JL, Lew AM (1998) Enhanced responses to a DNA vaccine encoding a fusion antigen that is directed to sites of immune induction. Nature 392:408–411

Brundler MA, Aichele P, Bachmann M, Kitamura D, Rajewsky K, Zinkernagel RM (1996) Immunity to viruses in B cell-deficient mice:influence of antibodies on virus persistence and on T cell memory. Eur J Immunol 26:2257–2262

Buchmeier MJ, Oldstone MBA (1979) Protein structure of lymphocytic choriomeningitis virus: evidence for a cell associated precursor of the virion glycopeptides. Virol 99:111–120

Childs JE, Glass GE, Korch GW, Ksiazek TG, Leduc JW (1992) Lymphocytic choriomeningitis virus infection and house mouse (*Mus musculus*) distribution in urban Baltimore. Am J Trop Med Hyg 47:27–34

Childs JE, Glass GE, Ksiazek TG, Rossi CA, Oro JG, Leduc JW (1991) Human-rodent contact and infection with lymphocytic choriomeningitis and Seoul viruses in an inner-city population. Am J Trop Med Hyg 44:117–121

Cummins D, Bennett D, Fisher-Hoch SP, Farrar B, Machin SJ, McCormick JB (1992) Lassa fever encephalopathy: clinical and laboratory findings. J Trop Med Hyg 95:197–201

Cummins D, Fisher-Hoch SP, Walshe KJ, Mackie IJ, McCormick JB, Bennett D, Perez G, Farrar B, Machin SJ (1989) A plasma inhibitor of platelet aggregation in patients with Lassa fever. Br J Haematol 72:543–548

Cummins D, McCormick JB, Bennett D, Samba JA, Farrar B, Machin SJ, Fisher-Hoch SP (1990) Acute sensorineural deafness in Lassa fever. JAMA 264:2093–2096

de Manzione N, Salas RA, Paredes H, Godoy O, Rojas L, Araoz F, Fulhorst CF, Ksiazek TG, Mills JN, Ellis BA, Peters CJ, Tesh RB (1998) Venezuelan hemorrhagic fever: clinical and epidemiological studies of 165 cases. Clin Infect Dis 26:308–313

Dekonenko EP, Ivanov AP, Andreeva LS, Tkachenko EA (1985) Appearance of antibodies to two viruses in cerebrospinal fluid of patients with aseptic meningitis. Acta Neurol Scand 71:146–149

Di Simone C, Zandonatti MA, Buchmeier MJ (1994) Acidic pH triggers LCMV membrane fusion activity and conformational change in the glycoprotein spike. Virol 198:455–465

Donnelly JJ, Ulmer JB, Liu MA (1997) DNA vaccines. Life Sci 60:163–172

Dykewicz CA, Dato VM, Fisher-Hoch SP, Howarth MV, Perez-Oronoz GI, Ostroff SM, Gary H Jr, Schonberger LB, McCormick JB (1992) Lymphocytic choriomeningitis outbreak associated with nude mice in a research institute. JAMA 267:1349–1353

Eddy GA, Wagner FS, Scott SK, Mahlandt BJ (1975) Protection of monkeys against Machupo virus by the passive administration of Bolivian haemorrhagic fever immunoglobulin (human origin) Bull World Health Organ 52:723–727

Enria DA, Maiztegui JI (1994) Antiviral treatment of Argentine hemorrhagic fever. Antiviral Res 23:23–31

Fisher-Hoch SP, McCormick JB, Auperin DD, Brown BG, Castor M, Perez G, Ruo S, Conaty A, Brammer L, Bauer S (1989) Protection of rhesus monkeys from fatal Lassa fever by vaccination with a recombinant vaccinia virus containing the Lassa virus glycoprotein gene. Proc Natl Acad Sci USA 86:317–321

Fisher-Hoch SP, Tomori O, Nasidi A, Perez-Oronoz GI, Fakile Y, Hutwagner L, McCormick JB (1995) Review of cases of nosocomial Lassa fever in Nigeria: the high price of poor medical practice. BMJ 311:857–859

Fulginiti VA, Eller JJ, Downie AW, Kempe CH (1967) Atypical measles in children previously immunized with inactivated measles virus vaccine. JAMA 202:1075–1080

Fulhorst CF, Bowen MD, Salas RA, De Manzione NM, Duno G, Utrera A, Ksiazek TG, Peters CJ, Nichol ST, De Miller E, Tovar D, Ramos B, Vasquez C, Tesh RB (1997) Isolation and characterization of Pirital virus, a newly discovered South American arenavirus. Am J Trop Med Hyg 56:548–553

Fulhorst CF, Bowen MD, Salas RA, Duno G, Utrera A, Ksiazek TG, De Manzione NM, De Miller E, Vasquez C, Peters CJ, Tesh RB (1999) Natural rodent host associations of Guanarito and Pirital viruses (Family Arenaviridae) in central Venezuela. Am J Trop Med Hyg 61:325–330

Gandsman EJ, Aaslestad HG, Ouimet TC, Rupp WD (1997) Sabia virus incident at Yale University. Am Ind Hyg Assoc J 58:51–53

Hany M, Oehen S, Schulz M, Hengartner H, Mackett M, Bishop DHL, Overton H, Zinkernagel RM (1989) Anti-viral protection and prevention of lymphocytic choriomeningitis or of the local footpad swelling reaction in mice by immunization with vaccinia-recombinant virus expressing LCMV-WE nucleoprotein or glycoprotein. Eur J Immunol 19:417–424

Hassett DE, Slifka MK, Zhang J, Whitton JL (2000a). Direct ex vivo kinetic and phenotypic analyses of $CD8^+$ T cell responses induced by DNA immunization. J Virol 74:8286–8291

Hassett DE, Whitton JL (1996) DNA Immunization. Trends in Microbiol 4:307–312

Hassett DE, Zhang J, Slifka MK, Whitton JL (2000b) Immune responses following neonatal DNA immunization are long-lived, abundant, and qualitatively similar to those induced by conventional vaccination. J Virol 74:2620–2627

Hassett DE, Zhang J, Whitton JL (1999) Induction of antiviral antibodies by DNA immunization requires neither perforin-mediated nor $CD8^+$-T-cell-mediated lysis of antigen-expressing cells. J Virol 73:7870–7873

Hinman AR, Fraser DW, Douglas RG, Bowen GS, Kraus AL, Winkler WG, Rhodes WW (1975) Outbreak of lymphocytic choriomeningitis virus infections in medical center personnel. Am J Epidemiol 101:103–110

Howard CR (1987) Neutralization of arenaviruses by antibody. Curr Top Microbiol Immunol 134:117–130

Jahrling PB (1983) Protection of Lassa virus-infected guinea pigs with Lassa-immune plasma of guinea pig, primate, and human origin. J Med Virol 12:93–102

Jahrling PB, Peters CJ (1984) Passive antibody therapy of Lassa fever in cynomolgus monkeys: importance of neutralizing antibody and Lassa virus strain. Infect Immun 44:528–533

Johnson KM, Mackenzie RB, Webb PA, Kuns ML (1965) Chronic infection of rodents by Machupo virus. Science 150:1618–1619

Kagi D, Ledermann B, Burki K, Seiler P, Odermatt B, Olsen KJ, Podack ER, Zinkernagel RM, Hengartner H (1994) Cytotoxicity mediated by T cells and natural killer cells is greatly impaired in perforin-deficient mice. Nature 369:31–37

Kapikian AZ, Mitchell RH, Chanock RM, Shvedoff RA, Stewart CE (1969) An epidemiologic study of altered clinical reactivity to respiratory syncitial (RS) virus infection in children previously vaccinated with an inactivated RS vaccine. Am J Epidemiol 89:404–421

Kenyon RH, Peters CJ (1986) Cytolysis of Junin infected target cells by immune guinea-pig spleen cells. Microb Pathog 1:453–464

Kilgore PE, Ksiazek TG, Rollin PE, Mills JN, Villagra MR, Montenegro MJ, Costales MA, Paredes LC, Peters CJ (1997) Treatment of Bolivian hemorrhagic fever with intravenous ribavirin. Clin Infect Dis 24:718–722

Klavinskis LS, Oldstone MBA, Whitton JL (1989b) Designing vaccines to induce cytotoxic T lymphocytes: protection from lethal viral infection. In: Brown F, Chanock R, Ginsberg H, Lerner R (eds) Vaccines 89. Modern Approaches to New Vaccines Including Prevention of AIDS. Cold Spring Harbor Laboratory, Cold Spring Harbor: pp 485–489.

Klavinskis LS, Whitton JL, Joly E, Oldstone MBA (1990) Vaccination and protection from a lethal viral infection: identification, incorporation, and use of a cytotoxic T lymphocyte glycoprotein epitope. Virol 178:393–400

Klavinskis LS, Whitton JL, Oldstone MBA (1989a) Molecularly engineered vaccine which expresses an immunodominant T-cell epitope induces cytotoxic T lymphocytes that confer protection from lethal virus infection. J Virol 63:4311–4316

Leifer E, Gocke DJ, Bourne H (1970) Lassa fever, a new virus disease of man from West Africa. II. Report of a laboratory-acquired infection treated with plasma from a person recently recovered from the disease. Am J Trop Med Hyg 19:677–679

Lisieux T, Coimbra M, Nassar ES, Burattini MN, de Souza LT, Ferreira I, Rocco IM, da Rosa AP, Vasconcelos PF, Pinheiro FP (1994) New arenavirus isolated in Brazil. Lancet 343:391–392

Lukashevich IS (1992) Generation of reassortants between African arenaviruses. Virol 188:600–605

Mackenzie RB (1965) Epidemiology of Machupo virus infection. I. Pattern of human infection, San Joaquin, Bolivia, 1962–1964. Am J Trop Med Hyg 14:808–813

Maetz HM, Sellers CA, Bailey WC, Hardy GE Jr (1976) Lymphocytic choriomeningitis from pet hamster exposure: a local public health experience. Am J Public Health 66:1082–1085

Maiztegui JI (1975) Clinical and epidemiological patterns of Argentine haemorrhagic fever. Bull World Health Organ 52:567–575

Maiztegui JI, Fernandez NJ, de Damilano AJ (1979) Efficacy of immune plasma in treatment of Argentine haemorrhagic fever and association between treatment and a late neurological syndrome. Lancet 2:1216–1217

Maiztegui JI, McKee KT Jr, Barrera Oro JG, Harrison LH, Gibbs PH, Feuillade MR, Enria DA, Briggiler AM, Levis SC, Ambrosio AM, Halsey NA, Peters CJ (1998) Protective efficacy of a live attenuated vaccine against Argentine hemorrhagic fever. AHF Study Group. J Infect Dis 177:277–283

McCormick JB (1987) Epidemiology and control of Lassa fever. In: Oldstone MBA (ed) Arenaviruses: biology and immunotherapy. Current Topics in Microbiology and Immunology. 134. Springer-Verlag, New York: pp 69–78

McCormick JB, King IJ, Webb PA, Scribner CL, Craven RB, Johnson KM, Elliott LH, Belmont-Williams R (1986) Lassa fever. Effective therapy with ribavirin. N Engl J Med 314:20–26

McCormick JB, Mitchell SW, Kiley MP, Ruo S, Fisher-Hoch SP (1992) Inactivated Lassa virus elicits a non protective immune response in rhesus monkeys. J Med Virol 37:1–7

McCormick JB, Webb PA, Krebs JW, Johnson KM, Smith ES (1987) A prospective study of the epidemiology and ecology of Lassa fever. J Infect Dis 155:437–444

McKee KT Jr, Huggins JW, Trahan CJ, Mahlandt BG (1988) Ribavirin prophylaxis and therapy for experimental Argentine hemorrhagic fever. Antimicrob Agents Chemother 32:1304–1309

McKee KT Jr, Oro JG, Kuehne AI, Spisso JA, Mahlandt BG (1992) Candid No. 1 Argentine hemorrhagic fever vaccine protects against lethal Junin virus challenge in rhesus macaques. Intervirology 34:154–163

McKee KT Jr, Oro JG, Kuehne AI, Spisso JA, Mahlandt BG (1993) Safety and immunogenicity of a live-attenuated Junin (Argentine hemorrhagic fever) vaccine in rhesus macaques. Am J Trop Med Hyg 48:403–411

Mercado R (1975) Rodent control programmes in areas affected by Bolivian haemorrhagic fever. Bull World Health Organ 52:691–696

Monath TP, Maher M, Casals J, Kissling RE, Cacciapuoti A (1974) Lassa fever in the Eastern Province of Sierra Leone, 1970–1972. II. Clinical observations and virological studies on selected hospital cases. Am J Trop Med Hyg 23:1140–1149

Morrison HG, Bauer SP, Lange JV, Esposito JJ, McCormick JB, Auperin DD (1989) Protection of guinea-pigs from Lassa fever by vaccinia virus recombinants expressing the nucleoprotein or the envelope glycoproteins of Lassa virus. Virol 171:179–188

Murali-Krishna K, Altman JD, Suresh M, Sourdive DJ, Zajac AJ, Miller JD, Slansky J, Ahmed R (1998) Counting antigen-specific CD8 T cells: a reevaluation of bystander activation during viral infection. Immunity 8:177–187

Oehen S, Hengartner H, Zinkernagel RM (1991) Vaccination for disease. Science 251:195–198

Oldstone MBA (1975) Virus neutralization and virus-induced immune complex disease. Virus-antibody union resulting in immunoprotection or immunologic injury–two sides of the same coin. Prog Med Virol 19:84–119

Park JY, Peters CJ, Rollin PE, Ksiazek TG, Katholi CR, Waites KB, Gray B, Maetz HM, Stephensen CB (1997) Age distribution of lymphocytic choriomeningitis virus serum antibody in Birmingham, Alabama: evidence of a decreased risk of infection. Am J Trop Med Hyg 57:37–41

Peters CJ, Jahrling PB, Liu CT, Kenyon RH, McKee KT Jr, Barrera OJG (1987) Experimental studies of arenaviral hemorrhagic fevers. In: Oldstone MBA (ed) Arenaviruses: biology and immunotherapy. Current Topics in Microbiology and Immunology. 134. pp 5–68

Planz O, Ehl S, Furrer E, Horvath E, Brundler MA, Hengartner H, Zinkernagel RM (1997) A critical role for neutralizing-antibody-producing B cells, CD4$^+$ T cells, and interferons in persistent and acute infections of mice with lymphocytic choriomeningitis virus: implications for adoptive immunotherapy of virus carriers. Proc Natl Acad Sci USA 94:6874–6879

Planz O, Seiler P, Hengartner H, Zinkernagel RM (1996) Specific cytotoxic T cells eliminate cells producing neutralizing antibodies. Nature 382:726–729

Puglielli MT, Browning JL, Brewer AW, Schreiber RD, Shieh WJ, Altman JD, Oldstone MB, Zaki SR, Ahmed R (1999) Reversal of virus-induced systemic shock and respiratory failure by blockade of the lymphotoxin pathway. Nat Med 5:1370–1374

Rai SK, Micales BK, Wu MS, Cheung DS, Pugh TD, Lyons GE, Salvato MS (1997) Timed appearance of lymphocytic choriomeningitis virus after gastric inoculation of mice. Am J Pathol 151:633–639

Riviere Y, Ahmed R, Southern PJ, Buchmeier MJ, Dutko FJ, Oldstone MBA (1985) The S RNA segment of lymphocytic choriomeningitis virus codes for the nucleoprotein and glycoproteins 1 and 2. J Virol 53:966–968

Robinson HL (1999) DNA vaccines: basic mechanism and immune responses. Int J Mol Med 4:549–555

Rodriguez F, An LL, Harkins S, Zhang J, Yokoyama M, Widera G, Fuller JT, Kincaid C, Campbell IL, Whitton JL (1998) DNA immunization with minigenes: low frequency of memory CTL and inefficient antiviral protection are rectified by ubiquitination. J Virol 72:5174–5181

Rodriguez F, Whitton JL (2000) Enhancing DNA immunization. Virol 268:233–238

Rodriguez F, Zhang J, Whitton JL (1997) DNA immunization: ubiquitination of a viral protein enhances CTL induction, and antiviral protection, but abrogates antibody induction. J Virol 71:8497–8503

Roebroek RM, Postma BH, Dijkstra UJ (1994) Aseptic meningitis caused by the lymphocytic choriomeningitis virus. Clin Neurol Neurosurg 96:178–180

Sabattini MS, Gonzalez DRLE, Diaz G, Vega VR (1977) Natural and experimental infection of rodents with Junin virus. Medicina (B Aires) 37:149–161

Salas R, de Manzione N, Tesh RB, Rico-Hesse R, Shope RE, Betancourt A, Godoy O, Bruzual R, Pacheco ME, Ramos B (1991) Venezuelan haemorrhagic fever. Lancet 338:1033–1036

Salvato MS, Shimomaye E, Oldstone MBA (1989) The primary structure of the lymphocytic choriomeningitis virus L gene encodes a putative RNA polymerase. Virol 169:377–384

Salvato MS, Shimomaye EM (1989) The completed sequence of lymphocytic choriomeningitis virus reveals a unique RNA structure and a gene for a zinc finger protein. Virol 173:1–10

Samoilovich SR, Carballal G, Weissenbacher MC (1983) Protection against a pathogenic strain of Junin virus by mucosal infection with an attenuated strain. Am J Trop Med Hyg 32:825–828

Samoilovich SR, Pecci Saavedra J, Frigerio MJ, Weissenbacher MC (1984) Nasal and intrathalamic inoculations of primates with Tacaribe virus: protection against Argentine hemorrhagic fever and absence of neurovirulence. Acta Virol (Prague) (Engl Ed) 28:277–281

Schulz M, Aichele P, Vollenweider M, Bobe FW, Cardinaux F, Hengartner H, Zinkernagel RM (1989) Major histocompatibility complex-dependent T cell epitopes of lymphocytic choriomeningitis virus nucleoprotein and their protective capacity against viral disease. Eur J Immunol 19:1657–1668

Selin LK, Varga SM, Wong IC, Welsh RM (1998) Protective heterologous antiviral immunity and enhanced immunopathogenesis mediated by memory T cell populations. J Exp Med 188:1705–1715

Sheinbergas MM (1976) Hydrocephalus due to prenatal infection with the lymphocytic choriomeningitis virus. Infection 4:185–191

Sheinbergas MM, Kilchavskiene VV, Tulevichiene JP (1984) Prenatal lymphocytic choriomeningitis (LCM): three new cases. Infection 12:105–106

Sheinbergas MM, Lewis VJ, Thacker WL, Verikiene VV (1981) Serological diagnosis in children infected prenatally with lymphocytic choriomeningitis virus. Infect Immun 31:837–838

Skinner HH, Knight EH (1973) Natural routes for post-natal transmission of murine lymphocytic choriomeningitis. Lab Anim 7:171–184

Slifka MK, Rodriguez F, Whitton JL (1999) Rapid on/off cycling of cytokine production by virus-specific $CD8^+$ T cells. Nature 401:76–79

Slifka MK, Whitton JL (2000a) Activated and memory $CD8^+$ T cells can be distinguished by their cytokine profiles and phenotypic markers. J Immunol 164:208–216

Slifka MK, Whitton JL (2000b). Clinical implications of dysregulated cytokine production. J Mol Med 78:74–80

Stephensen CB, Blount SR, Lanford RE, Holmes KV, Montali RJ, Fleenor ME, Shaw JF (1992) Prevalence of serum antibodies against lymphocytic choriomeningitis virus in selected populations from two US cities. J Med Virol 38:27–31

Tobery T, Siliciano RF (1999) Induction of enhanced CTL-dependent protective immunity in vivo by N-end rule targeting of a model tumor antigen. J Immunol 162:

Tobery TW, Siliciano RF (1997) Targeting of HIV-1 antigens for rapid intracellular degradation enhances cytotoxic T lymphocyte (CTL) recognition and the induction of de novo CTL responses in vivo after immunization. J Exp Med 185:909–920

Vanzee BE, Douglas RG, Betts RF, Bauman AW, Fraser DW, Hinman AR (1975) Lymphocytic choriomeningitis in university hospital personnel. Clinical features. Am J Med 58:803–809

Walker DH, Wulff H, Lange JV, Murphy FA (1975) Comparative pathology of Lassa virus infection in monkeys, guinea-pigs, and *Mastomys natalensis*. Bull World Health Organ 52:523–534

Walsh CM, Matloubian M, Liu CC, Ueda R, Kurahara CG, Christensen JL, Huang MT, Young JD, Ahmed R, Clark WR (1994) Immune function in mice lacking the perforin gene. Proc Natl Acad Sci USA 91:10854–10858

Weaver SC, Salas RA, de Manzione N, Fulhorst CF, Duno G, Utrera A, Mills JN, Ksiazek TG, Tovar D, Tesh RB (2000) Guanarito virus (Arenaviridae) isolates from endemic and outlying localities in Venezuela: sequence comparisons among and within strains isolated from Venezuelan hemorrhagic fever patients and rodents. Virol 266:189–195

Whitton JL (1990) Lymphocytic choriomeningitis virus CTL. Sem Virol 1:257–262
Whitton JL, Rodriguez F, Zhang J, Hassett DE (1999) DNA immunization: mechanistic studies. Vaccine 17:1612–1619
Whitton JL, Sheng N, Oldstone MBA, McKee TA (1993) A "string-of-beads" vaccine, comprising linked minigenes, confers protection from lethal-dose virus challenge. J Virol 67:348–352
Wilson SM, Clegg JC (1991) Sequence analysis of the S RNA of the African arenavirus Mopeia: an unusual secondary structure feature in the intergenic region. Virol 180:543–552
Xiang R, Lode HN, Chao TH, Ruehlmann JM, Dolman CS, Rodriguez F, Whitton JL, Overwijk WW, Restifo NP, Reisfeld RA (2000) An autologous oral DNA vaccine protects against murine melanoma. Proc Natl Acad Sci USA 97:5492–5497
Yokoyama M, Zhang J, Whitton JL (1995) DNA immunization confers protection against lethal lymphocytic choriomeningitis virus infection. J Virol 69:2684–2688

Junin Virus Vaccines

D.A. Enria[1] and J.G. Barrera Oro[2]

1	Introduction	240
2	Historical Background of Junin Virus Vaccine Development	242
2.1	Inactivated Vaccines	242
2.1.1	Formolized Mouse Brain	242
2.1.2	Formolized Antigen Produced in Cell Culture	243
2.1.3	Light Inactivated Antigens	243
2.1.4	Antigenic Subunit	243
2.2	Live Attenuated Vaccines	243
2.2.1	Heterologous Virus	243
2.2.2	Homologous Virus	244
3	Junin Virus Attenuated Candid #1 Vaccine	245
3.1	Development and Preclinical Studies	246
3.1.1	Virus Strain	246
3.1.2	Passage History	247
3.1.3	Characterization of Candid #1 Vaccine	247
3.1.3.1	Comparative Quantitative Virulence Testing of Candid #1	247
3.1.3.2	Candid #1 Attenuation Phenotype After Replication In Vivo	248
3.1.3.3	Long-Term Effect and Viral Persistence	249
3.1.3.4	Immunogenicity and Protective Efficacy of Candid #1	250
3.1.4	Testing of Candid #1 Secondary Seed and Vaccine	250
3.1.5	Molecular Characterization of Candid #1 Vaccine	251
3.2	Clinical Studies	251
3.2.1	Safety and Immunogenicity of Candid #1	251
3.2.2	Efficacy of Candid #1	252
3.2.2.1	Vaccine Reactogenicity	253
3.2.2.2	Vaccine Immunogenicity	253
3.2.2.3	Vaccine Efficacy	253
4	Current Studies with Candid #1 Vaccine	254
4.1	Immunogenicity and Duration of the Immune Response	255
4.2	Candid #1 Effectiveness for the 1992–99 Period	255
4.3	Impact of Vaccination of Selected High-Risk Groups	255
5	Candid #1 as an Orphan Vaccine	255
6	Concluding Remarks and Future Expectations	256
References		258

[1] Instituto Nacional de Enfermedades Virales Humanas (INEVH) "Dr. Julio I. Maiztegui", Monteagudo 2510, 2700 Pergamino, Argentina
[2] Instituto Malbran; CONICET, Argentina; The Salk Institute, USA

1 Introduction

Argentine hemorrhagic fever (AHF) was recognized as a new clinical entity from the richest farming region of Argentina in the 1950s (ARRIBALZAGA 1955). The etiologic agent of this disease, Junin virus (JUN), was isolated in 1958 (PARODI et al. 1958; PIROSKI et al. 1959). JUN belongs to the Arenaviridae family, which includes other rodent-borne pathogens which are important causes of hemorrhagic fever in Africa, (Lassa) and South America (Machupo, Guanarito, and Sabia viruses). The Arenaviridae comprises at least 20 recognized members. Arenaviruses are enveloped RNA viruses, and are divided into two groups, with low level antigenic relatedness: the Old World group, and New World group or Tacaribe complex. Two antigenic subgroups were defined within the New World arenaviruses. JUN is contained into the first group, together with Amapari, Latino, Machupo and Tacaribe viruses. Phylogenetic analysis have shown that Old World and New World arenaviruses occupied two distinct clades, and that New World arenaviruses comprise three evolutionary lineages, named A, B, and C. JUN is contained in lineage B, together with the other three agents causing South American hemorrhagic fevers (ENRIA et al. 1999a).

AHF is a systemic illness characterized by hematological, neurological, cardiovascular, renal and immunological alterations. Although subclinical cases occur, most infections are sufficiently severe to be recognized by the clinicians. This disease, like other arenaviral hemorrhagic fevers, is distinguished by the insidious onset of a prodromal phase lasting one week, characterized by fever and constitutional signs and symptoms, with minor neurologic and/or hemorrhagic signs. A severe hemorrhagic-neurologic phase, often fatal, is seen in 20%–30% of cases during the second week. The specific treatment consists of the transfusion of immune plasma with standardized dose of neutralizing antibodies. To be effective, this should be given within 8 days of onset. A late neurological syndrome, characterized mainly by truncal cerebellar ataxia with fever, is seen in around 10% of cases in association with this treatment (ENRIA et al. 1998).

The principal epidemiologic characteristics of AHF are determined by the natural cycle of JUN and by the behavior of its rodent reservoirs. *Calomys musculinus* (family Muridae, subfamily Sigmodontinae) has been identified as the principal reservoir of JUN, although the virus has also been isolated from the organs and body fluids of *Calomys laucha* and *Akodon azarae*, and occasionally from *Mus musculus*, *Necromys benefactus* and *Oligoryzomys flavescens* (SABATTINI and CONTIGIANI 1982). The exact mechanism of viral transmission from rodents to humans is unknown, but there is strong evidence to support the likelihood that these viruses are infectious as aerosols. AHF is not usually contagious among humans, although human-to-human transmission can occasionally occur (ENRIA et al. 1998).

AHF is endemic in the humid pampas, the most fertile region in Argentina. The most striking epidemiologic characteristic of AHF is its progressive geographic extension. In 1958, cases were limited to an area of approximately 16,000km^2, with a population at risk estimated to be 270,000 In 1963, cases of AHF were confirmed in the southeast of the province of Córdoba, and between 1964 and 1967 new areas were found in the province of Buenos Aires. Cases began to appear later in the south of Santa Fe province (MAIZTEGUI and SABATTINI 1977). At present, the endemic-epidemic region covers an area of approximately 150,000km^2, with a population at risk estimated at more than 5,000,000 (Fig. 1). Geographic extensions in the latest periods have been smaller than those seen in previous years, suggesting an autolimitation in the extension of the endemic area. Nevertheless, rodent studies indicate the possibility of a northward extension, as well as of the re-emergence of the disease in areas currently considered to be only of historical importance (MILLS et al. 1992; GARCÍA et al. 1996). The incidence of AHF varies in time. In general, it is higher during a period of 5–10 years in newly involved areas, and later declines. Nevertheless, cases continue to be reported from older locations (MAIZTEGUI et al. 1986).

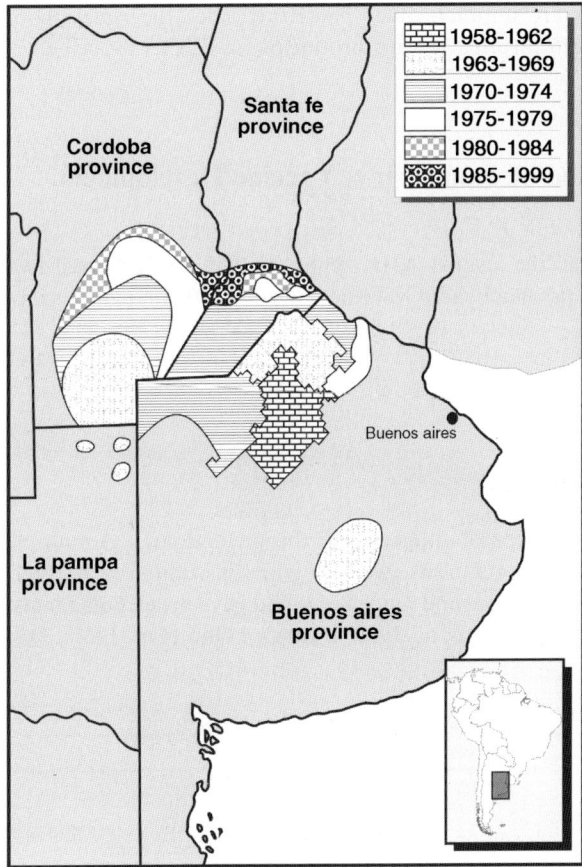

Fig. 1. AHF endemic area and progressive extension

Although attempts to quantify the social and economic impact of AHF have not been made, incidence rates as high as 140/100,000 in the general population and 355/100,000 in adult males have been registered in some rural areas, in a region whose prosperity is dependent on agriculture. The case fatality rate is as high as 30% in untreated individuals. Specific treatment consists of the administration of immune plasma with standardized doses of neutralizing antibodies (MAIZTEGUI et al. 1979). This treatment is only effective when it is administered during the first week of illness (ENRIA and MAIZTEGUI 1994). Thus, a surveillance system for all febrile syndromes of undetermined etiology should be kept permanently in the endemic area to identify suspected AHF cases for early treatment. The disease lasts from seven to 14 days, and is followed by a protracted convalescence that leaves people unable to work for up to 90 days. Approximately 10% of the patients treated with immune plasma develop a late neurological syndrome. (MAIZTEGUI et al. 1979; ENRIA et al. 1985). These considerations clearly indicate the substantial social impact and economic loss due to AHF.

For AHF prevention, rodent control and the control of human contact with infected rodent populations are impracticable. Therefore, since the isolation of JUN, several efforts have been directed toward producing an effective vaccine.

2 Historical Background of Junin Virus Vaccine Development

The attempts to obtain a vaccine against AHF begun in 1959 and followed two principal lines of research: (a) inactivated vaccines, (b) live attenuated vaccines (Table 1).

2.1 Inactivated Vaccines

2.1.1 Formolized Mouse Brain

Two studies reported the elaboration of antigens with this methodology. A multiple shot scheme of three to four inoculations, weekly, with or without a booster, induced neutralizing antibodies in mice and guinea pigs and gave protection against challenge with virulent JUN (PIROSKY et al. 1959; BARRERA ORO et al. 1967). One

Table 1. Historical background of Junin virus vaccine development

Inactivated vaccines	Live attenuated vaccines	
	Heterologous virus	Homologous virus
Formolized mouse brain		XJ 0 strain
Formolized antigen in cell culture	Tacaribe virus	XJ Clone 3 strain
Light inactivated antigens		Candid #1 strain
Antigenic subunit		

of these antigens was tested for protective efficacy in humans (AGNESE 1999), and then given to more than 15,000 persons of the AHF endemic area between 1959 and 1962, but neither the immunogenicity nor the protective efficacy were evaluated (METTLER 1969).

2.1.2 Formolized Antigen Produced in Cell Culture

This antigen induced neutralizing antibodies in experimental animals, but those immunized did not survive challenge with virulent JUN strains (VIDELA et al. 1989).

2.1.3 Light Inactivated Antigens

Several techniques were used to obtain inactivated JUN antigens, including exposition to photoactive colorants, ultraviolet radiation and acetone processing (PARODI et al. 1965; MARTINEZ SEGOVIA et al. 1980; D'AIUTOLO et al. 1979; CARBALLAL et al. 1985). In one experiment, mouse brain pools with two different JUN strains were inactivated with neutral red and ultraviolet light, and were inoculated in guinea pigs three times during a period of a month. Fifty percent of immunized animals survived challenge with virulent JUN strains. This antigen was used in 11 human volunteers, but only elicited complement fixing antibodies in two previously immune persons (GUERRERO 1977).

2.1.4 Antigenic Subunit

A JUN antigen subunit consisting of a purified capsular glycoprotein G38 was given with complete Freund's adjuvant to guinea pigs. This induced neutralizing antibodies and protected on challenge with virulent strains (CRESTA et al. 1980).

2.2 Live Attenuated Vaccines

Two lines of vaccine development were followed, one using homologous and the other heterologous virus.

2.2.1 Heterologous Virus

Tacaribe virus, the prototype member of New World arenaviruses, is biochemical and serologically closely related to JUN (DOWNS et al. 1963; DAMONTE et al. 1986; WEISSENBACHER et al. 1987). The virus is considered non-pathogenic or of very low pathogenicity for humans, as only one symptomatic laboratory acquired infection has been reported. Early studies done in several laboratories on the immunological relationships among Tacaribe-complex arenaviruses, showed that hamsters, mice and guinea-pigs hyperimmunized with Tacaribe virus frequently developed small amounts of neutralizing antibodies to JUN (PARODI and COTO 1964; TAURASO and

SHELOKOV 1965; WEISSENBACHER et al. 1975). In guinea pig and primate experimental models, Tacaribe virus showed cross-protection with JUN and the heterologous immunity induced was shown to persist (COTO et al. 1980; DAMONTE et al. 1981; WEISSENBACHER et al. 1982; SAMOILOVICH et al. 1984, 1988; WEISSENBACHER and DAMONTE 1983).

From this basis, Tacaribe virus was proposed as a candidate for an heterologous vaccine against AHF (WEISSENBACHER et al. 1987; SAMOILOVICH et al. 1988).

2.2.2 Homologous Virus

The guinea-pig adapted XJ prototype strain of JUN multiplies rapidly and widely in this animal, producing fever, viremia, weight loss, leucopenia, thrombocytopenia, and death in 10–15 days. Serendipitously, some investigators at the virus laboratories from the Rockefeller Foundation noticed that the XJ JUN strain lost virulence to guinea pigs after several passages in mouse brain (METTLER et al. 1963). Originally, this was attributed to a host's genetically-derived resistance (CASALS 1965). A plaque purified clone from a passage 14 in mouse-brain of XJ, originally designated as XJ Clone 3, was obtained at the National Institutes of Health, USA (BARRERA ORO and MCKEE 1992). The XJ Clone 3 JUN strain was subsequently grown in continuous cell lines MA-104 and MA-111. The product of one of the cycles in the continuous rabbit kidney line MA-111 was later grown in mouse-brain, giving the basis for the development of the XJ Clone 3 vaccine (RUGIERO et al. 1981).

XJ Clone 3 was found to replicate poorly in guinea pigs: no viremia, no weight loss, neither fever nor important hematological changes were detected in this experimental model. Antibodies were detectable by both complement fixation and neutralization tests 20 days after the administration of the vaccine. Different groups of animals survived challenge with the virulent XJ strain, at 10, 20, 30 and 60 days after vaccination (GUERRERO et al. 1969).

In 1968, the first seven human volunteers were inoculated with the XJ Clone 3 antigen (RUGIERO et al. 1969; WEISSENBACHER et al. 1969). Other three cohorts of 64, 159 and 406 volunteers, a total of 636 persons of both sexes, ranging from 11 to 77 years, followed this group in subsequent years. No serious illness occurred among vaccinees. Clinical studies established that among 213 vaccinees, 24.4% did not report any symptoms, 29.1% had headache, retro-ocular pain, and myalgia without fever, 44% had a febrile syndrome characterized by asthenia, headache, retro-ocular pain, and myalgia, appearing between day 3 and 10 and lasting no more than 4 days.In addition, 2.4% had a local reaction at the site of inoculation, which lasted less than 2 days. Clinical laboratory studies performed in 57 vaccinees showed no hematological alterations in 21%, 50% had leukopenia, and 9% had thrombocytopenia, and 20% had both leukopenia and thrombocytopenia. In all cases, recovery to normal values was achieved. JUN neutralizing antibodies were detected in 96% of 276 volunteers studied between 1 and 3 months after vaccination (RUGIERO et al. 1974).

Follow-up studies performed in 267 of these volunteers between 7 and 9 years after vaccination did not identify long-term adverse reactions, and 90% (153/165) still had detectable neutralizing antibodies against JUN (RUGIERO et al. 1981).

In 1971, human inoculations with the XJ Clone 3 vaccine candidate were interrupted. Reasons for this decision were the absence of a good documentation on the successive virus growth cycles, including the previous passages in heteroploid cell lines and mouse-brain substrate (WEISSENBACHER et al. 1987). However, the research with XJ Clone 3 continued, and other derivatives from new growth cycles were evaluated. The comparison with wild JUN strains showed that XJC clone 3 was very attenuated, although some evidences of residual peripheral virulence, neurotropism and neurovirulence was shown both in guinea pig and primates (CONTIGIANI and SABATTINI 1977; AVILA et al. 1979; CARBALLAL et al. 1982; LAGUENS et al. 1983). Viral persistence was demonstrated up to 3 months after infection in guinea pigs, and viral antigens were detected in infected primates up to 1 year after infection (AVILA et al. 1981, 1985; Malumbres et al. 1984).

The XJ0 JUN strain, a mouse-brain passage subline of the prototype XJ strain, proved to be immunogenic in guinea pigs, although a similar pattern to that observed with XJC clone 3 in persistence, neurotropism and neurovirulence was also described (GUERRERO and BOXACA 1980; BOXACA et al. 1981, 1984; GUERRERO et al. 1983, 1985).

3 Junin Virus Attenuated Candid #1 Vaccine

In 1976, the first International Seminar on Arenaviral Hemorrhagic Fevers was held in Buenos Aires, Argentina. Among the conclusions and recommendations of this seminar, the development of a vaccine against AHF was considered of high priority (EDDY and BARRERA ORO 1977). With this objective, an international collaborative project involving the government of Argentina, Pan American Health Organization (PAHO), the United Nation Development Program and the United States Army Medical Research Institute of Infectious Diseases (USAMRIID) was initiated.

In 1979, the goals of the program were: (1) to identify a virus with a well documented passage history, and at least as attenuated as XJ Clone 3, (2) to adapt the virus to diploid cell lines of certified vaccine substrate quality, (3) to select attenuated variants with minimum residual virulence, while keeping their immunogenicity and their capacity to protect experimental animals, and (4) to produce an experimental vaccine to be tested in human volunteers under the supervision of US, Argentine, and PAHO regulatory agencies. To achieve this objective, two approaches were followed simultaneously. The first attempted to identify a JUN strain already attenuated that could be used as the original virus for further modifications. The second approach was directed towards the isolation of a new natural JUN strain, that would be exposed to successive growth cycles in reliable cell substrates so as to obtain several attenuated strains. The latter approach yielded viruses with reduced but still significant virulence and was abandoned (BARRERA ORO and MCKEE 1992).

The first approach resulted in the development of the current Candid #1 AHF vaccine (Fig. 2).

3.1 Development and Preclinical Studies

3.1.1 Virus Strain

Candid #1 was derived from a 44th mouse brain passage (XJ #44) of the XJ prototype strain of JUN. XJ #44 was selected for further study because it had a well defined passage history; was as attenuated as XJ Clone 3 vaccine when tested in guinea pigs, and gave negative results when tested for murine viruses (BARRERA ORO and MCKEE 1992).

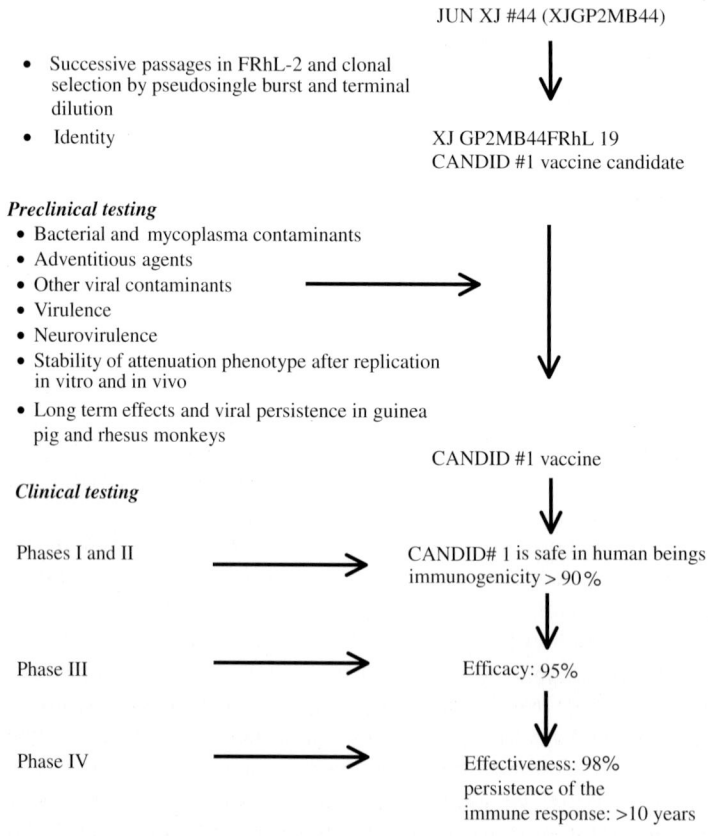

Fig. 2. Candid #1 outline of studies

3.1.2 Passage History

The XJ strain was isolated in guinea pigs from a human case in Argentina, and was subjected to another guinea pig passage and 11 mouse brain (MB) passages. The virus was then passed in MB 32 times at the Rockefeller Foundation virus laboratory. This material, $XJGP_2MB_{43}$ was received by the USAMRIID and passed once more in MB to prepare XJ #44, the starting material for the Candid #1 vaccine. At USAMRIID the virus was passed 12 times in certified FRhL-2. To purify and to attempt to select still further attenuated subpopulations in FRhL-2, the virus was subjected to the single burst technique (WALEN 1963), followed by two limiting dilution passages, and an expansion passage. This $XJGP_2$ $FRhL_{16}$ passage was named Candid #1. Candid #1 was passed three more times in FRhL-2 cells to produce master seed, secondary seed, and vaccine (XJ GP_2 MB_{44} $FRhL_{19}$) (Fig. 3).

3.1.3 Characterization of Candid #1 Vaccine

Extensive preclinical investigations were conducted to assure the safety, immunogenicity and protective efficacy of Candid #1. Animal studies were conducted to assure lack of neurovirulence, neurotropism, or hemorrhagic manifestations, the genetic stability of vaccine virus strain, and lack of viral persistence in monkeys. Comparison was frequently made to derivatives and/or progenitors of the Argentine XJ Clone 3 vaccine because of its common but less precise ancestry with the Candid #1 vaccine and its record of use in 636 volunteers without adverse effect.

3.1.3.1 Comparative Quantitative Virulence Testing of Candid #1

Comparative virulence of Candid #1 to XJ #44 parent, XJ Clone 3 vaccine, and passages of Candid #1 was studied quantitatively by conducting simultaneous titration of virus in mice, guinea pigs and Vero cell cultures, with the calculation of lethal indexes [plaque forming units (pfu) required to induce one LD50]. Outbred Hartley and inbred strain 13 guinea pigs inoculated intracerebrally with virus were monitored for death, paralysis and CNS lesions. From these observations, it was clear that Candid #1 and its passages derivatives were not lethal for guinea pigs. By contrast, 110 and 5,000pfu of XJ Clone 3 were necessary to obtain one LD50 in animals of 160 or 500g, respectively. Furthermore, all doses of XJ Clone 3 induced CNS lesion, whereas 200 to 700pfu of Candid #1 or its progeny were required to induce lesions in 50% of strain 13 guinea pigs. It was clear that Candid #1 was markedly less virulent and less neurovirulent for guinea pigs than other attenuated JUN strains. This attenuation remained stable and did not revert even after six cell culture passages.

In mice, it was apparent than Candid #1 and its progeny were more attenuated than XJ Clone 3. It took 100-fold more Candid #1 or Candid #1 cell-passages than XJ Clone 3 or XJ-44 to obtain one ICLD50. These experiments also appeared to

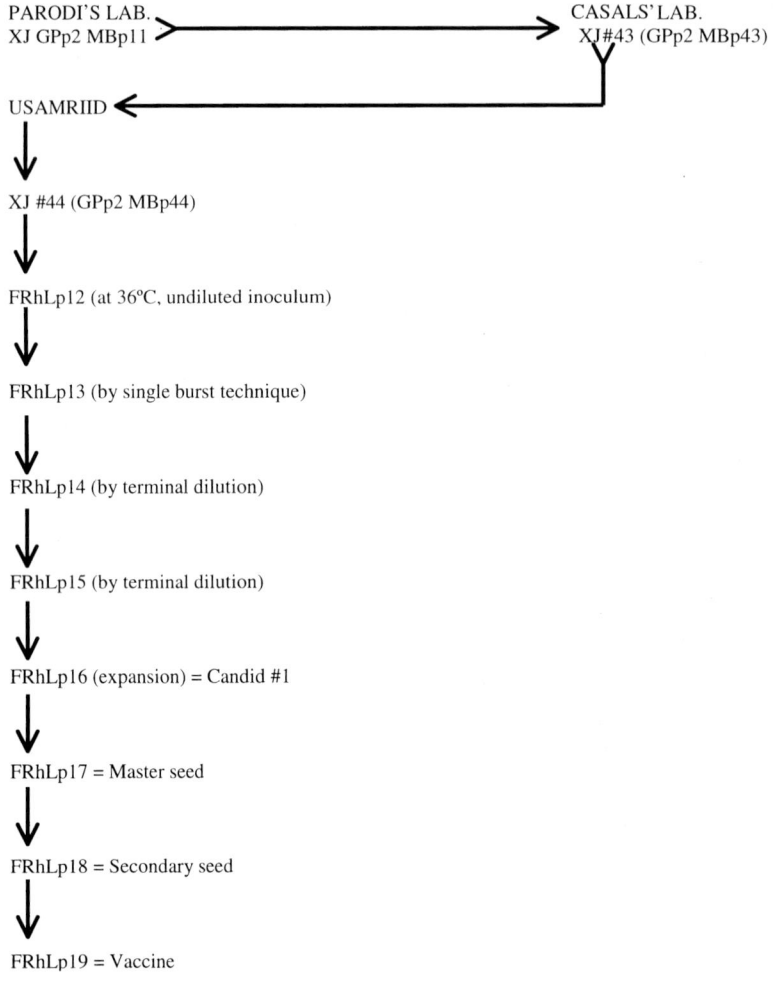

Fig. 3. Candid #1 passage history

confirm the retention of Candid #1 attenuation phenotype after passage in vitro (Table 2). (US Food and Drug Administration's Investigational New Drug # 2257).

3.1.3.2 Candid #1 Attenuation Phenotype After Replication In Vivo

The virulence of Candid #1 virus re-isolated from blood mononuclear cells of 11 vaccinated rhesus monkeys was compared quantitatively to the Candid #1 vaccine inoculum and XJ Clone 3 attenuated control virus, in 11–12-day-old mice. The lethal index of the Candid #1 vaccine inoculum was 3.62, but those of the blood mononuclear cell isolates showed a broad spectrum ranging from 1.98 to > 4.4. The

Table 2. Examples of comparative intracerebral virulence testing of Candid #1

Test animal	Virus tested	Passage level	Lethal index[a]
Guinea pig (160g)	Candid #1 p6	FRhlp22	> 5.9[b]
	XJ Clone 3	Vaccine MBp1	2.0
Mice (11 day old)	Candid #1 sec. seed	FRhlp18	4.1
	Candid #1 vaccine	FRhlp19	> 4.0
	XJ Clone 3	Vaccine MBp1FRhlp1	1.6
	XJ #44	MBp44	1.8

[a] Log 10pfu resulting in 1 intracerebral LD50.
[b] Maximum dose of undiluted virus.

lethal index of one isolate (1.98) approached the index of XJ Clone 3 (1.23); however the remaining ten isolates had indexes (2.82, 2.93, 3.13, 3.25, 3.51, 4.01, 4.29, > 4.14, > 4.20, > 4.40) 10 to 1,000 higher than the attenuated control. This reflected the instability of the Candid #1 attenuation phenotype after replication in vivo, but lent assurance that an attenuation equal to or higher than that of the XJ Clone 3 vaccine was retained (US Food and Drug Administration's Investigational New Drug # 2257).

The attenuation of Candid #1 was unrelated with plaque size or temperature sensitivity (BARRERA ORO and EDDY 1982). More recently, attempts have been made for the eventual identification of attenuation markers of Candid #1 (see Sect. 3.1.5.)

3.1.3.3 Long-Term Effect and Viral Persistence

These studies were performed using guinea pigs and rhesus monkeys. Guinea pigs inoculated intracerebrally with Candid #1 remained healthy throughout the observation period. Half of the animals killed 1–2 months after inoculation showed encephalitic lesions. No virus specific lesions were found at 4 and 8 months after inoculation. Spleen and CNS tissues, collected from animals at 8 months post-inoculation for virus isolation by co-cultivation techniques, yielded no virus. Guinea pigs inoculated subcutaneously with Candid #1p2 (secondary seed level) or Candid #1p7 (four passages beyond vaccine level) remained asymptomatic during a 649-day observation period. Neither JUN virus nor virus antigen were detected in tissues (cerebrum, medulla, cerebellum, spinal cord, spleen, liver, lymph nodes). No CNS lesions were found in 21 animals, microglial myelitis was observed in one animal 68 days after inoculation; fibrosis and colliculus in one animal 130 days after inoculation, and depletion of Purkinje cells in two animals, 577 and 631 days after inoculation, respectively.

Viral persistence and neurotropism were also studied in 24 rhesus monkeys inoculated intramuscularly with Candid #1 vaccine. Virus was isolated by co-cultivation from the liver of one animal and the spleen of another, 2 weeks after inoculation. No virus or virus antigen was detected in tissues (spleen, liver, lymph nodes, salivary gland, frontal cortex, temporal cortex, occipital cortex, thalamus, pons, medulla oblongata, cerebellum, and lumbar spinal cord) of the remaining animals, 1, 2, 3, and 7 months after inoculation. CNS lesions were observed in 19 of the 24 vaccinated, and in nine of the 12 control rhesus monkeys (there was no significant difference between the control and vaccinated groups). The CNS lesions were blindy graded as minimal both in control and vaccinated animals. However, in the control animals, lesions were limited to only one or two of the 16 CNS regions studied, whereas lesions distributed in a pattern similar to that observed in virulent JUN virus was found in four of the Candid #1 inoculated animals. No neuronal changes were found in vaccinated or control animals.

Lesions of lymphoid organs were observed with equal frequency and severity in control and principal macaques.

From these studies, it was concluded that Candid #1 appears to retain some minimal neurotropism (US Food and Drug Administration's Investigational New Drug Application # 2257).

3.1.3.4 Immunogenicity and Protective Efficacy of Candid #1

Guinea pigs and rhesus monkeys inoculated with graded doses of Candid #1 developed neutralizing antibodies and resisted parenteral challenge with highly virulent wild-type JUN virus strains (Ledesma and P3790, respectively) (BARRERA ORO and EDDY 1982; BARRERA ORO et al. 1985; MCKEE et al. 1992). The median protective dose of the vaccine was less than 34pfu for guinea pigs, and less than 16pfu for rhesus monkeys. Candid #1 also protected rhesus monkeys against aerosol challenge with P3790 (KENYON et al. 1989). Furthermore, Candid #1 protected guinea pigs and rhesus monkeys against challenge with Machupo virus, the agent of Bolivian hemorrhagic fever (BARRERA ORO and EDDY 1982; BARRERA ORO et al. 1988; LUPTON et al. 1988; JARHLING et al. 1988). A dose of 3pfu of the vaccine protected 6/6 rhesus against challenge with Machupo virus.

3.1.4 Testing of Candid #1 Secondary Seed and Vaccine

The testing of Candid #1 secondary seed and the first lot of vaccine were performed in compliance of FDA regulations for live attenuated virus vaccines. Secondary seed virus and vaccine were tested by appropriate cultural and fluorescent antibody techniques for bacterial and mycoplasma contaminants with negative results. Inoculation of adult and suckling mice, guinea pigs, rabbits, and four type of cell culture were used to demonstrate the absence of adventitious viral agents. In addition, the presence of other viral contaminants was ruled out by the murine antibody production test and reverse trancriptase test. Additional tests on young guinea pigs and on 11–12-day old mice demonstrated the decreased virulence of both secondary seed and vaccine, compared to parent virus or the XJ Clone 3 vaccine used in Argentina.

Breakthrough neutralization tests in mice < 24h old with seed and vaccine using specific reference rabbit antiserum, identified the virus as JUN.

The rhesus monkey neurovirulence safety test (intratalamic, bilateral; intraspinal, and intramuscular inoculation of undiluted virus) showed that the secondary seed and vaccine elicited no signs of neurological or other diseases, and no virus was isolated from brain or spinal cord. Histologic lesions were minimal or mild and significantly less severe than those seen in control monkeys that received wild JUN (Ledesma strain), parent virus, or the old XJ Clone 3 vaccine (US Food and Drug Administration's Investigational New Drug Application #2257).

Lots 2, 3 and 4 of Candid #1 vaccine, which were prepared to demonstrate lot to lot consistence of manufacturing, also passed the testing required by FDA.

3.1.5 Molecular Characterization of Candid #1 Vaccine

To characterize Candid #1 genome, the nucleotide sequences of its sRNA and those of its more virulent ancestors XJ #44, XJ prototype, XJ Clone 3, and other wild-type strain were analyzed. Comparison of the nucleotide and amino-acid sequences of N and GPC genes from Candid #1 and its progenitor strains revealed some changes that are unique to the vaccine strain. When Candid #1 was compared with the wild strain MC2, which is of intermediate virulence, one nucleotide insertion and four nucleotide substitutions were found at positions that do not affect the predicted secondary structure. The nucleotide sequence changes in the GPC open reading frame were concentrated in the amino-proximal and the carboxy-proximal regions. The comparison of the amino acid residues shows that the major changes were located in the amino-proximal region of GPC. These preliminary results suggested the involvement of the surface glycoprotein in attenuation of virulence, although they can not yet be definitively associated. On the other hand, it was also suggested that the overall positive charge of N protein had a tendency to decrease according to the attenuation of the viral strain (ALBARIÑO et al. 1997; GHIRINGHELLI et al. 1997).

3.2 Clinical studies

After the extensive preclinical testing, the regulatory authorities of both Argentina and the United States authorized the initiation of human trials. Candid #1 was considered as a live attenuated vaccine, that appears to be a phenotypically stable product, and that fulfilled or exceeded the requirements established for other live attenuated vaccines, such as measles, mumps, rubella and poliomyelitis.

3.2.1 Safety and Immunogenicity of Candid #1

Initial clinical studies were performed on both North American human volunteers at USAMRIID, USA, and Argentine human volunteers at INEVH, Argentina.

Phase I Trials. In these studies, Candid #1 was found to be safe and highly immunogenic. Volunteers were followed up with periodical clinical examinations. Laboratory parameters were extensively assessed, including routine hematological measurements, urinalyses, coagulation studies, measurements of muscle enzymes, amylase, hepatic and renal function assays, serum levels of immunoglobulins, complement, interferon, and T-lymphocytes and subsets. At USAMRIID, a total of 83 volunteers distributed in different cohorts using a variety of dosage and inoculation routes were immunized (MacDonald et al. 1987, 1988; Barrera Oro et al. 1987). In a concurrent study, 14 Argentine human volunteers males participated in a double blind placebo-controlled study. Four of them had had AHF and were immune for JUN, and all received vaccine. The other ten were negative for JUN by serology, and received either vaccine or placebo according to a double blind placebo-controlled design. Candid #1 was well tolerated and no clinical or laboratory abnormalities were detected during a follow-up period ranging from 1 month to 2 years (Maiztegui et al. 1988).

Viral isolation attempts from throat, plasma, PBMC, urine and semen in 46 vaccinees yielded positive results from PBMC only in 7 subjects during a window period of 7–13 days after vaccination. The virus could be recovered on only 1 day from six of seven subjects. The maximum duration of viremia was 3 days.

Among non-immune persons, up to 94% developed a positive response to one or more Junin virus-specific antibody tests (enzyme-linked immunoassay, indirect immunofluorescence assay, and/or neutralizing antibody tests), and 99% developed a JUN specific cellular immune response. In immune volunteers, Candid #1 induced an increase in the titers of JUN specific antibodies.

Phase II Trials. Subsequent studies validated safety and immunogenicity of Candid #1. In Argentina, 82 human volunteers at low risk for AHF participated in a trial under a double-blind design. Ultimately, 55 persons received 30,000pfu of Candid #1, and 27 received a placebo. None of the vaccine receptors developed significant clinical, hematological, biochemical or urinary abnormalities. Fifteen persons (27%) developed neutralizing antibodies by 15 days after inoculation [geometric mean titer (GMT) = 1:17], while 50 (91%) were positive at 30 days (GMT = 1:37), 60 days (GMT = 1:81), and 90 days (GMT = 1:87). At 6 months, 89% were found to have persistent antibody (GMT = 1:71) (Maiztegui et al. 1989). Similar findings were obtained among >50 additional volunteers inoculated at USAMRIID.

3.2.2 Efficacy of Candid #1

Phase III Trials. Protective efficacy of Candid #1 was shown in a prospective, randomized, double blind, placebo-controlled study performed in adult males from 41 localities in southern Santa Fe province, Argentina (Maiztegui et al. 1998). These individual counties were selected on the basis of their higher AHF incidence, in an endemic area where average annual incidence of AHF among males was 2.2/1000. Females were excluded because of their lower risk of contracting AHF and of the unknown effect of Candid #1 on pregnancy. Males were considered eligible if they were healthy, 15–64 years old, resided or worked in a rural agricultural area of the selected counties, had no history of AHF, had normal baseline blood values, were negative on screening for human immunodeficiency virus infection, and had no allergies to vaccine components. Eligible volunteers were randomly selected to receive either vaccine or placebo. A first group of participants was inoculated in 1988, with additional participants recruited in 1989. A single 1ml dose of either vaccine diluent or 40,000pfu of Candid #1 was injected. Following inoculation, volunteers were evaluated twice a year, before and after each AHF epidemic season. A passive case detection system for possible adverse events associated with inoculation and episodes of AHF was in place throughout the study. Volunteers hospitalized for suspected AHF were evaluated daily for clinical status, and clinical laboratory test were also monitored daily. Blood for JUN serology was collected every third day, and blood for virus isolation was obtained during the first 3 days of hospitalization. Volunteers with either a fourfold rise in antibody titer or the isolation of JUN from plasma were classified as having laboratory-confirmed JUN infection. Patients with a clinical case definition for AHF were treated with immune plasma.

A total of 7,450 were selected of 8,144 persons registered (the at risk population from which recruitment took place was estimated to be 21,000). Ultimately 6,500 volunteer were enrolled during the two recruitment periods. In 1990, when the study code was broken, it was revealed that 3,255 persons had received vaccine and 3,245 had received placebo. Pre-inoculation sera from 2.7% of the vaccine group and 2.5% of the placebo group were found to have JUN antibody titers of $\geq 1:4$ (MAIZTEGUI et al. 1998)

3.2.2.1 Vaccine Reactogenicity

Within the first 2 weeks after inoculation, 1.1% of the vaccine recipients had one or more reported adverse effects, versus 0.4% of the placebo recipients. The only adverse event that occurred significantly more often in vaccinees than in placebo recipients was headache with constitutional symptoms (MAIZTEGUI et al. 1998).

3.2.2.2 Vaccine Immunogenicity

Among vaccine recipients 91.1% had seroconverted (range $<1:4$ to ≥ 4096) during a mean interval period of 5 months between inoculation and first neutralizing antibody titer measurement, the GMT was 1:152 (MAIZTEGUI et al. 1998).

3.2.2.3 Vaccine Efficacy

Analysis of the entire study cohort revealed that among the 23 men with AHF, 22 had received placebo (35.5 cases/10,000 person-seasons of follow up), and 1 had received vaccine (1.6 cases/10,000 person-seasons); vaccine efficacy for protection against AHF was 95% (95% CI, 82%–99%; $P < 0.001$). A total of 29 volunteers were found to have laboratory evidence for JUN infection. Of these, 25 had received placebo (40.3 cases/10,000 person-seasons) and four had received vaccine (6.4 cases/10,000 person-seasons). Vaccine efficacy against infection with any clinical symptoms was 84% (95% CI, 60%–94%, $P < 0.001$) (MAIZTEGUI et al. 1998).

4 Current Studies with Candid #1 Vaccine

In 1991, with limited quantities of vaccine available produced at The Salk Institute, Swiftwater, PA, USA, INEVH initiated the vaccination of the adult population exposed to highest natural risk of acquiring AHF. Our aims in this phase of evaluation were:

1. To validate the protective efficacy of this vaccine in a large number of people at risk of acquiring AHF and to analyze the impact of this selected vaccination
2. To confirm the immunogenicity of Candid #1 vaccine
3. To observe a large group of volunteers and record their rare adverse clinical reactions
4. To determine the duration of protection against AHF

Areas and populations targeted were restricted according to vaccine availability, and were selected on the basis of their higher incidence during the previous 5 years. Incidence was calculated separately for men and women (FEUILLADE et al. 1997). The incidence rates from selected localities ranged between 0.16/1,000 and 26/1,000. Females participating in the study were required to have a negative blood test for pregnancy during the 2 days before vaccination and were requested to avoid pregnancy during the 3 months following immunization. Every participant was inoculated with a single dose of 10^4 plaque-forming units of Candid #1 virus, by intramuscular injection, as in previous phase I, II and III studies. A passive case-detection system for possible adverse effects associated with vaccination and for the detection of AHF was in place linked to surveillance carried out under the AHF National Control Program.

Between 1991 and 1999, a total of 202,972 persons (157,042 males and 45,930 females) coming from 206 selected counties were vaccinated. A total of 361 clinical events (0.17%) were reported among them within the first 2 weeks after vaccination with 188 vaccinees (0.1%) reported one or more adverse events. Fever, headaches with or without constitutional signs represented 76% (142/188) of these clinical manifestations. No severe adverse reactions attributed to vaccination were recorded during the study period.

Although strong recommendations were given to avoid pregnancy in the 3 months following vaccination, 104 pregnancies were reported among the 45,930 females vaccinated. Among the 103 that we could follow up, 82 resulted in a normal infant, one infant had anencephaly, one myelomeningocele, one aesophagus atresia, one hyperbilirubin, one had seizures associated with hypocalcemia, one experienced jaundice due to Rh-incompatibility, and one suffered from acute respiratory distress syndrome associated with a difficult delivery In addition, eight were electively aborted, and six were lost to miscarriages. Sixty-five mother-infant serum pairs were evaluated for neutralizing antibodies against JUN. Titers were similar for all of them (including the anencephalic infant). In 57 children evaluated at the age of 1 year, clinical examinations were normal, and the mother-children serum pairs revealed the persistence of antibodies in 54 mothers, with negative results in 3. The results for all the children were negative (including the child with myelomenigocele). The malformations observed were not judged to be a consequence of immunization.

4.1 Immunogenicity and Duration of the Immune Response

For the 1885 volunteers evaluated, immunogenicity during the study period ranged between 88.5% and 95.7%. Specific neutralizing antibodies at titers ≥ 1:10 persisted in 90% of a sample of 465 volunteers inoculated in 1988–89 up to 10 years after vaccination (GMT = 1:56).

4.2 Candid #1 Effectiveness for the 1992–99 Period

Eleven cases of AHF occurred during the period among the 202,972 vaccinees, and 276 cases of AHF in persons of ≥15 years with rural contact in the vaccination area. The estimated effectiveness for the 1992–99 period is thus 98.1% (IC 95%: 96.6%–99.0%) (Table 3). All AHF cases in vaccinated persons were mild.

4.3 Impact of Vaccination of Selected High-Risk Groups

To evaluate the impact of this selected vaccination, the epidemiological parameters of confirmed AHF cases studied at INEVH from 1987 to 1991, prior to the vaccination, and those from 1992 to 1997 were compared. A significant increase in the percentage of women (that comprised only 18% of the population targeted by vaccination), of children under 15 years (that were not vaccinated), and of cases without rural contact (only persons living or working in selected rural areas were eligible for vaccination) was noticed (ENRIA et al. 1999b). AHF epidemic outbreaks have been at their lowest recorded levels since the discovery of JUN (Fig. 4).

5 Candid #1 as an Orphan Vaccine

Candid #1 is destined to be used in a geographically limited region, with a total population at risk estimated in 5,000,000 persons. Candid #1 may also be effective

Table 3. Candid # 1 vaccine effectiveness (1992–1999)

Vaccine	AHF		Total
	Yes	No	
Yes	11	202,961	202,972
No	276	96,752	97,028
Total	287	299,713	300,000

Vaccine effectiveness: 98.1% (95% CI: 96.6%–99.0%)*.

*Vaccine effectiveness = $\frac{1-PCV}{1-PCV} \times \frac{1-PPV}{PPV}$.

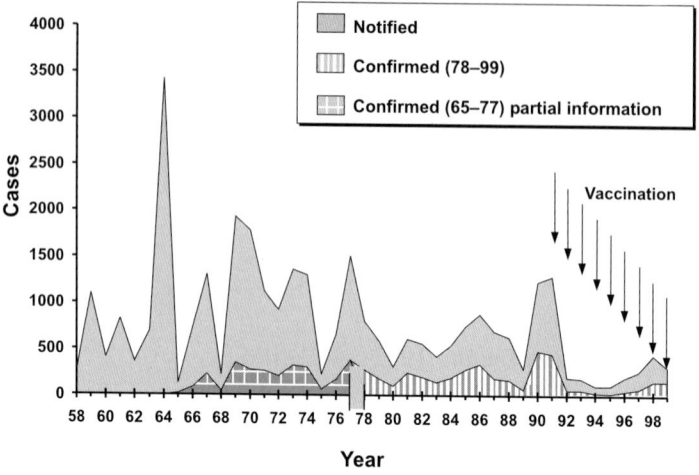

Fig. 4. Annual distribution of AHF cases (1958–1999)

for the prevention of Bolivian hemorrhagic fever, as suggested by studies in guinea pigs (BARRERA ORO and EDDY 1982), and rhesus monkeys (BARRERA ORO et al. 1988; LUPTON et al. 1988; JARHLING et al. 1988), but the vaccine does not cross-protect against infection with Guanarito virus in experimental studies (P.B. Jarhling, personal communication). For these reasons, Candid #1 is considered to be an orphan vaccine. The international collaborative project that developed Candid #1 envisioned the production problem, and proposed that the vaccine would be locally produced. To achieve this, INEVH laboratories would have to be adapted to fulfill Good Manufacture Procedures requirements and their personnel trained in the whole production and quality control process. The final phase of the project began to be financed by the Argentine Health Ministry during the last 3 years. It is expected that Candid #1 production in Argentina would be achieved in the near future, provided that the financial support is sustained. Until Argentina attains this goal, the vaccine available will only be sufficient to target people at a very high risk.

6 Concluding Remarks and Future Expectations

AHF has been a serious health problem in the richest areas of the Argentine pampas over the last five decades. The control of the rodent reservoir of JUN is impractical given the large geographic zones involved and the difficulties of intervening in local agriculture economies. Control of human contact with rodents is also not feasible. For this reason, almost since the discovery of the disease, all efforts for prevention have been directed towards the development of a vaccine.

The remarkable progress achieved with the discovery of the XJ Clone 3 vaccine candidate in the 1960s, prompted the development of a live attenuated JUN vaccine through the effort of an international collaborative project initiated in the late 1970s. Candid #1 vaccine was the final result of this project. Candid #1, the first vaccine against an arenavirus, was shown to be safe, highly immunogenic and effective for the prevention of AHF. Moreover, Candid #1 showed promise in preventing two natural arenaviral hemorrhagic fevers and thus potentially diminishing the medical significance of JUN and Machupo viruses (ALIBEK 1999).

Vaccination of high risk population corroborated findings in phases I, II and III trials, and indicated that the vaccine will also be immunogenic under field conditions. Studies of antibody persistence showed stability of antibody positivity for at least 10 years after immunization. The long-term protection afforded by only one dose of Candid #1 argues in favor of its excellence for the control of AHF.

The observed changes in the current composition of AHF cases, that are attributable to the impact of selected vaccination of high risk people, is indicative that a broader strategy of vaccination to protect the whole population at risk should be evaluated. This objective could only be accomplished with a full operation of vaccine production facilities at INEVH. With sufficient supplies of Candid #1 vaccine, the definitive control of AHF may be envisioned. However, the disease can not be eradicated because JUN reservoirs are rodents, and even with good vaccine coverage, small outbreaks and isolated AHF cases may be expected.

Candid #1 seems to be an excellent product, but does obviously not fulfill all the requirements for an ideal product. No currently available immunizing agent is both completely safe and totally effective. Contraindications to the administration of live viral vaccines include those conditions essentially associated with high risk for replication of viruses, and special considerations appear to be imperative in the case of immunocompromised patients. New system to deliver vaccines are being developed using biological engineering technology (KURSTAK 1994), and some projects are addressed to the search of vaccines for arenaviruses (CLEGG and LLOYD 1987; AUPERIN et al. 1988; MORRISON et al. 1989; FISHER-HOSCH et al. 1989; RICO-HESSE 1999; LOPEZ et al. 2000).

The question at present is whether we should replace this "conventional" vaccine in the future by a new one. As in other diseases, the answer is that we should replace "old" vaccines by "new" ones only if the latter demonstrate advantages in relation to those already used. In this respect, we should expect a new AHF vaccine to be safer, more effective, more stable during storage, and cheaper than Candid #1. More specifically, the design of new AHF vaccines should considered: (1) to eliminate the risk of reversion of virulence, (2) to eliminate the risk of persistent infections, (3) to allow the use without risk during reproductive ages, (4) to improve stability, and (5) to allow the use in the rodent reservoirs, opening in this way new possibilities for AHF control. An interesting consideration is the possibility of obtaining vaccine preparations that would protect against all arenaviruses, thus trying to overcome the difficulties involved with orphan status and the limited need for vaccine. If these lines of research continue to be financed and are successful, they could provide new products that ultimately might improve the current JUN vaccine. Meanwhile, Argentina should pursue the process with Candid #1 JUN live attenuated vaccine, that is currently providing the country with a great opportunity for a successful public health intervention against AHF.

References

Agnese G (1999) Historia de una enfermedad, imaginario y espacio rural: La fiebre hemorrágica argentina – 1943–1963. Tesis para optar al grado de Licenciatura en Historia. 5.2. Experiencias en el Hospital Neuropsiquiátrico. Pontificia Universidad Catolica Argentina. Facultad de derecho y Ciencias Sociales. Instituto de Historia. Rosario. Argentina.
Albariño CJ, Ghiringhelli PD, Posik DM, et al. (1997) Molecular characterization of attenuated Junin virus strains. J Gen Virol 78:1605–1610
Alibek K (1999) Biohazard. Biodefense (Junin virus: pp. 42, 113, 126, 202, 281; Machupo virus: pp. 18, 42, 113, 118, 126, 202, 234, 281). Random House, New York, NY
Arribalzaga RA (1955) Una nueva enfermedad epidémica a germen desconocido: hipertermia nefrotóxica, leucopénica y enantemática. Día Médico 27:1204–1210
Auperin DD, Esposito JJ, Lange JV, et al. (1988) Construction of a recombinant vaccinia virus expressing the Lassa virus glycoprotein gene and protection of guinea pigs from lethal Lassa virus infection. Virus Res 9:233–248
Avila MM, Samoilovich SR, Weissenbacher MC (1979) Infección del cobayo con la cepa atenuada del virus Junin $XJCl_3$. Medicina (B. Aires) 39:597–603
Avila MM, Laguens RM, Laguens RP, et al. (1981) Selectividad tisular e indicadores de virulencia de tres cepas del virus Junin. Medicina (B. Aires) 41:157–166
Avila MM, Frigerio MJ, Weber EL, et al. (1985) Attenuated Junin virus infection in *Callithrix jacchus*. J Med Virol 15:93–100
Barrera Oro JG, Girola RA, Frugone G (1967) Estudios inmunológicos con virus Junin: II, inmunidad adquirida por cobayos inoculados con virus inactivado por formol. Medicina (B. Aires) 27:279–282
Barrera Oro JG and Eddy GA (1982) Characteristics of Candidate live attenuated Junin Virus Vaccine. IV International Conference on Comparative Virology, Banff, Alberta, Canada. Abstract S4–10
Barrera Oro JG, McKee KT Jr, Kuehne AI, et al. (1985) Preclinical trials of a live, attenuated Junin virus vaccine in rhesus macaques. Institut Pasteur Symposium on Vaccines and Vaccinations, Paris, France. Abstract P28
Barrera Oro J, MacDonald C, Kenyon R, et al. (1987) Virus isolation and immune response in humans inoculated with a live-attenuated Junin virus (JV) vaccine. VIIth International Congress of Virology. Edmonton, Alberta, Canada. Abstract R3.45
Barrera Oro JG, Lupton HW, Jahrling PB, et al. (1988) Cross protection against Machupo virus with Candid #1 live-attenuated Junin virus vaccine. I. The post-vaccination prechallenge immune response. Second International Conference on Comparative Virology, Mar del Plata, Argentina. Abstract P.E.1
Barrera Oro JG, McKee KT Jr (1992) Hacia una vacuna contra la Fiebre Hemorrágica Argentina. Bol Of Sanit Panam 112:296–305.
Boxaca MC, Guerrero LB de, Weber L, Malumbres E (1981) Protección inducida en cobayo por la variente XJ0 del virus Junin. Medicina (B. Aires) 41:25–34
Boxaca MC, Guerrero LB de, Malumbres E (1984) Modification of Junin virus neurotropism in the guinea pig model. Acta Virol 28:198–203
Carballal G, Cossio PM, de la Vega MI, et al. (1982) Neurovirulencia de la cepa $XJCl_3$ de virus Junin en un primate sudamericano. Tercer Congreso de Microbiología, Córdoba, Argentina, Compendio 186
Carballal G, Videla C, Oubiña JR, et al. (1985) Antígenos inactivados de virus Junin. Medicina (B. Aires) 45:153–158
Casals J (1965) Serological studies on Junin and Tacaribe viruses. Am J Trop Med Hyg 14:794–795
Clegg JC, Lloyd G (1987) Vaccinia recombinant expressing Lassa-virus internal nucleocapsid protein protects guinea pigs against Lassa fever. Lancet 2 (8552):186–188
Contigiani MS, Sabattini MS (1977) Virulencia diferencial de cepas de virus Junin por marcadores biológicos en ratones y cobayos. Medicina (B. Aires) 37:244–251

Coto CE, Damonte EB, Calello MA, et al. (1980) Protection of guinea pigs inoculated with Tacaribe virus against lethal doses of Junin virus. J Infect Dis 141:389–393
Cresta B, Padula P, de Martínez Segovia ZM (1980) Biological properties of Junin virus proteins: I, identification of the immunogenic glycoprotein. Intervirology 13:284–288
D'Aiutolo AC, Lampuri JS, Coto CE (1979) Reactivación del virus Junin inactivado por la luz ultravioleta. Medicina (B. Aires) 39:801
Damonte EB, Calello MA, Coto CE, et al. (1981) Inmunización de cobayos contra la fiebre hemorrágica argentina con virus Tacaribe replicado en células diploides humanas. Medicina (B. Aires) 41:467–470
Damonte EB, Mersich SE, Candurra NA, et al. (1986) Cross reactivity between Junin and Tacaribe viruses as determined by neutralization test and immunoprecipitation. Med Microbiol Immunol 175:85–88
Downs WG, Anderson CR, Aitken THG, et al. (1963) Tacaribe virus, a new agent isolated from *Artibeus* bats and mosquitoes in Trinidad. Am J Trop Med Hyg 12:639–646
Eddy GA and Barrera Oro JG (1977) Discusión general. Medicina (B. Aires) 37 (Supl 3):257–259
Enria DA, Briggiler AM, Fernández NJ, et al. (1984) Importance of dose of neutralizing antibodies in treatment of Argentine hemorrhagic fever with immune plasma. Lancet 8397:255–256
Enria DA, Damilano AJ, Briggiler AM, et al. (1985) Síndrome neurológico tardío en enfermos de Fiebre Hemorrágica Argentina tratados con plasma inmune. Medicina (B. Aires) 45(6):615–620
Enria DA, Maiztegui JI (1994) Antiviral treatment of Argentine hemorrhagic fever. Antivir Res 23:23–31
Enria DA, Briggiler AM, Feuillade MR (1998) An overview of the epidemiological, ecological and preventive hallmarks of Argentine Hemorrhagic Fever (Junin virus). Bull Inst Pasteur 96:103–114
Enria D, Bowen M, Mills JN, et al. (1999a) I Arenaviruses. In: Guerrant RL, Walker DH, Weller PF, Saunders WB (eds) Tropical Infectious Diseases: Principles, Pathogen & Practice. New York, pp 1189–1212
Enria DA, Feuillade MR, Levis SC, et al. (1999b) Impact of vaccination of a high risk population for Argentine hemorrhagic fever with a live attenuated Junin virus vaccine. In: Saluzzo JF, Dodet B (eds) Factors in the Emergence and Control of Rodent-borne Viral Diseases. Elsevier SAS, Paris, France, pp 273–280
Farrington CP (1993) Estimation of vaccine effectiveness using the screening method. Int J Epidemiol 22:742–746
Feuillade MR, Sabattini M, Enria D (1997) Vacunación de una poblacion expuesta a riesgo natural de adquirir la fiebre hemorragica Argentina. Evolucion de la incidencia. V Congreso Latinoamericano de Medicina Tropical. 3–7 March 1997, Havana, Cuba, Abstracts:650
Fisher-Hoch SP, McCormick JB, Auperin DD, et al. (1989) Protection of rhesus monkeys from fatal Lassa fever by vaccination with a recombinant vaccinia virus containing the Lassa virus glycoprotein gene. Proc Nat Acad Sci USA 86:317–321
García J, Calderón G, Sabattini M, et al. (1996) Infección por virus Junin (VJ) de *Calomys musculinus* (Cm) en áreas con diferente situación epidemiológica para la Fiebre Hemorrágica Argentina (FHA). Medicina (B. Aires) 56:624
Ghiringhelli PD, Albarino CG, Piboul l, Romanowski V (1997) The glycoprotein precursor gene of the attenuated Junin Virus vaccine strain (Candid #1). Am J Trop Med Hyg 56(2):216–225
Guerrero LB de, Weissenbacher MC, Parodi AS (1969) Inmunización contra la fiebre hemorrágica argentina con una cepa atenuada del virus Junin; I: estudio de una cepa modificada del virus Junin, inmunización de cobayos. Medicina (B. Aires) 29:1–5
Guerrero LB (1977) Vacunas experimentales contra la fiebre hemorragica Argentina. Medicina (B. Aires) 37:252–259
Guerrero LB de, Boxaca MC (1980) Estudio preliminar de una variante atenuada del virus Junin derivada de la cepa prototipo XJ. Medicina (B. Aires) 40:267–274
Guerrero LB de, Boxaca MC, Rabinovich RD, et al. (1983) Evolución de la infección en cobayos infectados con la variante atenuada XJ_0 del virus Junin. Rev Argent Microbiol 15:205–212
Guerrero LB de, Boxaca MC, Malumbres E, et al. (1985) Pathogenesis of attenuated Junin virus in the guinea pig model. J Med Virol 15:197–202
Jahrling PB, Trotter RW, Barrera Oro JG, et al. (1988) Cross-protection against Machupo virus with Candid #1 Junin virus vaccine: III, post-challenge clinical findings. Second International Conference on the Impact of Viral Diseases on the Development of Latin American Countries and the Caribbean Region, Mar del Plata, Argentina, abstract PE3
Kenyon RH, McKee K, Barrera Oro JG, et al. (1989) Protective efficacies in rhesus monkeys of Candid 1 strain Junin virus (JV) and Tacaribe virus against aerosol challenge with virulent JV; 38th Annual Meeting of the American Society of Tropical Medicine and Hygiene, Honolulu. Abstract 136

Kurstak E (1994) Modern vaccinology: Progress towards the global control of Infectious Diseases. In: Kurstak E (ed) Modern Vaccinology, Plenun Medical Book Company, New York, London, pp 1–9

Laguens RM, Avila MM, Samoilovich SR, et al. (1983) Pathogenicity of an attenuated strain (XJCl$_3$) of Junin virus: morphological and virological studies in experimentally infected guinea pigs. Intervirology 20:195–201

López N, Scolaro L, Rossi C, et al. (2000) Homologous and heterologous glycoproteins induce protection against Junin virus challenge in guinea pigs. J Gen Vir:1273–1281

Lupton HW, Jahrling PB, Barrera Oro JG, et al. (1988) Cross-protection against Machupo virus with Candid #1 live-attenuated Junin virus vaccine: II, post-challenge virological and immunological findings. Second International Conference on the Impact of Viral Diseases on the Development of Latin American Countries and the Caribbean Region, Mar del Plata, Argentina, abstract PE2

MacDonald C, McKee K, Peters C, et al. (1987) Initial clinical assessment of humans inoculated with a live-attenuated Junin virus vaccine. In: Programs and abstracts of the VII International Congress of Virology (Edmonton, Canada). Abstract R3.27

MacDonald C, McKee K, Meegan J, et al. (1988) Initial evaluation in humans of a live-attenuated vaccine against Argentine hemorrhagic fever. In: Proceedings of the XVI Army Science Conference (Hampton, Virginia), US Army

McKee KT Jr, Barrera Oro JG, Kuehne AI, et al. (1992) Candid No 1 Argentine Hemorrhagic Fever vaccine Protects Against Lethal Junin Virus Challenge in Rhesus Macaques. Intervirology 34:154–163

McKee KT Jr, Green DE, Mahlandt BG, et al. (1985) Infection of Cebus monkeys with Junin virus. Medicina (B. Aires) 45:144–152

Maiztegui JI and Sabattini MS (1977) Extensión progresiva del área endémica de Fiebre hemorrrágica argentina. Medicina (B. Aires) 37:162–166

Maiztegui JI, Fernández NJ, Damilano AJ (1979) Efficacy of immune plasma in treatment of Argentine hemorrhagic fever and association between treatment and late neurological syndrome. Lancet 8154:1216–1217

Maiztegui JI, Feuillade MR, Briggiler AM (1986) Progressive extension of the endemic area and changing incidence of Argentine hemorrhagic fever. Med Microbiol Inmunol 175:149–152

Maiztegui J, Levis S, Enria D, et al. (1988) Inocuidad e inmunogenicidad en seres humanos de la cepa Candid 1 de virus Junin. Medicina (B. Aires) 6:660

Maiztegui JI, McKee KT Jr (1989) Inoculation of human volunteers with a vaccine against Argentine hemorrhagic fever. In: Program and abstracts of the VI International Conference on Comparative and Applied Virology (Banff, Canada). Abstract S4

Maiztegui JI, McKee KT Jr, Barrera Oro JG, et al. (1998) Protective efficacy of a live attenuated vaccine against Argentine Hemorrhagic Fever. J Infect Dis 177:277–283

Malumbres E, Boxaca MC, Guerrero LB de, et al. (1984) Persistence of attenuated Junin virus strains in guinea pigs infected by IM or IC routes. J Infect Dis 149:1022

Martínez Segovia ZM de, Arguelles G, Tokman A (1980) Capacidad antigénica del virus Junin inactivado mediante oxidación fotodinámica. Medicina (B. Aires) 40:156–160

Mettler NE, Casals J, Shope RE (1963) Study of the antigenic relationships between Junin virus, the etiological agent of Argentinean hemorrhagic fever, and other arthropod-borne viruses. Am J Trop Med Hyg 12:647–652

Mettler N (1969) Argentine hemorrhagic fever: current knowledge. PAHO Scientific Publication 183, Washington DC, Pan American Health Organization.

Mills JN, Ellis BA, McKee KT, et al. (1992) A longitudinal study of Junin virus activity in the rodent reservoir of Argentine hemorrhagic fever. Am J Trop Med Hyg 47:749–763

Morrison HG, Bauer SP, Lange JV, et al. (1989) Protection of guinea pigs from Lassa fever byvaccinia virus recombinants expressing the nucleoprotein or the envelope glycoproteins of Lassa virus. Virology 171:179–188

Parodi AS, Geenway DJ, Rugiero HR, et al. (1958) Sobre la etiología del brote epidémico de Junin. Día Médico 30:2300–2301

Parodi AS, Coto CE (1964) Inmunización de cobayos contra el virus Junin por inoculación del virus Tacaribe. Medicina (B. Aires) 24:151–153

Parodi AS, de Guerrero LB, Weissenbacher M (1965) Fiebre hemorrágica argentina: vacunación con virus Junin inactivado. Ciencia e Investigación 21:132–133

Pirosky I, Martini P, Zuccarini J, et al. (1959) Virosis hemorrágica del noroeste bonaerense (endemoepidémica, febril, exantemática y leucopénica): V, la vacuna específica y la vacunación. Orientación Médica 8:743–744

Rico-Hesse R (1999) Vaccines for emergent American arenaviruses. In: Saluzzo JB, Dodet B (eds) Factors in the emergence and control of rodent-borne viral diseases. Editions scientifiques et medicales Elsevier SAS, pp 267–272

Rugiero HA, Astarloa L, González Cambaceres C, et al. (1969) Inmunización contra la fiebre hemorrágica argentina con una cepa atenuada de virus Junin: II, inmunización de voluntarios, análisis clínico y del laboratorio. Medicina (B. Aires) 29:81–92

Rugiero HA, Magnoni C, Cintora FA, et al. (1974) Inmunización contra la fiebre hemorrágica argentina con una cepa atenuada de virus Junin: análisis de 636 vacunados. Pren Med Argent 61:231–240

Rugiero HA, Magnoni C, Guerrero LB de, et al. (1981) Persistence of antibodies and clinical evaluation in volunteers 7 to 9 years following the vaccination against Argentine hemorrhagic fever. J Med Virol 7:227–232

Sabattini MS and Contigiani MS (1982) Ecological and biological factors influencing the maintenance of arenaviruses in nature, with special reference to the agent of Argentine hemorrhagic fever. In: Pinheiro FD (ed) International Symposium on Tropical Arbovirus and Hemorrhagic Fevers, Academia Brasilera de Ciencias, Rio de Janeiro, pp 251–262

Samoilovich SR, Pecci Saavedra J, Frigerio MJ, et al. (1984) Nasal and intrathalamic inoculations of primates with Tacaribe virus: protection against Argentine hemorrhagic fever and absence of neurovirulence. Acta Virol 28:277–281

Samoilovich SR, Calello MA, Laguens RP, et al. (1988) Long-term protection against Argentine hemorrhagic fever in Tacaribe virus infected marmosets: virologic and histopathologic findings. J Med Virol 24:229–236

Tauraso N, Shelokov A (1965) Protection against Junin virus by immunization with live Tacaribe virus. Proc Soc Exp Biol Med 119:608–611

US Food and Drug Administration's Investigational New Drug Application #2257

Videla C, Carballal G, Remorini P, et al. (1989) Formalin inactivated Junin virus: immunogenicity and protection assays. J Med Virol 29:215–220

Wallen KH (1963) Demonstration of inapparent heterogeneity in a population of an animal virus by single-burst analyses. Virology 20:230–234

Weissenbacher M, Guerrero LB de, Help G, et al. (1969) Inmunización contra la fiebre hemorrágica argentina con una cepa atenuada de virus Junin: III, reacciones serológicas en voluntarios. Medicina (B. Aires) 29:88–92

Weissenbacher MC, Coto CE, Calello MA (1975) Cross-protection between Tacaribe complex viruses: presence of neutralizing antibodies against Junin virus (Argentine hemorrhagic fever) in guinea pigs infected with Tacaribe virus. Intervirology 6:42–49

Weissenbacher MC, Coto CE, Calello MA, et al. (1982) Cross-protection in non-human primates against Argentine hemorrhagic fever. Infect Inmun 35:425–430

Weissenbacher MC, Damonte EB (1983) Fiebre hemorrágica argentina. Ade Microbiol Enf Infecc 2:119–171

Weissenbacher MC, Laguens RP, Coto CE (1987) Argentine hemorrhagic fever. Curr Top Microbiol Immunol 134:79–116

Subject Index

A

aborted axon myelination 102
abrasion, transmission route 225
acetylase 101
acetylchloline
- degeneration 101
- synthesis 101

acute infection and its control 46–49, 84
- LCMV infection, CD4⁺ T cells 47–49

adoptive transfer 76
affinity
- affinity/avidity model 123
- maturation 73, 74

AHF (Argentine hemorrhagic fever) 240–245, 242–258
Akodon azarae 240
alpha
- α-dystroglycan (αDG) 93
- IFNα 8

alterations in behavior and learning 108
altered ligand hypothesis 122
Amapari 240
anergy 132
antibody
- antigen-antibody complexes 74
- antigen-independent model for maintaining antibody production: long-lived plasma cells 74–77
- antiviral (see there) 67, 69–70
- autoreactive 73
- environmental 73
- high-affinity 70
- in immunocytotherapy 58
- immunopathology, antibody-mediated 228
- neutralizing (see there) 68, 226–227, 231–232
- non-replicating 69, 73
- secreting cells 69
- virus-antibody immune complex 97–100

antigens
- antigen-antibody complexes 74
- antigen-dependent motels for maintaining antibody production 70–74
- antigen presenting cells (see APCs) 45, 150, 164–165
- antigen-specific T cell responses 84
- CD8⁺ T cell, antigen responsiveness 230
- environmental 73
- non-replicating 69
- self antigens 73

antiviral antibody 67, 69–70
- anatomical sites of antiviral antibody production 69–70
- immunity 44–46
- lymphocytes 162–164

APCs (antigen presenting cells) 45, 150, 164–165
- activated 150
- costimulation and infection 164–165

arenaviridae (arenavirus-arenaviruses) 240, 243, 245, 257

Argentine hemorrhagic fever (see AHF) 240–245, 242–258

autoimmunity, virally induced 145–169
- autoreactive lymphocytes, correlation with several autoimmune disease 152
- viral etiology of autoimmune diseases 146

autoreactive
- antibodies 73
- lymphocytes, correlation with several autoimmune disease 152

axon myelination, aborted 102

B

B cells
- in acute LCMV infection 49
- B cell-depleted mice 49
- in immunocytotherapy 58
- memory B cells (see there) 69, 70, 75

B7.1 153
behavior, alterations in behavior and learning 108
beta
- IFNβ 8
- TGFβ 156

blastogenesis, NK cells 10

Subject Index

blood 16–20
blood-brain barrier 178–179, 185, 187
bm mutant 123
Bolivian hemorrhagic fever 250, 256
bone marrow 69
booster vaccination 69, 73, 74
bottleneck transmissions 205, 215
BrdU (bromodeoxyuridine) 77
bromodeoxyuridine (BrdU) 77
bystander suppression 157

C

Calomys (C.)
- C. laucha 240
- C. musculinus 240
- - familiy muridae 240
- - subfamily sigmodontinae 240
Candid #1 242, 245–258
CD2 230
CD3 230
CD4+ T cells, LCMV infection 47–49
- antigen-specific CD4+ T cell response 84
- CD4-deficient mice 47, 48
- TRC transgenic 57
- virus-specific 55–57
CD8+ T cells
- against New World viruses 226
- against Old World viruses 226
- antigen responsiveness 230
- DNA vaccines 232
- immunopathology 229
- regulation 230
- TRC transgenic 57
- virus-specific 54–55
- - CD8+ CTL response 83
- - escape recognition by 89
CD11c+ 85
- splenic dentritic cells 95
CD28 130
CD40 131, 136
CD40:CD40L 48, 55
CD40L 48, 55
CD95 136
chemokines 165–166, 187–188
choriomeningitis virus, lymphocytic (see LCMV) 1–6, 8
chromosome 17 108
- region on chromosome 17 encompassing 2.5cM region 108
chronic or persistent infection 47, 72, 84, 92–109
- alters the function of differentiated cells 100–109
- immunocytotherapy of 50–59
- negative selection 92

clonal
- deletion 121
- diversity 54–55
- inactivation 124
- XJ clone 3 242, 244–251, 257
CNS persistent LCMV infection
- CNS physiology and virus-induced diseases 177–192
- immunocytotherapy and the CNS 53–54
compartmental issues 16–20
competition 206
complexity 201–202
consensus seuqences 213
cross-reactivity
- envirionmental antigens 73
- self antigens 73
CTL (cytotoxic T lymphocyte) 83, 178, 183–187, 189–192, 227–229
- against Old World viruses 227
- CD8+ CTL response, virus-specific 83, 227–229
- CTL-escape mutants 201
- immunopathology 229
- lower affinity CTL 149
- MHC class I-restricted CTLs 154
cytokines 7, 58, 165–166, 186–187
- inflammatory 154
cytomegalovirus (CMV) 8
- murine CMV (MCMV) 10
cytotoxic T lymphocytes (see CTL) 83, 178, 183–187, 189–192, 227–229
cytotoxicity, NK cells 10

D

DEC205+ 85
- splenic dentritic cells 95
deletion
- clonal 121
- of T cells 121, 136
dendritic cell (DC) 74, 152, 206
- DC transfer 59
- follicular (FDC) 74
- killer dendritic cells 167
- splenic 95
- - CD11c+ 95
- - DEC205+ 95
diabetes mellitus, insulin-dependent 145–169
diptheria 73
DNA
- immunization 160
- vaccines 167, 232–233
- - CD8+ T cell 232
draining lymph node 167
α-dystroglycan (αDG) 93

E

ELISPOT 69
ERK1/2 127
esterase 101
exocrine pancreas, infection of 151

F

FDC (follicular dendritic cells) 74
fitness 201, 213
follicular dendritic cells (FDC) 74
founder effects 205

G

gamma
- IFNγ 10, 58, 156, 186, 188–190
- – IFNγ-deficient mice 58–59
- – IFNγ-receptor 58
- IFNγ (Mig) 12
GAP-43 109
genetic factors 146
growth
- factor, tumor (see TGF)
- hormone (GH)
- – deficiency syndrome 102–108, 180, 197
- – GHF1 (Pit1) 107
- retardation 206
Guanarito 240, 256

H

³H-thymidine 77
hemorrhagic fever 221, 223–225, 240, 245
- Argentine (AHF) 240–245, 242–258
- Bolivian 250, 256
homeostasis 70
- of T cells 137
human-to-human (horizontal) transmission
- Lassa virus 223
- LCMV 223
- risk of 231
- Sabia 225
hydrocephalus 189–190
hypoglycemia 206

I

idiotypic networks 73
ignorance 150
ignorant 133
immune
- complexes 74
- – half-life 74
- privilege 178, 187, 190
immunocytotherapy 43–60
- B cells and antibodies 58
- CD4+ T cells, virus-specific 55–57
- CD8+ T cells, virus-specific 54–55

- persistent LCMV infection, immunocytotherapy and the CNS 53–54
- virus control after 58–59
immunological memory 67, 70, 84
immunomodulatory treatments 162
immunopathology/immunopathologic effects 163
- antibody-mediated 228
- CD8+ T cell 232
- CTL 229
- virally induced 145–169
immunosuppression/immunosuppressive
immunosuppressive
- generalized 204
- phenotype 205
- strains 93
- T cell-mediated 229
- vaccine-related 229
- variant 204
inducible nitric oxide synthase (see iNOS) 12–16
infection
- acute (see there) 46–49, 84
- chronic or persistent (see there) 42, 44, 47, 50–72, 72, 84, 92–109
- exocrine pancreas 151
- locally impaired ("sequestered") 47
- lymphocytic choriomeningitis virus (see LCMV) 1–6, 8, 47–50, 67–78
- protracted 47
- recrudescent 47
- recurrent 47
inflammatory cytokines 154
inhalation, transmission route 225
innate
- immunity 7–9
- responses and disease 23–25
iNOS (inducible nitric oxide synthase) 12
- endogenous 13–16
insulin B-chain 159
insulinomas 134
interferon (IFNs)
- IFNα 8
- IFNα/β/γ 44
- IFNα/β receptors, deficient for 59
- IFNβ 8
- IFNγ 10, 58–59, 156, 186, 188–190
- – IFNγ-deficient mice 58–59
- – IFNγMig (monokine induced IFNγ) 12
- – IFNγ-receptor 58
- – IFNγ-receptor-deficient recipients 59
- – T-cell IFNγ responses 15
- signaling pathways (see there) 20–23
- type I 7, 13–15
interleukin (IL)
- IL 2 156

- IL 4   156, 159
- IL 10   156
- IL 12   8, 13–15

J
Junin virus (JUN)   240–247, 249–258
- Akodon azarae   240
- Calomys laucha   240
- Calomys musculinus (*see there*)   240
- Guanarito virus   240
- hemorrhagic fever   240
- Machupo virus   240
- Mus musculus   240
- Necromys benefactus   240
- Oligoryzomys flavescens   240
- Sabia virus   240
- vaccines   222, 224

L
Lassa virus   223, 228, 240
- human-to-human (horizontal) transmission   223
- vaccines, experimental   223, 228
Latino   240
- Argentine hemorrhagic fever (AHF)   240–245, 242–258
- Bolivian hemorrhagic fever   250, 256
LCMV (lymphocytic choriomeningitis virus) infection   1–6, 8, 47–50, 67–78, 147–149, 223, 227
- acute   49, 67, 84
- - B cells   49
- biology and pathogenesis of   83–111
- CD4+ T cells   47–49
- human-to-human (horizontal) transmission   223
- LCMV model system   68
- mechanisms of humoral immunity explored through studies of   67–78
- persistent LCMV infection and tolerance   49–50
- RIP-LCMV mouse model   147–149
- vaccines   227
learning, alterations in behavior and learning   108
LFA-1   130
life attenuated vaccines   242, 423, 250, 251, 257, 258
ligand hypothesis, altered   122
lineage commitment thymocytes   126
liver   16–20
locally impaired ("sequestred") infection   47
luxory function   180
lymph node, draining   167
lymphocytes/lymphocytic
- anti-viral   162–164

- lymphocytic choriomeningitis virus (*see* LCMV)   1–6, 8, 47–50, 67–78, 147–149, 223, 227

M
Machupo   240, 250, 257
memory B cells   69, 70
- radiosensitivity   75
MHC (major histocompatibility complex)   178, 185–187, 190–191
- class I   153–154
- - blocking peptides   153
- - restricted CTLs   154
- class II-deficient mice   47
microsatellite mapping   107
minor lymphocyte stimulatory (Mls-1α)   121
mitomycin C   76
molecular mimicry   147, 191
multiple sclerosis   191
Muridae   240
murine CMV (MCMV)   10
mus musculus   240
mutation
- rates   198–199
- somatic   70, 73

N
natural killer (NK) cells   7–25
- blastogenesis   10
- cell responses and functions   9–13
- cytotoxicity   10
- killer dendritic cells   167
necromys benefactus   240
negative selection   121
- persistent infection with LCMV   92
neurobehavioral alterations   108
neurochemical abnormalities   108
neuroinvasion   190
neurotransmitters   108
neutralizing antibodies   68, 226–227, 231–232
- against New World viruses   226
- against Old World viruses   227
- vaccine-induced   231–232
nitric oxide synthase, inducible (*see* iNOS)   12
non-immunosuppressive strains   93

O
oligoryzomys flavescens   240
oral transmission   225
Orphan vaccine   255–257

P
P14 TCR transgenic   120
pancreas, infection of the exocrine pancreas   151
perforin   154

persistent infection (*see* chronic infection) 44, 47, 50–59, 72, 84, 92–109
pituitary 209
plasma cells 69, 70
- antigen-independent model for maintaining antibody production: long-lived plasma cells 74–77
- longevity 75, 77
- ovalbumin-specific 77
- radiolabeled 77
positive selection 121
presenting cells (PC) 211
- antigen presenting cells (*see* APCs) 45, 150, 164–165
protracted infection 47

Q
quasispecies 198–200, 205–206, 209, 213

R
RANK 131
RdRp (RNA-dependent RNA polymerases) 198
reassortant viruses 106
reassortment 199, 204
receptor
- binding 93
- editing 73
recombinants 199
recrudescent infection 47
recurrent infection 47
red pulp 199
regulatory cells 160, 168
RING finger 203
RIP-B7.1 153
RIP-gp 133
RIP-LCMV mouse model 147
RNA
- overview of RNA virus evolution: relevance to arenaviruses 197–201
- RNA-dependent RNA polymerases (RdRp) 198
- S RNA 106
- viruses 240
routes of transmission 225–226
- abrasion 225
- inhalation 225
- oral 225

S
S RNA 106
Sabia 225, 240
- human-to-human (horizontal) transmission 225
selection 205–206
Shannon entropy 201
sigmodontinae 240

signaling pathways IFNs 20–23
- STAT (signal transducer and activator of transcription) 1 and 2 20
somatic mutations 70, 73
spleen 16–20, 69
STAT (signal transducer and activator of transcription) 1 and 2 20
sterile immunity 44
SV40 large T antigen 134

T
T cell
- antigen-specific T cell responses 84
- cytotoxic T lymphocytes 178, 183–187, 189–192
- deletion 121, 136
- downregulation, T cell receptor 128
- homeostasis of 137
- IFNγ responses 15
- immunopathology, T cell-mediated 229
- oligomerization, T cell receptor 128
- persistent infection, T cell tolerance 54
- SV40 large T antigen 134
Tacaribe virus 240
- vaccine 242–244
tetanus toxoid 73
TGFβ 156
threshold levels 213
^{3}H-thymidine 77
thymocytes
- lineage commitment 126
- tuning of 125
thymus 150
thyroid-stimulating hormone β (TSHβ) 106
TNF (tumor necrosis factor)
- TNFα 58
- TNF receptor 1 136
- TNF receptor superfamily 136
tolerance 44, 49–54, 133
- persistent LCMV infection
- - immunocytotherapy and the CNS 53–54
- - and tolerance 49–50
TRANCE 131
transgenic animals 147
transmission
- bottleneck 205, 215
- human-to-human/horizontal (*see there*) 223, 225, 231
- routes of (*see there*) 225–226
TRC transgenic CD8$^+$ T cells 57
TSHβ (thyroid-stimulating hormone β) 106
tumor 59
- growth factor (*see* TGF) 156
- immunotherapy 134
- necrosis factor (*see* TNF) 58, 136

tuning of thymocytes 125
two-signal model 130, 135

U
unresponsiveness 150

V
vaccines/vaccination 73, 242–258
- booster 69, 73, 74
- candid #1 242, 250
- DNA vaccines 167, 232–233
- efficacy 253
- in humans 222
- inactivated 242–243
- *Junin* virus 222, 224
- *Lassa* virus, experimental 223, 228
- LCMV 227
- life attenuated vaccines 242–243, 250, 251, 257, 258
- Orphan 255–257
- Tacaribe virus 242
- vaccine-induced neutralizing antibodies 231–232
- vaccine-related immunopathology 229
- XJ clone 3 strain 242

vaccinia 73
variants 200
virus/viral
- etiology of autoimmune diseases 146
- persistence 44–46
- virus-antibody immune complex 97–100

W
white pulp 206

X
XJ clone 3 242, 244–251, 257

Y
yellow fever virus 68, 73

Z
ZAP-70 127

Printing (Computer to Film): Saladruck Berlin
Binding: Stürtz AG, Würzburg

Current Topics in Microbiology and Immunology

Volumes published since 1989 (and still available)

Vol. 219: **Gross, Uwe (Ed.):** Toxoplasma gondii. 1996. 31 figs. XI, 274 pp. ISBN 3-540-61300-5

Vol. 220: **Rauscher, Frank J. III; Vogt, Peter K. (Eds.):** Chromosomal Translocations and Oncogenic Transcription Factors. 1997. 28 figs. XI, 166 pp. ISBN 3-540-61402-8

Vol. 221: **Kastan, Michael B. (Ed.):** Genetic Instability and Tumorigenesis. 1997. 12 figs.VII, 180 pp. ISBN 3-540-61518-0

Vol. 222: **Olding, Lars B. (Ed.):** Reproductive Immunology. 1997. 17 figs. XII, 219 pp. ISBN 3-540-61888-0

Vol. 223: **Tracy, S.; Chapman, N. M.; Mahy, B. W. J. (Eds.):** The Coxsackie B Viruses. 1997. 37 figs. VIII, 336 pp. ISBN 3-540-62390-6

Vol. 224: **Potter, Michael; Melchers, Fritz (Eds.):** C-Myc in B-Cell Neoplasia. 1997. 94 figs. XII, 291 pp. ISBN 3-540-62892-4

Vol. 225: **Vogt, Peter K.; Mahan, Michael J. (Eds.):** Bacterial Infection: Close Encounters at the Host Pathogen Interface. 1998. 15 figs. IX, 169 pp. ISBN 3-540-63260-3

Vol. 226: **Koprowski, Hilary; Weiner, David B. (Eds.):** DNA Vaccination/Genetic Vaccination. 1998. 31 figs. XVIII, 198 pp. ISBN 3-540-63392-8

Vol. 227: **Vogt, Peter K.; Reed, Steven I. (Eds.):** Cyclin Dependent Kinase (CDK) Inhibitors. 1998. 15 figs. XII, 169 pp. ISBN 3-540-63429-0

Vol. 228: **Pawson, Anthony I. (Ed.):** Protein Modules in Signal Transduction. 1998. 42 figs. IX, 368 pp. ISBN 3-540-63396-0

Vol. 229: **Kelsoe, Garnett; Flajnik, Martin (Eds.):** Somatic Diversification of Immune Responses. 1998. 38 figs. IX, 221 pp. ISBN 3-540-63608-0

Vol. 230: **Kärre, Klas; Colonna, Marco (Eds.):** Specificity, Function, and Development of NK Cells. 1998. 22 figs. IX, 248 pp. ISBN 3-540-63941-1

Vol. 231: **Holzmann, Bernhard; Wagner, Hermann (Eds.):** Leukocyte Integrins in the Immune System and Malignant Disease. 1998. 40 figs. XIII, 189 pp. ISBN 3-540-63609-9

Vol. 232: **Whitton, J. Lindsay (Ed.):** Antigen Presentation. 1998. 11 figs. IX, 244 pp. ISBN 3-540-63813-X

Vol. 233/I: **Tyler, Kenneth L.; Oldstone, Michael B. A. (Eds.):** Reoviruses I. 1998. 29 figs. XVIII, 223 pp. ISBN 3-540-63946-2

Vol. 233/II: **Tyler, Kenneth L.; Oldstone, Michael B. A. (Eds.):** Reoviruses II. 1998. 45 figs. XVI, 187 pp. ISBN 3-540-63947-0

Vol. 234: **Frankel, Arthur E. (Ed.):** Clinical Applications of Immunotoxins. 1999. 16 figs. IX, 122 pp. ISBN 3-540-64097-5

Vol. 235: **Klenk, Hans-Dieter (Ed.):** Marburg and Ebola Viruses. 1999. 34 figs. XI, 225 pp. ISBN 3-540-64729-5

Vol. 236: **Kraehenbuhl, Jean-Pierre; Neutra, Marian R. (Eds.):** Defense of Mucosal Surfaces: Pathogenesis, Immunity and Vaccines. 1999. 30 figs. IX, 296 pp. ISBN 3-540-64730-9

Vol. 237: **Claesson-Welsh, Lena (Ed.):** Vascular Growth Factors and Angiogenesis. 1999. 36 figs. X, 189 pp. ISBN 3-540-64731-7

Vol. 238: **Coffman, Robert L.; Romagnani, Sergio (Eds.):** Redirection of Th1 and Th2 Responses. 1999. 6 figs. IX, 148 pp. ISBN 3-540-65048-2

Vol. 239: **Vogt, Peter K.; Jackson, Andrew O. (Eds.):** Satellites and Defective Viral RNAs. 1999. 39 figs. XVI, 179 pp. ISBN 3-540-65049-0

Vol. 240: **Hammond, John; McGarvey, Peter; Yusibov, Vidadi (Eds.):** Plant Biotechnology. 1999. 12 figs. XII, 196 pp. ISBN 3-540-65104-7

Vol. 241: **Westblom, Tore U.; Czinn, Steven J.; Nedrud, John G. (Eds.):** Gastroduodenal Disease and Helicobacter pylori. 1999. 35 figs. XI, 313 pp. ISBN 3-540-65084-9

Vol. 242: **Hagedorn, Curt H.; Rice, Charles M. (Eds.):** The Hepatitis C Viruses. 2000. 47 figs. IX, 379 pp. ISBN 3-540-65358-9

Vol. 243: **Famulok, Michael; Winnacker, Ernst-L.; Wong, Chi-Huey (Eds.):** Combinatorial Chemistry in Biology. 1999. 48 figs. IX, 189 pp. ISBN 3-540-65704-5

Vol. 244: **Daëron, Marc; Vivier, Eric (Eds.):** Immunoreceptor Tyrosine-Based Inhibition Motifs. 1999. 20 figs. VIII, 179 pp. ISBN 3-540-65789-4

Vol. 245/I: **Justement, Louis B.; Siminovitch, Katherine A. (Eds.):** Signal Transduction and the Coordination of B Lymphocyte Development and Function I. 2000. 22 figs. XVI, 274 pp. ISBN 3-540-66002-X

Vol. 245/II: **Justement, Louis B.; Siminovitch, Katherine A. (Eds.):** Signal Transduction on the Coordination of B Lymphocyte Development and Function II. 2000. 13 figs. XV, 172 pp. ISBN 3-540-66003-8

Vol. 246: **Melchers, Fritz; Potter, Michael (Eds.):** Mechanisms of B Cell Neoplasia 1998. 1999. 111 figs. XXIX, 415 pp. ISBN 3-540-65759-2

Vol. 247: **Wagner, Hermann (Ed.):** Immunobiology of Bacterial CpG-DNA. 2000. 34 figs. IX, 246 pp. ISBN 3-540-66400-9

Vol. 248: **du Pasquier, Louis; Litman, Gary W. (Eds.):** Origin and Evolution of the Vertebrate Immune System. 2000. 81 figs. IX, 324 pp. ISBN 3-540-66414-9

Vol. 249: **Jones, Peter A.; Vogt, Peter K. (Eds.):** DNA Methylation and Cancer. 2000. 16 figs. IX, 169 pp. ISBN 3-540-66608-7

Vol. 250: **Aktories, Klaus; Wilkins, Tracy, D. (Eds.):** Clostridium difficile. 2000. 20 figs. IX, 143 pp. ISBN 3-540-67291-5

Vol. 251: **Melchers, Fritz (Ed.):** Lymphoid Organogenesis. 2000. 62 figs. XII, 215 pp. ISBN 3-540-67569-8

Vol. 252: **Potter, Michael; Melchers, Fritz (Eds.):** B1 Lymphocytes in B Cell Neoplasia. 2000. XIII, 326 pp. ISBN 3-540-67567-1

Vol. 253: **Gosztonyi, Georg (Ed.):** The Mechanisms of Neuronal Damage in Virus Infections of the Nervous System. 2001. approx. XVI, 270 pp. ISBN 3-540-67617-1

Vol. 254: **Privalsky, Martin L. (Ed.):** Transcriptional Corepressors. 2001. 25 figs. XIV, 190 pp. ISBN 3-540-67569-8

Vol. 255: **Hirai, Kanji (Ed.):** Marek's Disease. 2001. 22 figs. XII, 294 pp. ISBN 3-540-67798-4

Vol. 256: **Schmaljohn, Connie S.; Nichol, Stuart T. (Eds.):** Hantaviruses . 2001, 24 figs. XI, 196 pp. ISBN 3-540-41045-7

Vol. 257: **van der Goot, Gisou (Ed.):** Pore-Forming Toxins, 2001. 19 figs. IX, 166 pp. ISBN 3-540-41386-3

Vol. 258: **Takada, Kenzo (Ed.):** Epstein-Barr Virus and Human Cancer. 2001. 38 figs. IX, 233 pp. ISBN 3-540-41506-8

Vol. 259: **Hauber, Joachim, Vogt, Peter K. (Eds.):** Nuclear Export of Viral RNAs. 2001. 19 figs. IX, 131 pp. ISBN 3-540-41278-6

Vol. 260: **Burton, Didier R. (Ed.):** Antibodies in Viral Infection. 2001. 51 figs. IX, 309 pp. ISBN 3-540-41611-0

Vol. 261: **Trono, Didier (Ed.):** Lentiviral Vectors. 2002. 32 figs. X, 258 pp. ISBN 3-540-42190-4

Vol. 262: **Oldstone, Michael B.A. (Ed.):** Arenaviruses I. 2002, 30 figs. XVIII, 197 pp. ISBN 3-540-42244-7